Isotope Production and Applications in the 21st Century

TRIUMF

Third International Conference on Isotopes
3ICI
Vancouver Canada '99

Proceedings of the 3rd International Conference on Isotopes

ISOTOPE PRODUCTION AND APPLICATIONS IN THE 21ST CENTURY

Vancouver, Canada 6 – 10 September 1999

Editor

Nigel R. Stevenson
TRIUMF, Canada

World Scientific
Singapore • New Jersey • London • Hong Kong

Published by

World Scientific Publishing Co. Pte. Ltd.
P O Box 128, Farrer Road, Singapore 912805
USA office: Suite 1B, 1060 Main Street, River Edge, NJ 07661
UK office: 57 Shelton Street, Covent Garden, London WC2H 9HE

British Library Cataloguing-in-Publication Data
A catalogue record for this book is available from the British Library.

ISOTOPE PRODUCTION AND APPLICATIONS IN THE 21ST CENTURY
Proceedings of the 3rd International Conference on Isotopes

ISBN 981-02-4200-X

Printed in Singapore by World Scientific Printers

This book is printed on acid and chlorine free paper.

FOREWORD

The Third International Conference on Isotopes (3ICI) was held at the Renaissance Hotel, Vancouver, Canada, from September 6-10, 1999 and followed successful meetings in Beijing, China (1995) and Sydney, Australia (1997). 3ICI was hosted by TRIUMF, Canada's national accelerator facility and a major center for isotope production and applications in research and medicine. The conference was co-organized by the Canadian Chapter of the International Isotope Society.

The meeting had the theme "Isotope Production and Applications in the 21st Century" and emphasized the new and exciting developments that are imminent in this growing field. In line with the previous meetings the conference was structured to focus on the three main fields of isotope applications: in medicine, industry and the environment. The subject material of the papers that were presented spanned a wide range of interests reflecting the diversity of the attendees. Over 300 people attended from 23 countries and 120 oral and 50 poster papers were submitted for presentation. It was particularly pleasing that representatives from both the Federal and Provincial governments of Canada and British Columbia were able to address the conference delegates during 3ICI.

One of the highlights of this meeting was the technical tour of TRIUMF, MDS Nordion (Vancouver Operations) and the University of British Columbia's Positron Emission Tomography (PET) facilities. The tour included visits to the new radioactive beam facility (ISAC) at TRIUMF and the MDS Nordion isotope production cyclotrons. The tour was successfully organized by Ms. Maria Freeman with the assistance of numerous TRIUMF, MDS Nordion and UBC staff.

Generous support from several Principal and Major Sponsors allowed the Organizing Committee to sponsor many delegates from developing countries to attend 3ICI. The concerns of the sponsored delegates were looked after by Ms. Ann Fong. The Technical Exhibition was coordinated by Mr. Jozef Orzechowski and it was particularly pleasing that this aspect of the conference was fully subscribed. Thanks are also due to Mr. Rick Bloemhard for producing the very informative and useful web pages and to Mr. Rene van den Elzen for arranging the AV and office computers that delegates were able to use during the conference.

The Banquet dinner afforded the opportunity to recognize the organizers of ICI (Prof. Jinrong Zhang) and 2ICI (Dr. Clarence Hardy) with special awards. At this time, it was a pleasure to also acknowledge the efforts (often beyond the "call of duty") of the 3ICI Conference Coordinator – Ms. Elly Driessen. Her willingness to deal with every sort of issue and problem was a major contributing factor to the success of this meeting. Finally, the role of the Conference Program Chairman, Dr. John Rolston, in arranging many of the details of the scientific program was recognized.

As this series of conferences has grown and become established, it has become appropriate to produce these published proceedings in the hope of achieving wider circulation. In this regard we wish to acknowledge the work of Ms. Jana Thomson (TRIUMF) in her role of editing and organizing several aspects of these proceedings.
The need for this series of conferences has been demonstrated by the sustained support of attending delegates from all corners of the world. The subjects discussed at these meetings were obviously topical and relevant – as demonstrated by the fact that attending members of international research laboratories and universities were also joined by their colleagues from industry. The unique mix of people attending 3ICI reflects well on the future health of this community and the outlook for future such meetings. In this regard, it is a pleasure to announce that the next meeting (4ICI) will be held in Cape Town, South Africa in early 2002. On behalf of the Organizing Committee and the International Advisory Committee I would like to thank all attendees for their participation and contributions to the success of 3ICI.

Nigel R. Stevenson
Chairman, 3ICI

CONFERENCE ORGANIZATION

International Advisory Committee

Dr. Nigel R. Stevenson (TRIUMF)
Dr. John Rolston (International Isotope Society)
Dr. S. Machi (The International Atomic Energy Agency, Austria)
Prof. Quian Gaoyun* (The Chinese Nuclear Society, China)
Prof. Jinrong Zhang (The Isotope Society of China, China)
Dr. K.K.S. Pillay (ANS)
Dr. J.M. Cuttler (The Canadian Nuclear Society)
Dr. Heiko Gerstenberg (The German Nuclear Society, Germany)
Dr. Dewi Lewis (The British Nuclear Energy Society, England)
Dr. Hirokazu Umezawa (The Japan Radioisotope Associate, Japan)
Prof. Chizuo Mori* (Atomic Energy Society of Japan, Japan)
Prof. Chang Hyo Kim* (The Korean Nuclear Society, Korea)
Dr. William P. Duncan* (The International Isotope Society, USA)
Dr. Sergei V. Kushnarev* (The Nuclear Society of Russia, Russia)
Dr. Peter L. Airey (The Australian Nuclear Association, Australia)
Dr. Clarence J. Hardy (The Australian Nuclear Association, Australia)
Dr. S.M. Rao (Indian Nuclear Society, India)
Dr. Konrad Hadener* (The European Nuclear Society, Switzerland)
Dr. Peter Louw (SAFARI-1 Reactor, South Africa)

Organizing Committee

R. Bloemhard
E. Driessen
A. Fong
M. Freeman
P. Gardner
J. Orzechowski
J. Rolston
N. Stevenson (Chairman)
J. Thomson
I. Tsui
R. van den Elzen

*or Alternate

Principal Sponsors

MDS Nordion Inc.
Atomic Energy Corporation of South Africa
Institut National Des Radioelements, Belgium
The Information, Sciences and Technology Agency of B.C.

Major Sponsors

TRACE Sciences International
Chemgas
Reviss Services (UK) Ltd.
Isoflex

Advertising Sponsors

International Isotope Clearing House Inc.
Australian Radioisotopes
Canadian Nuclear Society

Organizers

TRIUMF
International Isotope Society, Canadian Chapter

Industrial Exhibitors

Booth	Company
1	MDS Nordion
2	American Radiolabeled Chemicals, Inc.
3	Tenex
4	Reviss Services Ltd.
5	Bioscan
6	AEC
7	IRE
8	Dehnel Consulting Ltd.
9	Trace Sciences Int.
10	Amersham Pharmacia Biotech Inc.
11	Isoflex
12	NEN Life Science Products, Inc.
13	Isotec Inc.
14	C/D/N Isotopes Inc.
15	Chemgas
16	Cambridge Laboratories
17	TRIUMF
18	The Information, Sciences and Technology Agency

Booths 1-13 are in Ballroom 3 (see map)
Booths 14-18 are in the Foyer and/or Ballrooms 1& 2

CONTENTS

Session A

Opening Plenary Session

PROSPECTS FOR THE APPLICATION OF ISOTOPES AND RADIATION AND THE ACTIVITIES OF THE IAEA

SUEO MACHI

Deputy Director General, International Atomic Energy Agency, Wagramer Strasse 5, A-1400 Vienna, Austria, E-mail: S.Machi@iaea.org

Applications of isotopes and radiation for improved technologies in the fields of food and agriculture, industry, health care, water resource management, and environmental protection have been benefiting people. The IAEA promotes these applications through internationally co-ordinated research programmes and technical cooperation programmes.

Applications in food and agriculture are in particular important for mutation breeding in improving crop production, bio-fertilizer development, control of insect pests by sterile insect technique, and food irradiation. In the field of human health, nuclear medicine and radiation therapy of cancer are major applications achieving remarkable progress. Radiation processing has been used in industry for production of materials with improved properties and sterilization of medical products NDT by radiography and nucleonic control systems are well established and indispensable technologies for modern industry with improved safety and energy efficiency. Radiation technology has been studied for its application in environmental protection such as treatment of flue gases, waste water and sewage sludge. The activities of the IAEA in these applications are also reviewed

Isotopes are being used in industry, health care and agriculture. The IAEA is the World's international forum for the promotion of research and technology transfer of these isotope applications. This paper focuses on the application of isotopes, not including electron beams.

1 Industrial Applications

1.1 Nucleonic gauges

To maintain competitive edges in a changing economy, nuclear or isotope-based techniques are extensively used in developed countries. The IAEA catalyses transferring technology, such as nucleonic gauges and radiotracing that enhance efficiency and improve product quality, to developing countries.

For example, the Guanzhou Paper Mill operates seven machines using nucleonic control gauges for quality control. Through saving energy and materials yearly savings are at a level of US$ 500,000, while maintenance costs of gauges are US$ 60,000.

1.2 Radiation processing and sterilization

Sterilization of single use medical products continues to be the major industrial application of Co-60 sources. About 40% of single use medical products are currently sterilized by radiation; this amount is expected to be increased up to 80% in future.

New applications are emerging such as blood irradiation and sterilization of tissue graft and pharmaceuticals. Tissue graft projects have been implemented by the IAEA in Sri Lanka, Malaysia, Indonesia, Zambia and Singapore. To prevent graft versus host disease (GVHD) in patients, blood irradiation is practised using Cs-137 in developed and a few developing countries. In India and Argentina disinfection of sewage sludges by Co-60 is under development for improving environment and agriculture.

2 Human Health Applications

Nuclear technologies are contributing considerably to human health, and many IAEA Member States have established medical care programmes that include essential nuclear elements. Examples of nuclear technologies are those involved in radiation treatment of cancer. Another example is provided by nuclear medicine.

2.1 Radiation therapy of cancer

The number of new cancer cases per year is about 10 million (5.5 million deaths); it is expected to increase to 15 million by the year 2015, and about 2/3 of these cases will occur in developing countries. The IAEA devotes considerable resources to extending radiation therapy technology to developing Member States. Assistance with radiotherapy in Ghana is a recent example. Thanks to the help of the IAEA and to extrabudgetary contributions from China and USA, the first radiotherapy centre was established in Ghana in 1997.

Novel treatment modalities in brachytherapy by means of micro sources of Ir-192 are also being explored by the Agency. This new approach better confines the beam to the tumour thereby reducing undesirable side effects, and even allows outpatients to be treated in fewer visits. BNCT (boron neutron capture therapy) another promising technology is now being intensively studied in the USA (Brookhaven N.L.), Japan (JAERI, Kyoto University etc.) and the EU (PETEN) and will soon be addressed under the Agency's programme.

2.2 Nuclear medicine

Classical methodologies employing radioimmunoassays have been transferred for many years by the Agency to developing Member States. RIAs are attractive even for fragile infrastructures because they are usually cheap, robust and allow fast and accurate diagnosis of conditions such as hepatitis, insufficient growth hormones and hypothyroidism. Some RIAs can detect “tumour marker molecules” of certain types of cancer.

Mass screening of hypothyroidism in new born babies, which, if untreated, results in permanent physical and mental disability, has been made by RIA, followed by prompt treatment with thyroid hormones in some developing countries such as Tunisia and Thailand, by the IAEA assistance.

One application in nuclear medicine supported by the Agency is cancer treatment by unsealed radioactive sources where the Agency has been quite successful. Thus in the 90s one International Research Programme co-ordinated by the Agency developed and evaluated therapeutic radiopharmaceuticals for bone pain palliation; bone seeking agents labelled with Sm-153 and Re-188 were developed. They are now attaining widespread use and have clearly greatly improved cancer therapy.

The most comprehensive element of nuclear medicine is in vivo procedures by gamma camera imaging, including SPECT (single photon emission tomography). These procedures allow clinicians to detect and treat abnormalities, often long before changes can be revealed by other tests. Thus SPECT is being used to improve the diagnosis of Alzheimer and Parkinson diseases, cerebrovascular disease and even trauma. In the World of today, a total of 18,000 gamma cameras are in use. Only some 2000 of these are used in the developing countries.

In assisting developing countries, more than 150 existing analogue gamma cameras have been upgraded during the last couple of years by the IAEA’s personal computer-based nuclear medicine computer system. The IAEA assisted Argentina by providing equipment and experts for the Nuclear Medicine School in Mendoza.

3 Food and Agriculture

About 600 million people, mainly in Africa and Asia, suffer from chronical malnutrition and hunger. “Food for all” was the stated goal of the 1996 Declaration of the World Food Summit. To achieve this goal countries will need increased resources, suitable technologies and the strategies to use them effectively. The IAEA’s food and agriculture activities, operated jointly with FAO, are greatly contributing to this objective.

3.1 Isotope techniques for better soil management

Soil nutrients and water are essential to plant growth. Isotopes can play a crucial role in understanding soil-plant interaction and to improve soil management.

Biological nitrogen fixation is an essential process in growing legumes such as peas, beans, soybeans and alfalfa. N-15 is used to measure the efficiency of the nitrogen fixation. Many different kinds of rhizubian bacteria are known by scientists to stimulate the production of root modules which act as a nitrogen fixation catalyst from atmospheric nitrogen. Rhizobia is, therefore, called upon as a "biofertilizer" to increase the yield of grain legumes. In Zimbabwe, in co-operation with the IAEA, this technology is being used in some commercial farms. In field trial the yield of soybeans has been increased by a factor of 1.5 ~ 2.0. The biofertilizer is much cheaper than the imported chemical fertilizer and is also environmentally friendly. The IAEA is implementing a project to set up a demonstration plant in Bangladesh to commercially produce the biofertilizer. The N-15 isotope technique is essential in identifying the best suited strain of Rhizobia for specific grain legume crops. Biofertilizer is used in quite a few countries such as Thailand and Brazil.

3.2 Food Irradiation

Post harvest losses due to damage by insects and bacteria are estimated to be 25~40%. Irradiation reduces such losses, improves food hygiene and extends shelf-life without using chemical fumigation.

In the USA, for example, food derived pathogenic bacteria such as Salmonella, E.Coli and other parasites claim an estimated 9000 lives annually and from 24 to 81 million cases of diarrhoea. The related economic losses are enormous. Food irradiation has been recognized as an efficient way to decrease these food borne diseases.

Global production of irradiated food, while still small in volume, has been increasing steadily in the past five years. In about 30 countries there are one or more irradiated food products such as spices, potatoes, onions, herbs, garlic, strawberries, frozen shrimps, frozen chicken meat, etc. mainly for their own requirements. Irradiated spices and dried vegetable seasonings are traded widely among countries.

In international trade, irradiation is beginning to play an important role as a sanitary and phytosanitary treatment applied to meet trade requirements. Radiation provides an excellent method of ensuring the quality of more solid food, especially that of animal origin, certain fresh fruits and vegetables to be consumed raw, minimally processed food to be consumed without cooking, spices and dried vegetable seasonings. Co-60 is a major radiation source while electron accelerators are slowly being introduced.

3.3 Increasing production of meat and milk

Isotopes provide unique information on the nutrition of farm animals to find better ways of converting animal feeds into milk and meat. IAEA and FAO have helped to develop supplemental animal feeds, UMB (Urea-Molasses Block) that is rich in calories, nourishing and balanced in fostering growth, milk production, reproduction and work. UMB is manufactured from local products and labour. For example, in West Java, Indonesia, UMBs have increased milk production by 10-23% and cattle growth rates by up to 200%, leading to a doubling in farmers' incomes.

RIA is used to measure hormones; this determines when animals are ready for breeding, diagnoses pregnancy earlier and improves artificial insemination. In Mexico, by the assistance of IAEA and FAO, RIA has recently been introduced to improve the production of cows and milk. Milk production increased from 4.5 lit. to 7.2 lit per cow per day - equivalent to 10 million lit. of milk per country each year.

3.4 Insect and pest control by nuclear techniques

The sterile insect technique (SIT) using radiation from Co-60 to Cs-137 has been successfully used to control insect pest. In Chile the medfly has recently been eradicated by SIT with IAEA's support, yielding considerable economic and environmental benefits. The tsetse fly on Zanzibar island of Tanzania has been eradicated through the IAEA's project in 1997. A larger project having the same purpose in Ethiopia is under field study.

4 Fresh Water Resource Management

The global demand for fresh water is doubling every 20 years. Growing scarcity is compounded by pollution wherever fresh water is used for agriculture, industry and domestic purposes. Isotope hydrology provides an important tool for studying groundwater dynamics, including recharge rates which are crucial for improved water resource management.

In Damascus, Syria, when new wells were needed as population increased to more than two millions, the IAEA helped to investigate the groundwater recharge at proposed well sites by measuring ^{3}H at various depths. The investigation confirmed that the new reservoir is sufficiently large and recharges quickly enough to meet the growing needs. The IAEA has promoted isotope techniques for more than 30 years and supported projects in 63 countries by building up local capacities in isotope hydrology.

5 Demand and Supply of Radioisotopes

The OECD Nuclear Energy Agency (NEA), in co-operation with the IAEA, conducted a world-wide comprehensive survey of isotopes uses and production capabilities during 1997-1998, including an analysis of the trends in isotope demand

and supply in order to identify issues of interest to governments. The study focused on topics relevant to governments; the following conclusions were reached:

At present, most of the isotope production facilities are operated in OECD countries, which are also the main users. Demand is increasing in non-OECD countries, and their production capabilities may not increase as fast as their demand.

Isotope demand is evolving owing to the development of new applications. As far as the production is concerned, most production facilities require several years to be built and commissioned, so that it is essential to monitor the projected demand and the planned production capacities at world level in order to alleviate the risk of inadequate supply in the future.

- Recognizing the great potential of isotopes in their beneficial uses for medicine, industry and scientific applications, governments should consider policy measures to ensure adequate supply.

In terms of uses, it is recognized that medicine and industry are the two major sectors of applications. 30 million medical procedures involving the use of isotopes are carried out every year. Gamma imaging represents a turn-over of around US$ 1 billion in the world for a total of about 17 000 gamma cameras in use. The demand in this sector is estimated to grow by 5% per year. Co-60 radiotherapy represents an annual turn-over (in terms of the value of cobalt sources) of around US$ 35 million, but demand appears to be declining since Co-60 is being replaced by electron accelerators.

More quantitative information on the production capacities will be available at the end of the year 2000, after a survey to be conducted in 1999.

THE MAIN TENDENCIES OF MODERN RADIOCHEMISTRY DEVELOPMENT AND RADIONUCLIDES PRODUCTION IN RUSSIA

B.F. MYASOEDOV

V.I. Vernadsky Institute of Geochemistry and Analytical Chemistry, Russian Academy of Sciences,
Kosygin str.,19, 117975 Moscow, Russia
E-mail: bf@ncrc.msk.ru

Several periods in radiochemistry development history is observed. In the report the sources of radionuclide income into environment are examined including long-lived transuranic elements on the different stages of full nuclear fuel cycle. Radioactive substance contamination analysis is given for different regions of Russia from natural and man-caused sources. The problems of radioactive contamination of the territories around the radiochemical plant "Mayak" are discussed. Isotope production in Russia began more than 50 years ago in Biophysics Institute. Industrial production of radioisotopes has been started at “Mayak” Production Association and then at The Institute of Nuclear Reactors in Melikess. The main characteristics of isotopes produced in reactors, cyclotrons and also from fission products isolation are shown in. the report. At that special attention was given to isotopes of transuranic elements.

1 Introduction

A hundred years ago Henri Becquerel discovered the phenomenon of radioactivity. Seventy years are passed also from publishing of the pioneer work by George von Hevesy and Hilde Levi opened using of activation analysis.

In the history of radiochemistry it is possible to distinguish few periods [3]. The first period of radiochemistry formation as an independent scientific direction, was closely related with the discovery of new radioactive elements, understanding of the main laws of radioactive substance behavior.

During the second period (40-50-th years) the radiochemical investigations were focused on the practical utilization of nuclear energy, studying of the chemical properties of artificially obtained elements, development of technology of processing of the irradiated nuclear fuel, resolving of the problem of radioactive wastes burial.

The intensive development of activation analysis in the 50-th years also was definitely stimulated by needs in qualitatively new materials for nuclear technology. Later years the method begins widely to be applied for the analysis of environmental samples, for investigations in medicine and biology. Therefore, one can trace the similar tendencies in priorities like in the case of radiochemistry: the exclusive work for purposes of nuclear technology is replaced by maintenance of diverse needs of the society.

From the 60-th the society step by step began realize the global character of consequences of the contemporary activity of mankind. The main attention shifted to the problems of maintenance of the sustainable development, including such aspects as: remediation of the polluted territories, study of the radionuclides behavior in nature, reduction of the amount of unavoidable (for the current nuclear technology) radioactive wastes, development of technology of long-term radioactive wastes storage and many other things.

Today radioactive isotopes are less often used for the elements determination, but they find more and more wide application in medicine, environmental science, biology for the purposes of diagnostics, elements speciation and migration, studying of fine biochemical processes.

Environmental problems are known to be among the most important issues of humankind. Solutions to these problems in Russia are considerably aggravated by heavy nuclear pollution in many regions of the country.

2 The main sources of nuclear pollution in some regions of Russia

2.1 Nuclear tests

In the USSR, nuclear tests in the atmosphere were carried out from 1949 to 1962 in the Semipalatinsk area (124 explosions with a total capacity of 6,3 Mt releasing 0,1MCi ^{90}Sr, and 0,2MCi ^{137}Cs), and on the North Land (87 explosions, including 3 submarine ones with a total capacity of 250Mt releasing 4,2 and 8,0MCi of ^{90}Sr and ^{137}Cs, respectively).

In addition, Russia performed 81 commercial nuclear explosions, including those aimed at intensifying oil production, and for seismic surveying, and construction.

The tests of nuclear weapons have resulted in the emission into the environment of a considerable amount of uranium, plutonium, and their fission products, which were distributed as aerosols and gases across large distances and produced the anthropogenic radioactive background on the earth's surface, especially in the northern hemisphere.

The density of pollution (Bq/m^2) with the most dangerous radionuclide, plutonium, varies in different countries as follows [4]: Germany, 67 -- 148; Ireland, 33 -- 127; Great Britain, 33-122; Russia: Leningrad Nuclear Power Plant region, 14 -- 262, Beloyarka Nuclear Power Plant region, 116 -- 183; Mayak, up to 1400; Ukraine: Chernobyl Nuclear Power Plant (30-km zone), 3700; and Japan, 90.

2.2 Nuclear power plants

The commercial use of atomic energy was launched in 1954 by the installation of the first 5-MW nuclear power station in Obninsk. By the end of 1998, 29 nuclear power units of Russia's nine nuclear power plants with an established electric capacity of 21242 MW accounted for 11,8% of total Russia's electrical energy supply. The work of a nuclear power plant is environmentally less hazardous than that of a power station that uses natural coal. Nuclear power plants release a certain amount of long-lived radionuclides (^{85}Kr, ^{3}H, ^{14}C, and ^{129}I) into the atmosphere. Overall, the contribution of nuclear power plants to the dose of natural background radiation was less than 0.1% in the early 1980s [1] and will be less than 1% in 2000, if the expectations of its development are justified.

The main problems in the development of nuclear energy are to ensure its safety and to dispose of the high-level radioactive waste safely. In this context, potential use of the thorium fuel cycle with fluoride melts, utilization of highly enriched (up to 99,9%) uranium-235, new technological approaches to the reprocessing of spent nuclear fuel, and transmutations of nuclear waste are now discussed.

2.3 Plutonium production for military purposes and reprocessing of irradiated nuclear fuel

The Production Association "Mayak" was established in late 40's for production of plutonium for military purposes and for processing fissile materials to achieve a parity in the nuclear arms race [5]. Regretfully, at the early stage of running in a number of unique and complicated technologies some serious technical mistakes were made. Significant territories around "Mayak" have undergone radioactive contamination as result of the 1957 and 1967 accidents and technological discharges into atmosphere. Discharges of liquid wastes to the river Techa at early stages of "Mayak" operation have resulted in contamination of its floodplane and radiation exposure of the local population inhabiting the shore area. At total a huge amount of radioactive wastes have been accumulated and stored "Mayak" (about 1 billion Curies), which represent a great ecological hazard.

Nowadays, the reactor division of "Mayak" includes two operating reactors producing radionuclides both for military purpose and general applications and five uranium graphite reactors, which were shut down in 1987 -- 1991. Production of the weapon grade plutonium at "Mayak" was stopped in 1987. The new plant operation started in 1977 and since then the staff of it has been involved in spent fuel reprocessing from different power reactors, as well as from transport and research reactors. For the whole operation period 2380 tons of spent fuel was received from a number of domestic and the foreign power plants. The plant comprises a spent fuel storage pool, three chopping-dissolution process lines, a PUREX extraction with separated plutonium, technetium streams output. In January 1996 the waste partitioning facility has been put into operation after a prolonged preparatory period. Currently the work is under way on development of the cementation and bituminization facilities for the medium level wastes utilization.

3 Radionuclides production in Russia

Isotopes production in Russia began more than 50 years ago at the Biophysics Institute. Industrial production of radioisotopes has been started at Production Assoiciation "Mayak" and then at The Institute of Atomic Reactors in Melikess. They are leaders among rodioisotopes producers in Russia.

The manufacturing of radionuclides for sources and preparations at Production Association "Mayak" goes back into 60s, at first on industrial uranium-graphite reactors and heavy-water reactors simultaneously with plutonium production. After interruption of the work of these reactors isotope production now continued on heavy-water reactor L-2 and light-water reactor "Ruslan".

The production of transplutonium elements (TPE) and technological items on their base as well as investigations of isolation, separation and purification technology of these elements, their individual physico-chemical properties and properties of their compounds are carried out at special research structure organized in The Research Institute of Atomic Reactors (RIAR). RIAR provides other scientific centers of Russia for preparations of ^{242}Pu, ^{243}Am, ^{244}Cm, ^{248}Cm, ^{249}Bk, ^{249}Cf, ^{252}Cf. The technology of manufacturing of ionizing radiation sources on the base of TPE so as gamma- (^{241}Am, ^{243}Am), alpha- (^{244}Cm), fission- ($^{249,252}Cf$), neutron- (^{248}Cm,^{252}Cf) ones have been developed. These sources are applied in various branches of industry and science: in geology, medicine, analytical control and so on.

The original technology for obtaining of radionuclides of high specific activity are elaborated at various organizations in Russia (Table.1). All their products are in wide demand in the world market.

Table 1. Preparations of high radioactive nuclides producing by The Research Institute of Atomic Reactors [2].

Radionuclide	Specific activity, Ci/g	Chemical compound
^{63}Ni	10-12	Nickel chloride
^{55}Fe	55-60	Ferrous (III) chloride
^{59}Fe	80-100	Ferrous (III) citrate
^{188}W	4-5	Sodium tungstate
^{89}Sr	>300 (without carriers)	Strontium chloride
^{51}Cr	>500	Sodium chromate, chromic chloride
^{113}Sn	30-50	Stannic (IV) chloride
^{119m}Sn	0,5-0,7	Stannic (IV) chloride
^{109}Cd	4-5	Cadmium chloride
^{54}Mn	without carriers	Manganese chloride

Reserves of “aged” for 40 years Ra-226 available to Russian producers of isotopes make possible daughter Pb-210 in amount satisfying the requirements of the world economy. Technique elaborated in The V.G.Khlopin Radium Institute enables to remove 99,95% of radioactive impurities from the purposeful product.

Russian producers along with the biggest foreign companies occupy a prominent place in the production of such isotopes as: carbon-14, phosphorus-33, cobalt-57, iridium-192, cesium-137, gadolinium-153 e.c. Radiopharmacentics on the base of isotopes ^{133}Xe, ^{201}Tl, ^{67}Ga are discharged for diagnosis of various illness. Radiopharmacentics with other isotopes are elaborated as well. The technology of creation, isolation and deep purification of ^{125}I, ^{153}Sm,^{127}Xe is made. At The Physico-energetic Institute main advantages of fast neutron reactors for production ^{32}P, ^{33}P, ^{35}S, ^{89}Sr under reaction (n,p), ^{117m}Sn under reaction (n,n’) are shown. They are: high output and radiochemical purity. Isotopes – ^{32}P, ^{89}Sr and ^{117m}Sn are mainly applied for therapy of oncological illness. The production of bioactive compounds with ^{32}P, ^{33}P, ^{35}S tracers is realized. Further investigations on development of nomenclature of preparation are continued. Isotopes ^{106}Ru, ^{90}Sr, ^{137}Cs, ^{241}Am are isolated from fuel rod. They are used as needle-type sources and applicators for medical exposure of limited surface or individual organs.

Fundamental studies on the mechanisms of chemical reactions between organic compounds and hydrogen isotopes made it possible to develop the methodology and techniques to obtain physiologically active molecules labeled with tritium or deuterium. (The Institute of Molecular Genetics of the Russian Academy of Sciences). Over 200 tritium labeled preparations made in the laboratory are provided with standard documentation. During the last five years these preparations were delivered abroad (to IICH) on a regular basis.

References

1. Babaev N.S., Dermin V.F., Il'in L.A., et al. In Nuclear Energy: Hurnans and the Environment, (Moscow: Atomizdat, 1984) pp.1-25.
2. Karelin Ye.A., Gordeev Ya.N., Filimona V.T. et al. *Appl. Radiat. Isoi.*, **48** (1997) pp.1585-1589.
3. Myasoedov B.F. Problem of nuclear pollution in some regions of Russia. *Environmental Science* **1** (1998) pp.3-13.
4. Pavlotskaya F.I., Fedorova Z.M., Emel'yanov V.V. et al., Plutonium Abundance in Soils of the Soviet Union, *At. Energ.*, **59** (1985), pp. 381-384.
5. Fetisov V.I. *Radiat. Safety Prob.* **1** (1996) p.5.

Session B

Isotope Facilities and Programs

AECL'S MAPLE RESEARCH AND ISOTOPE REACTORS

J-.P. LABRIE

General Manager, Research and Isotope Reactor Business, AECL, 2251 Speakman Drive, Mississauga, ON L5K 1B2 Canada E-mail: labriej@aecl.ca

P. GIROUARD

Engineering Manager., MMIR[1] Project, AECL, 2251 Speakman Drive, Mississauga, ON, L5K 1B2 Canada E-mail: girouardp@aecl.ca

A. LEE

Analysis Manager, MMIR Project, AECL, 2251 Speakman Drive, Mississauga, ON L5K 1B2 Canada E-mail: leea@aecl.ca

Atomic Energy of Canada Limited (AECL) develops, designs and markets CANDU power reactors, MAPLE research reactors and MACSTOR waste storage facilities and manages the construction of nuclear reactor projects worldwide.

AECL is also the main producer of medical isotopes, including Molybdenum-99 (Mo-99), Iodine-125, Iodine-131 and Xenon-133, for the global market. The National Research Universal (NRU) reactor at Chalk Rivers Laboratories (CRL) produces approximately two thirds of the world's market demand for Mo-99 - the isotope most widely used in hospitals and clinics to diagnose diseases.

AECL is currently building and commissioning for MDS Nordion, two new MAPLE reactors at CRL that will be dedicated to producing medical isotopes and secure the long-term supply of isotopes. The start-up for the first MAPLE reactor is scheduled for 1999 and start-up of the second reactor for the year 2000.

MAPLE reactors are among the most advanced multipurpose research reactors available today. The reactors are an open-tank-type reactor assembly within a light-water-filled pool based on AECL's well-established MAPLE technology. The presentation will highlight the features of AECL's MAPLE research and isotope reactors.

The first demonstration of MAPLE technology is Korea's HANARO 30 MW_{th} research reactor, which started up in 1995. Two MAPLE 10 MW_{th} isotope reactors are being completed at CRL for MDS Nordion. The National Research Council of Canada (NRC) and AECL - in partnership with universities and industry - are jointly proposing a new Canadian Neutron Facility, a 40 MW_{th} MAPLE research reactor, to support next-generation neutron-based materials research and innovation in Canada for the twenty-first century.

---

1 MMIR - MDS Nordion Medical Isotope Reactor

1 INTRODUCTION

Atomic Energy of Canada Limited (AECL) was established in 1952 as a Crown Corporation and has since developed into a global corporation. The flagship product, the CANDU reactor, supplies about 15 percent of Canada's electricity and is an important component of energy programs on four continents. Twenty-two CANDU reactors have been built in Canada and twelve overseas.

In addition to power reactors, AECL has extensive experience with the engineering, design, construction and operation of research and isotope reactors and associated facilities. The following table lists the non-power reactors designed by AECL.

Table 1. Non-power Reactors Designed by AECL

Reactor	Date in Service	Power
ZEEP	1945	10 W_{th}
NRX	1947	42 MW_{th}
PTR	1957	10 kW_{th}
NRU	1957	200 MW_{th}
ZED-2	1960	250 W_{th}
WR-1	1965	60 MW_{th}
SLOW POKE-2	since 1968 (8 produced)	20 kW_{th}
HANARO	1995	30 MW_{th}
MAPLE 1 & 2	1999 - 2000	10 MW_{th} each
CNF	proposed	40 MW_{th}

AECL is also the main producer of medical isotopes for the global market. The NRU reactor at AECL's Chalk Rivers Laboratories (CRL) currently produces about two thirds of the world's market demand for Molybdenum-99 which is the most effective radioisotope for diagnosis and treatment of patients with life-threatening illnesses, including cancer, lung disease, heart disease and kidney disease. More than 50,000 medical diagnostic procedures are performed daily using Technetium-99m, the daughter nucleus of Molybdenum-99, produced in the NRU reactor.

The NRU reactor also produces about 80% of the Cobalt-60 used in therapy machines. Cobalt-60 used in industrial irradiators for applications such as medical product sterilization is produced in CANDU reactors.

2 RESEARCH AND ISOTOPE REACTORS DEMANDS

Research and isotope reactors are a key element for developing nuclear technologies contributing to improving health care, developing industry and national economies. In Canada, the source of neutrons that research reactors provide has allowed:

- duplicating conditions in power reactors and therefore supports the CANDU technology
- molecular structure-neutron scattering supporting the research into new materials
- neutron activation for material analysis
- neutron radiography
- radioisotope production for industrial and medical applications
- cancer treatment
- training of the next generation of nuclear experts.

The same needs exist globally. There are some 318 Research Reactors, of which 45 operate above 10 MW_{th}. Of these 33 are over 30 years old. Many will need to be replaced over the next decade. AECL has the experience and capability to design, build and operate, research and isotope reactors and associated facilities meetings global demand.

3 MAPLE REACTOR TECHNOLOGY

MAPLE reactors are among the most advanced multipurpose research reactors available today. The reactors are an open-tank-type reactor assembly within a light-water-filed pool based on AECL's well-established MAPLE technology.

MAPLE reactors can be designed for various operating power levels, core configurations, beam port arrangements, hot and cold neutron sources and isotope production capabilities. All are pool type, light water moderated and cooled reactor. The MAPLE reactor core is located above a grid plate and inlet plenum structure at the base of the pool.

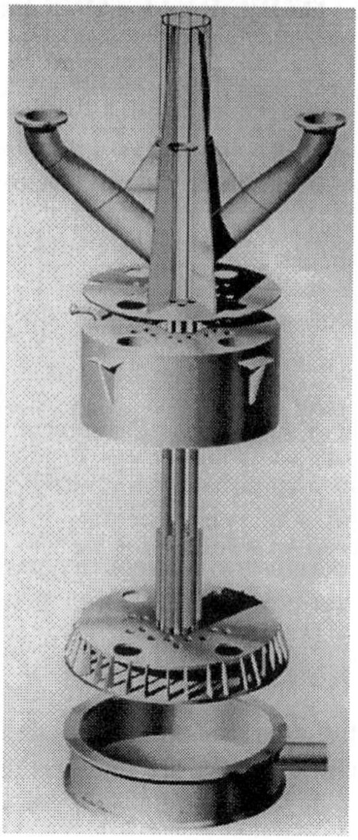

Figure 1: Typical MAPLE configuration

The core consists of low enrichment uranium (LEU) driver fuel bundles accessible via an open chimney. The core is surrounded by an annular reflector tank filled with heavy water that also contains a number of vertical tubes for isotope irradiation.

The core is cooled by low-temperature and low-pressure (~30°C and <0.4 MPa) light water pumped through the core. The primary coolant is piped from a heat exchanger located outside the pool, down to the inlet plenum, where it is directed upward through the grid plate, the flow tubes and the fuel, and exits into the chimney. Two nozzles attached to opposite sides of the chimney conduct the coolant into the core outlet piping leading to the pump inlet. Approximately 10% of the primary flow is released from the core inlet piping at the bottom of the pool. This bypass flow returns down the chimney to provide suppression of the core outlet flow to the lower part of the chimney. Decay heat removal is by natural circulation of the pool water through the core. Heat is removed from the heavy water in the reflector tank via the Reflector Cooling System. Pool water chemistry and purity are controlled by the Pool Purification and the Skimming Systems.

The reactor power level is controlled using Control Absorber Rods (CARs), consisting of annular hafnium absorbers, which slide over the extended flow tubes in the outer ring of the core. The CARs are controlled via an advanced dual redundant computer system. This allows automated computer control from shutdown to full power.

The core is under-moderated and has a negative reactivity coefficient. There are also two separate, independent and diverse shutdown systems for added safety.

The core design offers the maximum flexibility available and is easily accessible from the top. Its small size permits achieving higher neutron fluxes per unit of power then other multipurpose reactors. High burn-ups can be achieved resulting in low fuel cost. A high ratio of peak thermal flux in the reflector versus the total power results from the use of a heavy water reflector.

The pool offers a reliable heat sink in all cases. As the pool is below grade it cannot fail. The control and safety systems meet or exceed all international safety guides, i.e., International Atomic Energy Agency Safety Requirements for Research Reactors. Only a modest staff is required for operation.

4 MAPLE REACTOR EXPERIENCE

The first demonstration of MAPLE technology is Korea's HANARO 30 MW_{th} research reactor, which started up in 1995. Two MAPLE 10 MW_{th} isotope reactors are being completed at CRL for MDS Nordion (MMIR project). The National Research Council of Canada (NRC) and AECL - in partnership with universities and industry - are jointly proposing a new Canadian Neutron Facility (CNF), a 40 MW_{th} MAPLE research reactor.

The NRU reactor began operations in 1957, and reliance on the reactor for the majority of the supply of medical isotopes has been of concern in the nuclear medicine community. MDS Nordion contracted AECL in 1996 to design, build and operate two new MAPLE reactors and an isotope processing facility which are dedicated to producing Mo-99, I-125, I-131 and Xe-133 and will ensure a secure supply of medical isotopes well into the next century. Construction of these new facilities is currently being completed and their commissioning is underway. The MMIR Project started in September 1996. The environmental assessment was completed in April 1997, Construction Approval was granted in December 1997 and the Atomic Energy Control Board awarded an Operating License in August 1999. The start-up for the first MAPLE reactor is scheduled for 1999 and start-up of the second reactor for the year 2000.

NRU is both a source of radioisotopes and a research tool for fuel, fuel channel development and CANDU research. A new reactor called the Canadian Neutron Facility (CNF) is being proposed to support next-generation, neutron-based materials research and innovation in Canada. The purpose of the CNF is:

- to provide an advanced materials research capability to meet the needs of Canadian universities and industry

- to provide an essential testing facility to advance the CANDU power reactor design and ensure the future competitiveness of the Canadian nuclear industry

Construction of the CNF is planned to begin in Year 2000, with projected start-up in 2005.

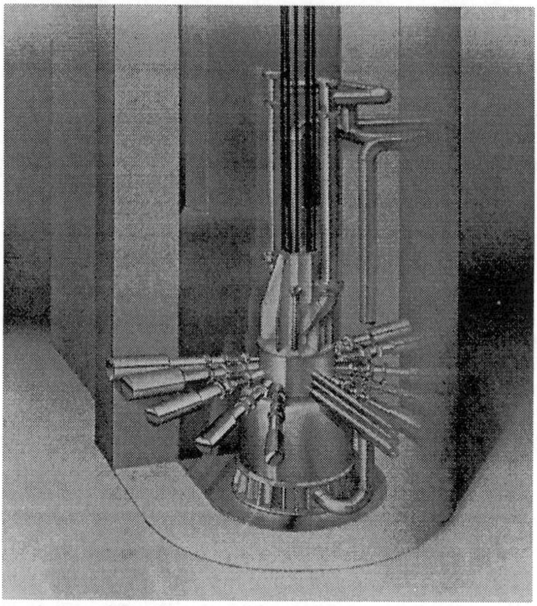

Figure 2: CNF, experimental facility and beam ports.

Test loops will traverse the CNF reactor core. The reactor will have extensive neutron beam facilities. One beam tube will be equipped with a cold neutron source and six cold neutron guides, five cold neutron guides to deliver cold neutrons to an adjacent guide hall and one cold neutron guide to deliver cold neutrons to an instrument within the reactor building. One beam tube will also be equipped with two thermal neutron guides to deliver neutrons to instruments in the guide hall. Six other beam tubes with thermal guides will deliver thermal neutrons to instruments within the reactor building. There will also be two hydraulic rabbit systems and four pneumatic rabbit systems to transfer irradiation capsules between the reactor and a shielded facility.

5 CONCLUSION

AECL's MAPLE reactors are the most advanced multipurpose research reactors currently available. AECL is one of the few suppliers actively designing, operating and selling research reactors and associated facilities. AECL with CANDU power reactors in operation in Asia, Europe, South America and North America and MAPLE reactors in Asia and North America has the tools and capabilities to address global demands for the installation of new research reactors, their operation, maintenance and service.

THE DEVELOPMENT AND PROGRESS OF RADIOPHARMACEUTICALS AT CIAE

HONG-YAN ZHANG, JIN-RONG ZHANG AND YOU-FENG HE

National Research Center of Isotope Engineering and Technology, Isotope Department, China Institute of Atomic Energy (CIAE), P.O.Box 275-12, Beijing 102413, P.R.China

This paper intends to present the current status of development and progress of radiopharmaceuticals at CIAE, P.R.China. As the main radiopharmaceutical center in China, CIAE produces and supplies radiopharmaceuticals for both in vivo and in vitro applications, including $^{99}Tc^{m}$ generator and $^{99}Tc^{m}$ labelled radiopharmaceuticals, reactor and cyclotron–produced radiopharmaceuticals, radioimmunoassay kits and some special radiation sources for therapy. Some future programs will be also described.

China Institute of Atomic Energy (CIAE), as a multidisciplinary institute and the main radiopharmaceutical center in China, situated some 40 Km southwest of Beijing, started producing radioisotopes after the first nuclear reactor was put into operation in 1958[1]. The technology for producing some types of radiopharmaceuticals as one of the major applications of radioisotopes, was developed in the late 1960's, such as ^{131}I from irradiated tellurium, ^{198}Au in form of colloid, and ^{32}P. Research and development work in the field of radioimmunoassay (RIA) started in the mid 1970's. Since then the large scale production of radiopharmaceuticals steadily increased both in quantity and types. For the development and production of radiopharmaceuticals, CIAE has been equipped with two reactors (10 MW heavy water research reactor and 3.5MW light water swimming pool reactor), one cyclotron (Cyclone-30), hot cells and other related facilities as well. The Isotope Department at CIAE produces and supplies radiopharmaceuticals for both in vivo and in vitro applications, including. $^{99}Tc^{m}$ generator and $^{99}Tc^{m}$ based radiopharmaceuticals, cyclotron-produced radiopharmaceuttcals, radiopharmaceuticals for therapy, radioimmunoassay reagents, some special radiation sources for therapy etc.

In view of the fact that more than 80% of the radiopharmaceuticals used in nuclear medicine are $^{99}Tc^{m}$ radiopharmaceuticals, we have therefore placed our emphasis on the development of ^{99}Mo extraction, $^{99}Tc^{m}$ generator and $^{99}Tc^{m}$ based radiopharmaceuticals. CIAE developed technology for producing ^{99}Mo by neutron capture of natural molybdenum oxide in 1970's. Since the production capability of that technology was not enough to meet with domestic needs, we decided to develop new technology for extracting ^{99}Mo from the uranium fission products. By now, CIAE has developed the technology of ^{99}Mo extraction from irradiated highly enriched uranium target, established $^{99}Tc^{m}$ generator production line and supplied to market over 10 different kinds of cold kits based on $^{99}Tc^{m}$, such as MIBI, HMPAO, DTPA-HSA, MDP, DTPA, PYP, ECD etc. To promote nuclear medicine in China, CIAE has set up regional radiopharmaceutical centers in Beijing, Shanghai, Guangzhou and Shenyang to provide ready-to-use $^{99}Tc^{m}$ labelled radiopharmaceuticals for hospitals in the respective regions, and more centers will be established in other regions in China.

Although there are several accelerators at CIAE, those are especially dedicated to fundamental research. The construction of Cyclone-30 was completed by the end of 1994 and made it possible to Cyclotron-produced radiopharmaceuticals on a large scale. Soon after the operation of Cyclone-30, CIAE offered many kinds of products to meet the demands for ^{201}Tl, ^{111}In, ^{67}Ga, ^{57}Co, ^{68}Ge and ^{18}FDG which is an important reagent for PET users. A plan has been made to develop ^{123}I and ^{123}I labelled radiopharmaceuticals.

Among the applications of radiopharmaceuticals in nuclear medicine, the therapeutic use of radionuclides is very important. Apart from some conventional radiotherapeuticals, such as ^{131}I (sodium iodine in capsule and solution), ^{131}I-MIBG and ^{32}P etc, the work in recent years is in progress to extend the range of radiotherapeuticals based on ^{153}Sm, $^{186,188}Re$ and ^{90}Y. CIAE has developed a radiotherapeutical agent ^{153}Sm-EDTMP for the treatment of pain associated with metastic bone cancer, which is now in the final stage of human clinical trial in China. Meanwhile a type of ^{90}Y glass microspheres was developed as a therapeutic agent for tumor. The yttrium-doped glass microspheres 30-50μm in diameter can be activated by neutron in the reactor and then concentrated in liver tumor after injection via intrahepatic artery catheter. The preliminary clinical results have shown good promise.

Due to the easy availability of ready-to-use RIA kits, the relatively simple instrumentation needed and the comparatively low cost, there is a fast and steady growth of RIA laboratories. Radioimmunoassays are widely used in China now. The laboratory of CIAE is the earliest and the biggest for the development, production and supply of RIA reagents in China [2]. A relatively complete system has been established for the development and preparations of RIA and Immunoradiometric assay (IRMA) reagents, including antigen extraction, binding preparations of polyclonal/monoclonal antibodies; solid phase separation techniques; tracer preparation and purification; methodologies and quality control [2]. There are more than 70 kinds of RIA kits in routine production, including those for thyroid function and newborn screening, tumor markers, renal function, the diabetes-related, fertility and pregnancy monitoring and diagnosis of virus hepatitis, etc.

CIAE has developed the technology of powder metallurgy to manufacture a variety of $^{90}Sr/^{90}Y$ surface applicators and $^{90}Sr/^{90}Y$ prostate hyperplasia applicator [3]. The clinical results indicate that the $^{90}Sr/^{90}Y$ prostate hyperplasia applicator is an effective and safe method for intracavitary therapy of benign prostate hyperplasia.

The Isotope Department at CIAE has finished two IAEA technical cooperation projects on " Fission ^{99}Mo production technology" and " Radiopharmaceuticals for cancer therapy" . Under these projects we received fellowships, expert missions and equipment (multichannel pulse height analyzer, liquid scintillation counter and HPLC system etc) from IAEA in the last few years. Some other research contracts are awarded to us. The support from IAEA is very effective and beneficial in strengthening our facilities and updating processing technology related to the development of radiopharmacenticals. We are very grateful for all the assistance from the Agency. On the other hand, the Isotope Department of CIAE also hosted training courses, expert workshops, scientific visits and fellowships from other countries. Moreover, our technical support and the related reagents are supplied to more than 20 Asia and Latin–American countries under the IAEA projects: RAS/6/018 and RLA/6/028.

In view of recent advances in molecular biology and the development of synthetic peptides and hormones, which provide new concepts for the design of potential high affinity biological carriers, the major emphasis now placed on the research and development of radiolabelled peptides and receptor imaging agents. Some receptor-positive tumor imaging agents, octreotide, octreotide derivative and VIP labeled with ^{125}I, ^{111}In and Tc-99m, have been studied at CIAE, and development of agents for imaging central neural system receptor based on ^{99}Tc^{m} is also in progress.

REFERENCES

1. XIAO L., Production and Application of Isotopes in China, J. Radioanal. and Nucl. Chem. **205**(1996) pp.9-20.
2. He Y. F., Research and Development Activities on Radioimmunoassay at CIAE. J. Radioanal. and Nucl. Chem. **206**(1996) pp.185-188.
3. Cai S.Y., Li Z. and Chen F., Sr-90/Y-90 Prostate Hyperplasia Applicators and its Clinical Effects. 10th Pacific Basin Nuclear Conference, Kobe, Japan, pp.20-25 October 1996.

AUSTRALIAN REPLACEMENT RESEARCH REACTOR

HELEN M. GARNETT
Australian Nuclear Science and Technology Organisation,
Private Mail Bag 1 Menai, NSW 2234 Australia
E-mail: hmg@ansto.gov.au

1 Introduction

On 3 September 1997, the Australian Government announced that, subject to the outcomes of an environmental assessment process, it would fund the construction of a research reactor to replace the Australian Nuclear Science and Technology Organisation's (ANSTO) existing HIFAR research reactor. On 26 August 1999, Parliament accepted the report of the Public Works Committee clearing the way for ANSTO to proceed to tender.

HIFAR is not capable of satisfying the growing demand for medical radiopharmaceuticals beyond 2004/2005. Its neutron beams are low in flux and its beam-lines are radial to the core giving excessive levels of gamma radiation. These factors seriously restrict the range of research projects. Moreover, HIFAR does not have a cold-source to produce the slow neutron beams required for investigating the structure of biological and other materials containing light elements.

The replacement reactor will be multi-purpose: producing isotopes for medical and industrial applications; providing neutron beams for scientific research and industrial applications; and providing irradiation services including silicon transmutation doping and neutron activation analysis. The state-of-the-art facility, to be commissioned in 2005, will keep Australia virtually self-sufficient in nuclear medicines and permit the development of new therapeutic and diagnostic substances in association with national and international teams. Demand for radiopharmaceuticals is predicted to double in the next five years and it is expected that almost every Australian will require a nuclear medical procedure in their lifetime. The replacement reactor, complemented by ANSTO's National Medical Cyclotron facility, will enable this demand to be met. In addition, the replacement reactor will support ANSTO's nationally important work in such areas as environmental studies, agriculture and in assisting industry. It will support leading-edge scientific research and provide students and researchers from universities in Australia and New Zealand with improved access to a modern, versatile neutron source.

2 Replacement Research Reactor

The main characteristics of the replacement reactor will be:

- it will be a pool type in which the core of the reactor is at the bottom of a pool of water at least six metres deep;
- it will produce a higher neutron flux than available from HIFAR;
- neutron beams for research experiments will have a wider range of energies and be cleaner and of higher intensity than HIFAR;
- it will use low-enriched uranium fuel;
- it will have flexible core and irradiation facilities;
- its isotope production and other irradiation facilities will have greater capacity than HIFAR and be designed to meet current and future needs.

The replacement reactor will be built on a site adjacent to HIFAR.

3 Radioisotope Production Capabilities

ANSTO has been providing radiopharmaceuticals to the Australian health and medical sector for over 35 years. ANSTO currently produces about 350,000 patient doses of reactor-produced radiopharmaceuticals annually (mainly molybdenum-99 for generation of technetium-99m). ANSTO routinely produces iodine-131, Quadramet (a samarium-153 labelled agent for the relief of pain caused by metastatic bone cancer) and both yttrium-90 and holmium-166 containing microspheres. Tetley Medical Limited, a private company based at the ANSTO Business and Technology Park, has developed and marketed worldwide the technetium-99m based agent 'Technegas' widely regarded as the world's best system for lung ventilation diagnostic imaging. In addition, ANSTO produces a range of industrial radioisotopes including tracers (gold-198, copper-64, chromium-51, iridium-192, bromium-82, cobalt-60 and sodium-24). The production of these industrial radioisotopes enables ANSTO to provide a range of services to industry. Examples include the investigation of the release of sewage from ocean outfalls in Sydney and Hong Kong.

4 Environmental Impact Assessment

As part of the stringent environmental assessment process, the project's Draft Environmental Impact Statement (EIS) was issued for a 3-month public comment period in August 1998. Some 935 submissions were received and a Supplement to the Draft Environmental Impact Statement, which addressed the comments and issues raised in those submissions, was prepared and lodged with Environment Australia on 18 January 1999.

Of the 935 submissions received, 776 were pro-forma submissions and a further 50 were either pro-forma based or repeated issues contained in the proformas. Substantive submissions included technical submissions from the local Sutherland Shire Council, from Greenpeace Australia, from the three Peer Review Consultants and from the NSW Government (via the Department of Urban Affairs and Planning). In all, 205 distinct issues requiring responses were extracted by the submission process. They ranged from simple editorial comments to discussion of detailed technical issues. The most frequently raised issues included: concern about a Chernobyl type incident; concern about the effect on the ozone layer; concern about the effects on health; belief that reactor funding could be better utilised; concern that cyclotron production of isotopes was not sufficiently addressed in the Draft EIS; and concerns that no solution or site exists for the disposal of radioactive waste in Australia.

In addition, three separate independent reviews were commissioned by the Minister for the Environment to assure the public that the process was rigorous and complied with world best practice.

- The firm CH2M HILL reviewed the safety and handling of nuclear wastes and emissions in accordance with international best practice.

- The IAEA conducted a hazard and risk analysis of the EIS, looking at the siting of the facility and possible risks to the community.

- Parkman Safety Management provided a further independent review of hazards and risks.

Environment Australia prepared an assessment of the documents and submissions for the Minister for the Environment and Heritage, who is responsible for the decision from this process. The Minister subsequently decided that there were no environmental reasons, including on safety, health, hazard or risk grounds, to prevent construction of a replacement reactor at Lucas Heights, but recommended that the proposal be implemented in accordance with twenty nine specified conditions. These recommendations were aimed at ensuring that the replacement reactor is built and operated in accordance with best international practice and covered the construction and operation of the replacement reactor, management of wastes, monitoring and containment of site emissions, management of hazards and risks, emergency management plans, and community consultation. In early-May, the Minister for Industry, Science and Resources announced that he accepted the Minister for the Environment's recommendations.

Like all large-scale public works in Australia, the replacement reactor was referred to the Parliamentary Public Works Committee. The Public Works Committee's report was accepted by Parliament on 26 August 1999.

5 Tender Process

A two-stage tender process is being used. Advertisements were placed in the worldwide press in June/July 1998 seeking expressions of interest from reactor vendors. Stage one of the tender selection process was completed in December 1998 with the prequalification of four consortia led by AECL, INVAP, Siemens and Technicatome respectively. We are now in the second stage and these prequalified reactor vendors have been invited to tender for the design and construction contract in their own right or as members of joint ventures.

An application for a site licence for the replacement research reactor facility has been submitted to the Australian Radiation Protection and Nuclear Safety Agency (ARPANSA), which is the new regulatory body established at the beginning of this year. ARPANSA is responsible for nuclear regulatory issues relating to the replacement reactor project. In particular, the construction will require separate authorisation from ARPANSA, and in addition ARPANSA will oversee the implementation of the twenty nine recommended environmental conditions.

Design is expected to commence in 2000 with construction commencing in 2001 and with commissioning being completed during 2005. It will have an operational life of at least 40 years. HIFAR and the replacement reactor will operate simultaneously for a short period of up to about six months during the commissioning of the replacement reactor. Once the replacement reactor is operational, HIFAR will be shut down and decommissioned.

PORTRAIT OF ISAC
THE NEW RADIOACTIVE BEAM FACILITY AT TRIUMF

J.-M. POUTISSOU

TRIUMF, 4004 Wesbrook Mall, Vancouver, B.C. Canada V6T 2A3
E-mail: jmp@triumf.ca

TRIUMF is building a new radioactive beam facility of the ISOL type, ISAC; beams of many radioisotopes at low energy (60 keV) have been available since November 1998, while beams with energies up to 1.5 MeV/u will be available by the fall of 2000. With further energy upgrades anticipated in the next five-year plan, Canada is positioning itself as a leader in the next generation of radioactive beam facilities. A general description of the facilities being built, will follow.

1 The Facilities

TRIUMF is constructing a radioactive beam facility, ISAC, which will provide intense beams of accelerated isotope beam by the fall of year 2000 with energies up to 1.5 MeV/u with a possible extension towards higher energies (6.5 MeV/u) in the future. The artist's conceptual view of the new ISAC facility at TRIUMF is presented in the attached Fig. 1.

ISAC will produce radioactive beams by the ISOL method using very intense beams (up to 100 μA) of protons at 500 MeV to bombard thick production targets. Although, initially, the proton current will be limited by the production targets, the facility is being constructed with a capability to handle proton beam intensities on target of up to 100 μA.

Protons from the TRIUMF main cyclotron have been extracted into a new beam line (2A) which feeds the new ISAC buildings. The new extraction system has been tested and ion beams were delivered to the new ISAC experimental hall.

In the target hall, two target stations will be receiving the proton beam alternatively to increase the efficiency of ion beams delivery. Typically, such targets present up to 20 g/cm^2 to the proton beam and are maintained at high temperature to release the products of fission, spallation or fragmentation reactions to an ion source. As the proton beam intensity will be increased, the mode of operation will change from keeping the operating temperature by supplying extra heat to removing the beam induced heating by cooling.

The target is coupled to the ion source via a small transfer tube. Several types of ion sources will be available; initially surface ionization is used to produce beams of alkali elements. Beam intensities of such alkalis (Li, Na, K,

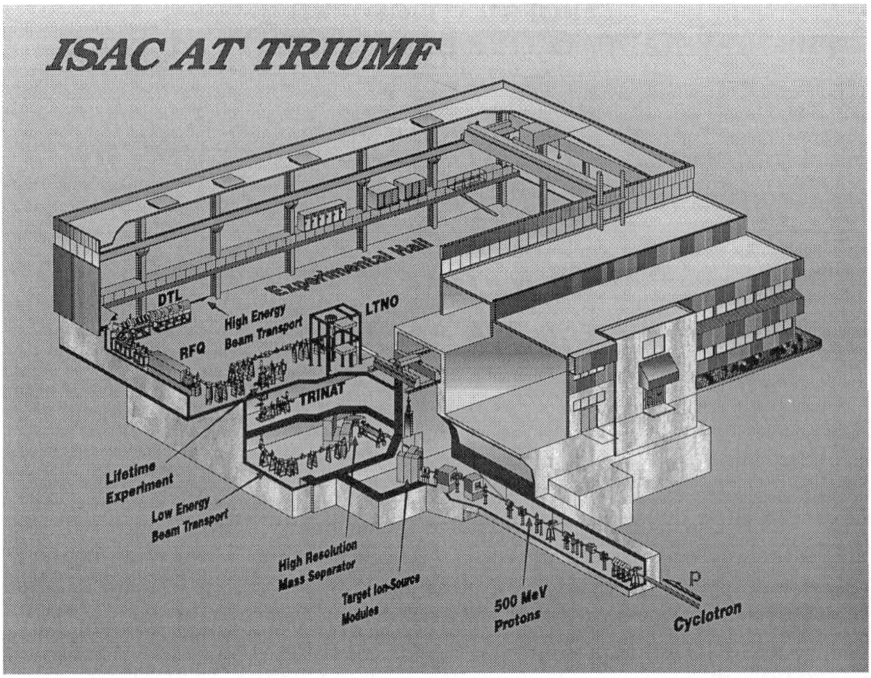

Figure 1. General ISAC Site Layout: The ISAC complex at TRIUMF includes the new ISAC buildings (from back to front – Accelerators and Experimental Areas, production target handling cave and Service Annex) and the new BL2A tunnel.

Rb, Cs, Fr) of the order of 10^{11}/sec. or higher at the peak of the mass yield curve could be expected. In addition, this ion source should also produce beams of elements such as Sr, Ga, Yb, and others. Intensities decrease as one approaches the limits of stability. ECR ion sources are being developed to produce beams of volatile elements such as He, O, N, F, Ne, C (as CO), Ar, and Kr. There are plans to develop other sources, such as laser ion sources which could ionize a wider selection of atoms.

After exiting the ion source, the ion beam is mass analyzed, first by a low resolution preseparator where most of the radioactivity is removed on to removable slits. The preselected beam is directed to a high resolution, high acceptance mass analyzer which is designed to reach a resolving power of $\frac{\Delta M}{M} \simeq \frac{1}{10000}$. The whole assembly sits on a HV platform to allow the operation of the mass separator at constant momentum, independent of the actual beam energy.

The mass separated beam is then directed to the experimental hall located at the grade level. Alternatively, the beam can be deflected to the TRINAT (TRIUMF Neutral Atom Trap) facility, located on a well-shielded intermediate floor. In the new ISAC experimental hall, the beam is directed either to the RFQ/DTL accelerators for experiments with accelerated beam (HE), or to the experimental areas (LE) planning to use the unaccelerated radioactive beams. A cw RFQ LINAC accelerator will take a singly charged ion beam of with $q/A > \frac{1}{30}$ from 2 keV/u to 150 keV/u. After the RFQ, the beam will be stripped, and the selected charge directed to a cw DTL which will accelerate an ion beam with $q/A > \frac{1}{6}$ to some desired energy in the range from 0.15 to 1.5 MeV/u. The RFQ will be operated in continuous duty cycle mode at 35 MHz. The entire acceleration system including matching sections between the RFQ and DTL are expected to be ready for commissioning by the beginning of 2000. It is anticipated that radioactive beams at full energy will be available in the HE experimental area by the fall of 2000.

2 Experimental Programme

In the experimental hall, there is space for at least 5 experimental stations in the LE area using unaccelerated beams in addition to the TRINAT facilities located on the mezzanine floor. Facilities being developed are a standard nuclear spectroscopy counting area, a dilution refrigerator system for low temperature nuclear orientation studies, a material studies facility using polarized beams, a beta-NMR system, and a penning trap system for mass measurements. The HE area for experiments with the accelerated beam will house a specially designed spectrometer to study reactions with the radioactive beam (DRAGON), a general purpose scattering chamber for particle reaction studies (TUDA), and possibly a Coulomb excitation system. Figure 2 gives a general overview of the planned experimental facilities in the ISAC hall.

These facilities have been conceived to support an experimental program covering nuclear physics including nuclear astrophysics, tests of symmetries, mass and life-time studies, etc., condensed matter research using polarized ion beams via β-NMR techniques, and a variety of life science related projects.

For the latter, isotope separators that are located at the source of isotope production offer the opportunity to produce and separate desired radiotoxic species in high yield and purity. At ISAC we will investigate the production and separation of radiotoxic nuclides for use in therapy as well as exploring the feasibility of using the post accelerator for generator production via ion implantation. As an example the implantation work can take the form of investigating a number of generator systems that could be more efficient if

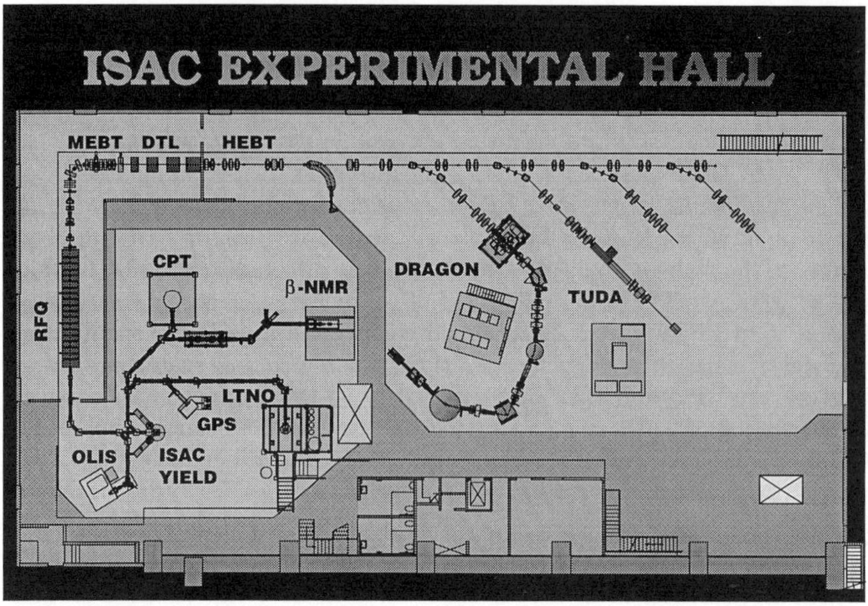

Figure 2. ISAC Experimental Hall Layout: On the left side are the off-line ion source, RFQ accelerator, medium energy beam transport system and drift tube linac. The low energy area features some of the initial facilities (lifetime measurement, low temperature nuclear orientation refrigerator and polarized ^{8}Li beams). In the high energy area the location of the recoil spectrometer (DRAGON) is indicated as well as a general purpose scattering facility and an implantation station.

the parent nuclide is held in matrix more suitable for the separation of the two species, that is daughter from the parent.

3 Conclusion

In conclusion, ISAC holds the promise to provide a unique scientific opportunity at TRIUMF supporting a strong, diversified program in nuclear physics, nuclear astrophysics, condensed matter research and biosciences. The first ^{38m}K beam was obtained in November 1998 and accelerated beams should be available later in year 2000; ISAC will be the premier radioactive beam facility in North America for years to come. Further upgrades to increase the ion energies to $\sim$ 6.5 MeV/u and extend the mass reach of ISAC are being considered as part of the next five-year plan.

Session C1

Radiochemical Synthesis — I

RADIOIODINATION VIA SOLID PHASE ORGANIC CHEMISTRY

DUNCAN H. HUNTER, MUSTAFA JANABI, BRIAR MANNING AND XIZHEN ZHU
Department of Chemistry, *The* University *of* Western Ontario
London, Ontario, Canada N6A 5B7

GIAN GOBBI
Dow Chemical Canada, Inc.
Vidal Street, P.O. Box 3030
Sarnia, Ontario, Canada N7T 7M1

In principle, insoluble polymer-supported aryl stannanes provide a potentially convenient route to radioiodinated compounds for several reasons. Radioiododestannylation can proceed under mild conditions to provide no-carrier-added material that can be isolated by simple filtration from any insoluble unreacted starting material and from insoluble by-products. The final product should be free of tin impurities since the tin remains attached to the insoluble polymer. To convert principle to practice, we have been exploring the generality and usefulness of this approach.

Through the use of appropriate protecting groups, several substituted phenyl rings have been bound through tin to the solid support. The meta and para substituents include hydroxy, methoxy, carboxyl, formyl and aminomethyl. Subsequent conversions of some of these groups have been attempted so that the aryl-stannanes can be used as precursors to libraries of radiopharmaceuticals.

Conditions for radioiododestannylation have been investigated. In the specific case of the guanidinomethyl group, radioiodination leads to the release of MIBG (meta-iodobenzyl guanidinium) which has found use both diagnostically and therapeutically. $[^{131}I]$MIBG is produced in ∃90% radiochemical yield and high chemical purity with a reaction time of 5 minutes at room temperature and a total preparation time of about 10 minutes. $[^{123}I]$MIBG requires a temperature of –65 ^{o} for 5 minutes and again results in a product of ∃90% radiochemical yield and high chemical purity.

1 Introduction

A very convenient route to no-carrier-added iodinated radiopharmacueticals is through the corresponding tri-n-butylstannyl or trimethylstannyl precursors [1]. Iododestannylation with radioiodide and an oxidant typically proceeds rapidly at room temperature in good radiochemical yield. However there are concerns associated with this approach. The potential toxicity of tin compounds makes it imperative to remove unreacted starting material and any tin containing by-products. One way to circumvent these problems is to attach the stannyl precursor to an insoluble polymeric support [2]. Thus, after radioiododestannylation, all tin materials should remain insoluble and should be readily removed with only the desired radiopharmacetical remaining dissolved. As well, use of an insoluble polymeric support will provide all the advantages associated with Solid Phase Organic Chemistry techniques so in vogue at present.

2 Results and Discussion

A polymer-supported chlorostannane has been prepared using a modification [3] of an existing method [4,5] such that about 50 g of material can be prepared in 3-4 days commencing with commercially available divinylbenzenes. Treatment of the polymeric chlorostannane with an appropriate aryllithium, prepared *in situ* from the corresponding arylhalide , has led to polymer-supported arylstannanes (step 1 in figure 1).

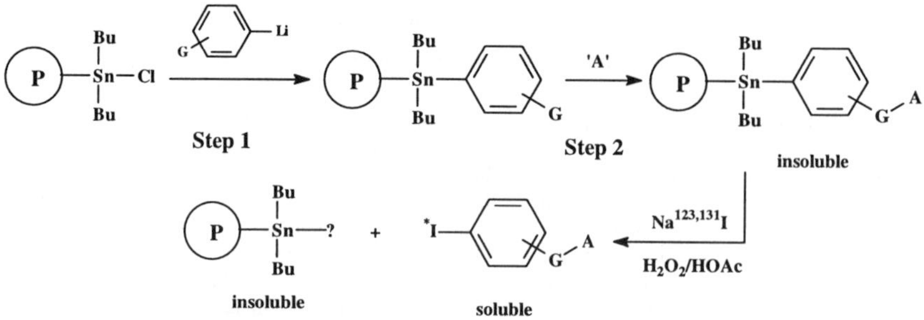

In some cases the organolithium could be prepared directly from the organohalide: 4-bromoanisole and 4-bromo-N-isopropylamphetamine. Because certain groups are sensitive to treatment with n-butyllithium, protecting group procedures had to be developed such that the protecting group could be removed without cleavage of the polymer-supported arylstannanes and subsequent loss of the aryl group from the polymer support. Thus the amino group of 3-bromobenzylamine was protected as the 'stabase' [6] and removed by treatment with dil. HCl/MeOH yielding polymer-supported 3-benzylamine. The 3- and 4-bromophenols were protected with the triisopropylsilyl group that was removed with TBAF/THF producing polymer-supported phenols. The aldehyde groups were protected as oxazolidines by treatment with pseudoephedrine [7] which was subsequently removed with dil. HOAc/EtOH/H_2O yielding the polymer-supported benzaldehydes.
These polymer-supported arylstannanes were characterized by MAS ^{119}Sn NMR spectroscopy and, in several cases, showed complete conversion of the chlorostannane to arylstannane. The amount of bound aryl moiety was determined by reaction with iodine that liberated the corresponding aryl iodide giving loading capacities in the range of 0.5-1.1 mmol/g of polymer.
In a subsequent step 2, the benzylammonium group was converted quantitatively to a benzylguanidinium group by reaction with guanidine [3]. The benzaldehydes were converted to benzoic acids by treatment with 3-chloroperbenzoic acid. In these cases, iodination was used to assess the extent of conversion and was shown to be complete (> 95 %).
In a final radioiodination step, the polymer supported 3-benzylguanidinium group was cleaved using Na ^{123}I and Na ^{131}I with peracetic, from H_2O_2/HOAc, as oxidant. In the case of $^{131}I^-$, the radiolabelling was complete in 5 minutes at room

temperature and gave [^{131}I]MIBG in 95 % radiochemical yield and > 99.5 % chemical and isotopic purity after filtration through a Sep-Pak. The more dilute $^{123}I^-$ still required only 5 minutes but at 65 °C with a radiochemical yield of 85 % and again of chemical and isotopic purity > 98 %. This seems to represent the most convenient route to no-carrier-added [*I]MIBG that is presently available.
The polymer-supported benzoic acids provide an opportunity for creating a library of radioiodinated benzamides as well as a convenient precursor for their preparation. There are several iodobenzamides which are being touted for the value as imaging agents for functions such as melanoma , breast cancer , MAO receptors and kidney function.

3 Acknowledgements

We wish to thank the Natural Science and Engineering Council of Canada and Merck-Frosst, Canada, Inc. for financial support via an IOR grant.

References

1. Ali, H. and van Lier, J.E., Synthesis of radiopharmaceuticals via organotin intermediates, *Synthesis*, (1996) pp. 423-445
2. Culbert, P.A. and Hunter, D.H., Polymer supported radiopharmaceuticals: ^{123}I and ^{131}I-Labelled N-Isopropyl-4-Iodoamphetamine, *Reactive Polymers*, **19**, (1993) pp. 247-253
3. Hunter, D.H. and Zhu, X., Polymer supported radiopharmaceuticals: [^{131}I]MIBG and [^{123}I]MIBG, *J. Labelled Compd. Radiopharm*, **42**, (1999), pp. 653-661
4. Gerigk, U., Gerlach, M., Neumann, W.P., Veiler, R. and Weintritt, Polymer-supported organotin hydrides as immobilised reagents for free-radical synthesis, *Synthesis*, (1990) pp 448-
5. Gerlach, M., Jordens, F., Kuhn, H., Neumann, W.P., and Peterseim, M., A polymer-supported organotin hydride and its multipurpose applications in radical organic chemistry, *J. Org. Chem.*, **5** (1991) pp. 5971-5972
6. Djuric, S., Venit, J. and Magnus, P., Silicon in synthesis: stabase adducts- a new primary amine protecting group: alkylation of ethyl glycinate, *Tetrahedron Lett.*, **22**, (1981) pp 1787-1790
7. Agami, C. and Rizk, T., Kinetic control of asymmetric induction during oxazolidine formation form (-)-ephedrine and aromatic aldehydes, *Tetrahedron*, **41**, (1985) pp537-540

DESIGN AND SYNTHESIS OF NEW NEUTRAL RE(V)O COMPLEX DERIVED FROM BENZAMIDE

JI HU , YOUFENG HE , YISHANG WANG AND LUN XIAO
Isotope Department, China Institute of Atomic Energy,
P. O. Box 275(103), Beijing 102413 , P. R. China

Due to excellent nuclide properties and wide availability of Tc-99m, development of Tc-99m labeled central nervous system (CNS) receptor imaging agent is expected to have a significant impact in the diagnosis and management of patients with brain disease. Based on the structure-activity relationship in a variety of substituted benzamide derivatives that exhibit high affinity for CNS dopamine D2/D3 subreceptors, new benzamide derivatives labeled with Tc-99m have been designed. In this presentation, a new neutral Re(v)O complex, derived from benzamide by introducing a mercapto group at 5-position of benzene ring to mimic sulpiride ligand, has been synthesized using the "3+1"mixed ligand approach. A reliable and reproducible synthetic route of MBZM ligand [5-mercapto-N-((1-ethyl-2-pyrrolidinyl) methyl)-2-methoxy benzamide] starting from commercially available 2-methoxy benzoic acid was developed. Tridentate ligand SNS (N-methyl di (2-mercaptoethyl) amine was synthesized as in the literature. Re(v)O(SNS)(MBZM) complex was obtained by reacting MBZM and SNS with $Re(V)OCl_3(PPh_3)_2$ or Re-gluconate in ethanol aqueous solution, and identified by IR, ^{1}H-NMR, MS and elemental analysis.

1. Introduction

In vivo imaging techniques of central nervous system (CNS) receptor with PET and SPECT provide important information about the change in receptor concentration or the appearance of receptors as a function of a specific pathological state[1]. Due to excellent nuclide properties and wide availability of Tc-99m, development of Tc-99m labeled CNS receptor imaging agent, which may facilitate use of SPECT imaging of CNS receptor as a cost-effective clinical tool, has been received considerable interest. However, apart from Tc-99m TRODAT1 for the dopamine transporter, previous attempts have been undertaken with little success[2,3].

Based on the structure-activity relationship in a variety of substituted benzamide derivatives[4,5] that exhibit high affinity for CNS dopamine D2/D3 subreceptors, new benzamide derivatives labeled with Tc-99m have been designed. In this presentation, a new neutral Re(v)O complex as surrogate for Tc-99m complex , derived from benzamide by introducing a mercapto group at 5-position of benzene ring to mimic Sulpiride ligand, has been synthesized using the "3+1"mixed ligand approach[6].

2. Experimental

A tridentate chelator, methyl di-(2-mercaptoethyl) amine, was synthesized according to the standard methods. Racemic 1-ethyl-2- amino methyl pyrrolidine was resolved by precipitation of di-tartrate salts from 50% aqueous ethanol as in the literature[7].

Synthesis of MBZM ligand [5-mercapto-N-((1-ethyl-2-pyrrolidinyl) methyl)-2-methoxy benzamide] was carried out as shown in Scheme1. The sulphonyl chloride, **2**, was obtained by the action of chlorosulphonic acid on commercially available 2-methoxy benzoic acid. Reduction of the above product using zinic dust and conc. HCl acid, followed by reaction with acetic anhydride under Schotten-Baumann condition, produced the acid, **4**. The acid, **4**, was activated by converted to acyl chloroide, and then coupled with the previous resolved pyrrolidine, to form the intermediate, **5**. The removal of acetyl moiety of **5** was accomplished through reaction with diethylamine to form disulfur bridge, and followed by reduction, to yield the deprotected benzamide, **6**.

Preparation of Re(v)O(SNS)(MBZM) complex was achieved by reacting the free ligang, monodentate, **6** and tridentate, SNS with $Re(V)OCl_3(PPh_3)_2$ in alkaline ethanol aqueous solution. The chemical structure of **7** was identified by IR, ^{1}H-NMR, fast-atom bombardment MS and elemental analysis, respectively.

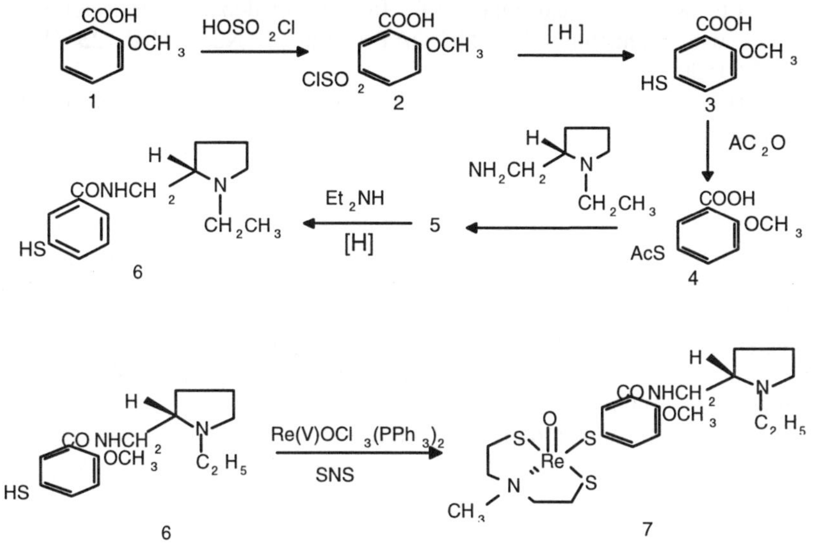

Scheme 1. Syntheses of 5-mercapto-N-((1-ethyl-2-pyrrolidinyl) methyl) -2-methoxy benzamide and Re(v)O(SNS)(MBZM)

3. Results and discussion

A reliable and reproducible synthetic route of MBZM ligand starting from commercially available 2-methoxy benzoic acid was developed. As expected, the new neutral Re(V)(SNS(MBZM) could be conveniently prepared using the "3+1"mixed ligand approach. Radiolabelling with 99mTc and in vitro receptor binding assay are currently in progress.

4. Acknowledgement

This work was a part of CRP project under auspices of IAEA.

References

1. Hu J., He Y.F., Wang Y.S. and Xiao L., Development of ^{99m}Tc labelled central nervous system receptor imaging agents. Isotopes. **11**(1998) pp.184-188.
2. Hom R.K. and Katzenellenbogen J.A., Technetium-99m-labeled receptor-specific small-molecule radiopharmaceuticals: recent developments and encouraging results. Nucl. Med & Biol. **23** (1997) pp.485- 498.
3. Kung H.F., Kim H.J., and Kung M.P., Imaging of dopamine transporters in humans with technetium-99m TRODAT-1. Eur. J. Nucl. Med. **23** (1996) pp.1527-1530.
4. Kung H.F., Kasliwal R. and Pan S., Dopamine D2 receptor imaging radiopharmaceuticals: synthesis, radiolabelling and in vitro binding of (R)(+)- and (S)(-)-3-iodo-2-hydroxy-6-methoxy-N-[(1-ethyl-2-pyrolidinyl)methyl] benzamide. J. Med. Chem. **31** (1988) pp.1039-1043.
5. Hogberg T., Strom P. And Hall H., Synthesis of [^{125}I], [^{123}I] and unlabelled (S)3-iodo-5,6-dimethoxy-N-[(1-ethyl-2-pyrolidinyl)methyl] salicylamide (NCQ298), a selective ligand for dopamine D-2 receptors. Acta. Pharm. Nord. **2** (1990) pp. 53-60.
6. Spies H., Fietz TH. and Pietzsch H.J., Neutral oxorhenium(V) complexes with tridentate and monodentate alkane- or arenethiolate coligands. J. Chem. Soc., Dalton Trans. (1995) pp. 2277-2280.
7. Bulteau G.S., African Pat. 6802593 (1968).

SCANDIUM-47: A REPLACEMENT FOR CU-67 IN NUCLEAR MEDICINE THERAPY WITH β/γ EMITTERS

L.F. MAUSNER, K.L. KOLSKY, V. JOSHI, M.P. SWEET, G.E.MEINKEN AND S.C. SRIVASTAVA

Medical Department, Brookhaven National Laboratory
Upton, NY 11973, USA
E-mail: mausner@bnl.gov

Cu-67, initially developed at BNL, is an excellent □/□ therapeutic label. However, its availability has been limited. Because of its very similar nuclear properties, we are proposing Sc-47 (t ½ 3.35d; β_{max} 441 keV, 68%, 601 keV, 32%; γ 159 keV, 68%) as a replacement for Cu-67. Sc-47 was produced using the $^{47}TiO_2$ (n,p)^{47}Sc fast neutron reaction (saturation yield at HFBR 25MBq/mg of target). Radiochemical processing utilized two cation exchange columns. Sc-47 was labeled to anti-CEA F(ab')$_2$ MAb with four chelating agents: 4-isothiocyanatocyclohexyl EDTA (4-ICE); mono-NHS ester of DOTA (DOTA-NHS); 2-(p-SCN-Bz)-6 methyl DTPA (1B4MDTPA); and the conventional DTPA dianhydride (DTPA-DA). Labeling yields were between 75-95%, and stability in serum (96 h) was >95% for all preparations, except for DTPA-DA (83%). Immunoreactivity ranged between 70-80%. These preparations were administered i.v. into human tumor (LS-174T cells) xenografted nude mice to study biodistribution (% ID per g). 4-ICE, DOTA-NHS and 1B4M-DTPA all showed high tumor uptake at 48 h (32.2, 34.7, 27.9). Kidney uptake of 4-ICE was higher than DOTA-NHS or 1B4MDTPA (15.4, 7.0, 7.0) but the uptake into liver was higher for DOTA-NHS than 4-ICE or 1B4MDTPA (9.4, 5.8, 5.6). The DTPA-DA conjugate gave poor tumor uptake (12.4), and high kidney (14.8) and liver (17.1) uptakes. Work done under US DOE, Contract #DE-AC02-98CH10886.

1 Introduction

The use of beta emitters attached to monoclonal antibody conjugates for radioimmunotherapy has been an active area of research. The desirable properties of the radiolabel for this application are a) intermediate to high energy beta emission; b) some gamma emission for low dose imaging and biodistribution determination; c) a physical half life of 1 to 14 days; d) a stable daughter product; e) chemical properties allowing stable in-vivo binding to a carrier molecule; f)and rapid excretion of free radionuclide. Copper-67, initially developed at BNL, meets these criteria very well. It has a 2.58d half life, a weighted average beta energy of 141 keV, and an imageable photon of 187 keV (49%). However, its availability has been too limited to support wide spread experimental or clinical use. Because of its very similar nuclear properties, we are proposing ^{47}Sc as a replacement for ^{67}Cu. Scandium-47 has a 3.35 d half life, a weighted average beta energy of 163 keV, and an imageable photon of 159 keV (68%). Scandium also displays favorable chemistry for conjugation to monoclonal antibody-chelate systems. It is chemically similar to ^{90}Y and ^{111}In. Thus, ligands already developed for these isotopes would be adaptable to ^{47}Sc. We have optimized the production of no-carrier added ^{47}Sc and investigated its labeling chemistry and the in-vivo distribution of labeled bioconjugates.

2 Methods

Scandium-47 was produced using the $^{47}TiO_2(n,p)^{47}Sc$ fast neutron reaction in the High Flux Beam Reactor (HFBR) and the High Flux isotope Reactor (HFIR). Radiochemical processing was based on two separations with cation exchange using Dowex AG 50W-X4 and scandium elution with 4N HCl/0.1N HF [1]. The expensive enriched ^{47}Ti was recovered by precipitation with NH_4OH, and then ignited back to the starting form, $^{47}TiO2$. The scandium was labeled to anti-CEA $F(ab')_2$ MAb with four chelating agents: 4-isothiocyanatocyclohexyl EDTA (4-ICE); mono-NHS ester of DOTA (DOTA-NHS); 2-(p-SCN-Bz)-6 methyl DTPA (1B4MDTPA); and the conventional DTPA dianhydride (DTPA-DA). These preparations were administered i.v. into human tumor (LS-174T cells) xenografted nude mice to study biodistribution.

3 Results

The saturation yield of ^{47}SC at the HFBR was measured as 13.1MBq/mg of TiO_2 target (94.5 % enriched in ^{47}Ti)., and at the HFIR (PTP position) we achieved 62.4 MBq/mg. Only traces of radionuclidic impurities were detected (0.012% ^{46}Sc at both reactors, and 0.075% at HFBR and 0.15% at HFIR of ^{48}Sc). Recovery of enriched ^{47}Ti was nearly quantitative at 98%. The labeling yields for the DOTA, IB4MDTPA, 4-ICE, and DTPA-DA chelates were 73%, 98%, 88%, and 62% respectively. Stability in serum at 96h was >95% for all preparations, except DTPA (83%). After linking to anti-CEA F(ab')2Mab the immunoreactivity ranged between 70-80%.

The biodistribution of 4-ICE, DOTA-NHS and 1B4M-DTPA all showed high tumor uptake at 48 h (32.2, 34.7, 27.9%ID/g). Kidney uptake of 4-ICE was higher than DOTA-NHS or 1B4MDTPA (15.4, 7.0, 7.0) but the uptake into liver was higher for DOTA-NHS than 4-ICE or 1B4MDTPA (9.4, 5.8, 5.6). The DTPA-DA conjugate gave poor tumor uptake (12.4), and high kidney (14.8) and liver (17.1) uptakes, as expected due to its relatively lower serum stability. In conclusion we have developed the large-scale production of no-carrier-added 47Sc, and demonstrated its potential for labeling antibodies for radioimmunotherapy. These methods can be used to label peptides and other bioengineered molecules as tumor associated carriers.

Reference

1. Kolsky K.L., Joshi V., Mausner L.F. and Srivastava S.C., Radiochemical purification of no-carrier-added Scandium-47 for radioimmunotherapy. *Appl. Radiat. Isot.* **49** (1998) pp.1541-1549.

SYNTHESIS OF ANTI-CMV RPR111423 LABELED WITH CARBON-14, DEUTERIUM AND TRITIUM

BO-CHAN KAO

Rhone-Poulenc Rorer Pharmaceuticals

Department of Drug Metabolism and Pharmacokinetics

500 Arcola Road, Collegeville, PA 19426, USA

RPR111423, 2-chloro-3-pyridin-3-yl-5,6,7,8-tetrahydroindolizine-1-carboxamide (CAS Registry Number: 186829-19-6) is a potent, orally active anti-CMV compound under investigation as a potential therapeutic agent for the treatment of cytomegalovirus infections. It is a very potent inhibitor of CMV replication *in vitro* (IC_{50} in the range of 0.3 to 3 ng/ml). It is 50 to 90,000 times more active than the reference drugs (ganciclovir, cidofovir, foscarnet). C-14, H-3 and H-2 -labeled RPR111423 were required for ADME, protein binding and mass spectometric assay. This paper will present the synthesis of the [C-14], [H-3], and [H-2] of RPR111423 in Scheme 1, 2, and 3.

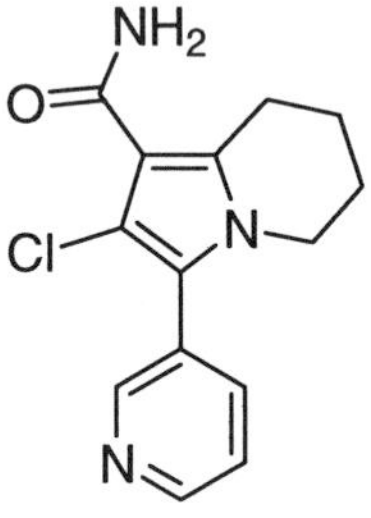

1 Introduction

RPR111423, 2-chloro-3-pyridin-3-yl-5,6,7,8-tetrahydroindolizine-1-carboxamide is an anti-CMV drug candidate under investigation as a potential therapeutic agent for the treatment of cytomegalovirus infections. As part of the continuing development program, a number of isotopically labeled forms of the compound were required: C-14 for ADME studies. H-3 for assay development and protein binding studies and H-2 as an internal standard for mass spectrometric assay of the compound. 30 mCi of RPR111423[C-14] (specific activity, 56 mCi/mmol) for drug metabolism studies was obtained using a 6-step process (Scheme 1) derived from the medicinal chemistry synthesis with an overall 12% radiochemical yield from nicotinic acid [3-carbonyl-C-14] and 99.2% radiochemical purity. 20 mCi of RPR 111423 [H-3] (specific activity, 28 Ci/mmol) was prepared through catalytic reduction of RPR120126 with tritium gas (scheme 2) with a 5% yield and radiochemical purity >99%. 500 mg of RPR111423[H-2] was obtained using a 6-step process (Scheme

3) derived from the medicinal chemistry synthesis with a 27% overall yield (from ethyl pipecolinate hydrochloride). The chemical purity was 99%.

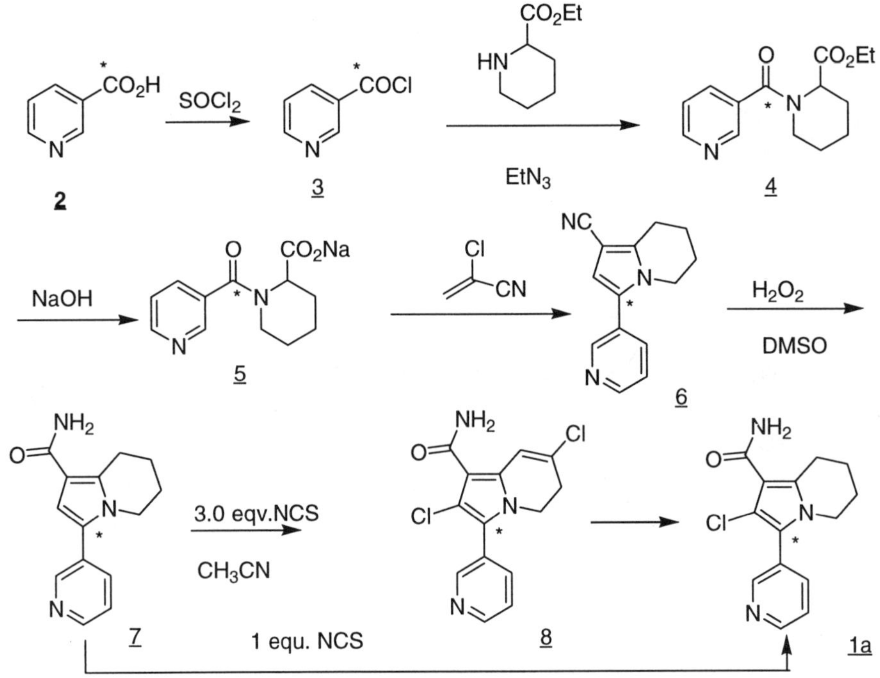

Scheme 1: Preparation of RPR111423[C-14]

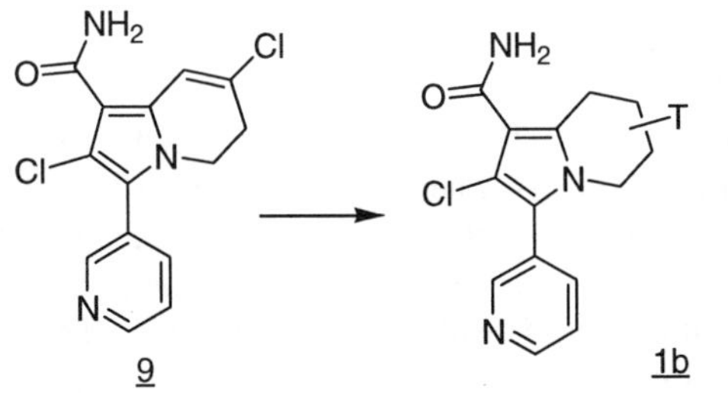

Scheme 2: Preparation of RPR111243[H-3]

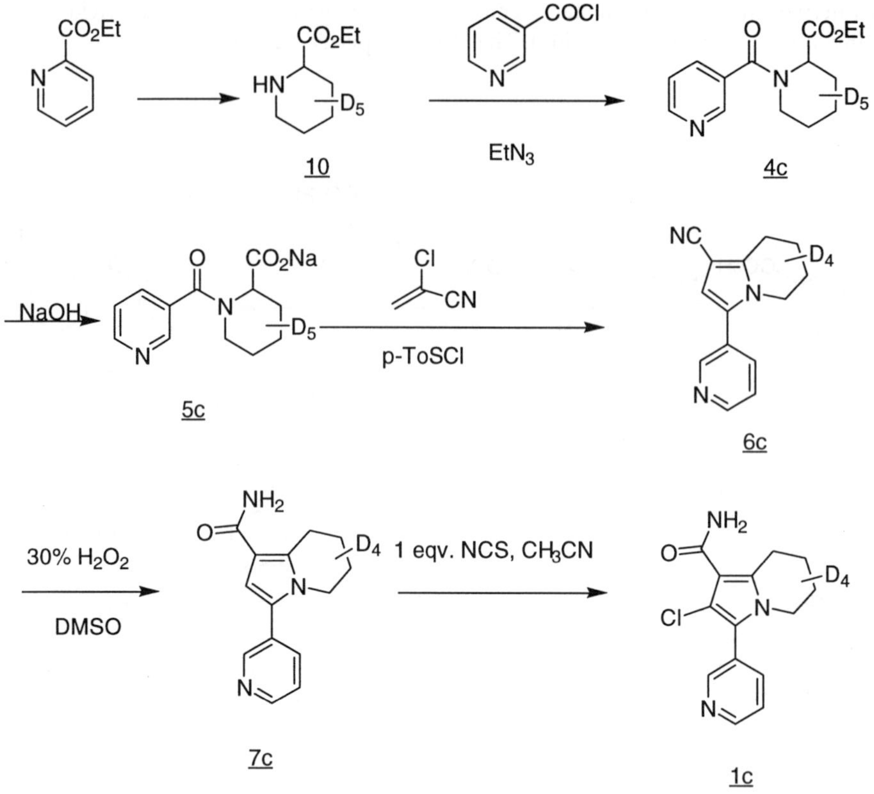

Scheme 3: Preparation of RPR111423[D-4]

2 Results and Discussion

RPR111423[C-14] Synthesis. The synthesis was based on the route developed by Rhone-Poulenc Rorer Discovery Chemistry[1] (in France by using nicotinic acid[C-14] (2) to introduce the C-14 label. The reaction of nicotinic acid chloride[C-14] (3) with ethyl pipecolinate proceeded with 90% yield of (4) followed by the conversion to the sodium salt (5) with sodium hydroxide. The resulting sodium salt was treated with 2-chloroacrylonitrile in acetonitrile at room temperature to give the indolizine (6) with 85% yield. In order to avoid the problems associated with the original methods for carrying out the acidic hydrolysis of the nitrile (6) to the amide (7) and the acid by-product. Hydrolysis was performed using 30% hydrogen peroxide in DMSO with catalytic amounts of potassium carbonate to prevent further hydrolysis of the amide (7) to the corresponding acid[2]. Instead of using an excess of N-chlorosuccinimide to synthesize the dichloro compound (8), the resulting amide (7) was treated with one equivalent of N-chlorosuccinimide in acetonitrile to give RPR111423[C-14] (1a) with an overall 12% radiochemical yield from nicotinic acid [3-carbonyl-C-14] and a 99.2% radiochemical purity.

RPR111423[H-3] Synthesis. 20 mCi of RPR 111423 [H-3](1b)(specific activity, 28 Ci/mmol) was prepared through catalytic reduction of RPR120126(9) with tritium gas (scheme 2) with a 5% yield and radiochemical purity >99%. Direct catalytic tritium exchange with RPR111423 failed. Instead of aliphatic tritium labeling (scheme 2), pyridine ring tritium labeled RPR111423[H-3] was therefore synthesized via bromination and tritiation.

RPR111423[H-2] Synthesis. RPR111423[d_4](1c) was required for use as internal standard in a mass spectrometric bioanalytical assay[3]. An enrichment of at least three deuterium atoms per molecule was required to prevent interference from the C-13 and Cl isotope. The catalytic reduction of ethyl pipecolinate with deuterium and platinum oxide on carbon yielded isotopically labeled d_5-pipecolic acid (10)and other minor products, i.e. d_4 and d_6-pipecolic acid. The same procedures for RPR111423[C-14] was followed and 500 mg of RPR111423[d_4] (1c)was obtained using a 6-step process (Scheme 3) derived from the medicinal chemistry synthesis with a 27% overall yield (from ethyl pipecolinate hydrochloride) and a 99% chemical purity.

3 Acknowledgments

The author thanks Rhone-Poulenc Rorer Medicinal Chemistry, Process Chemistry in France and also Ken Clow at Dangenham , UK for providing the original synthetic procedures of RPR111423.

References

1. Carry J.C., Mignai S., and Nemecek C., Rhone-Poulenc Rorer S.A. France,. PCT Int. Appl. WO 9825925 AL 19980618, pp 50 (1998).
2. Katritzky A., Pilarski B., and Urogdi L., Synthesis (1989), pp 949.
3. Thomas, A.F., "Deuterium labeling in Organic Chemistry", Appleton Century Croftd, New York, 1971.

Session C2

Nuclear Analytical Applications of Radioisotopes — I

THE UNAVOIDABLE USE OF RADIOISOTOPES AT RMC

K.S. NIELSEN, R.G.V. HANCOCK, L.G.I. BENNETT, W.J. LEWIS,
W.S. ANDREWS, B.J. LEWIS, AND H.W. BONIN

*SLOWPOKE-2 Facility & Department of Chemistry and Chemical Engineering
Royal Military College of Canada/Collège Militaire Royal du Canada
Kingston, Ontario, Canada*

At RMC/CMR, the SLOWPOKE-2 Facility provides a focus where various activities occur, with most centred around the research reactor as a source of neutrons. Instrumental Neutron Activation Analysis (INAA) is the main application in which short-lived radioisotopes are produced to identify elements in samples and to determine their concentrations. Neutron radiography is a nondestructive testing (NDT) technique whereby the presence of hydrogen-containing materials is determined in assembled structures such as aircraft components. A third application is the study of the effects of the reactor's neutron and gamma radiation on polymers including adhesives and nitrocelluloses. A 0.2 TBq Am-Be source is available for calibration of several low-level radiation detectors used for assessing the cosmic radiation exposure of aircrew at jet altitudes. Neutron-sensitive bubble detectors, a Tissue Equivalent Proportional Counter (TEPC), a portable ion chamber and a Microspec 2 are all checked with their own or other calibration sources in the Facility. Multi-element radioisotopes in a Marinelli configuration are used for calibration of gamma-ray spectroscopy systems located in the Facility as well as in laboratories near approved berths for visiting nuclear submarines.

The SLOWPOKE-2 Facility at RMC/CMR consists of a small, Canadian-designed research reactor surrounded by a number of laboratory rooms (Figure 1) developed for its support and applications. In the Reactor Room, the small reactor core is located near the bottom of a vertical, cylindrical reactor container which is suspended from two off-centred I-beams on top of a reactor pool (Figure 2). The reactor is controlled from a reactor console, operated simply with keys, located in an adjacent Control Room. In a manner not unlike a modern car, turning the key to an 'auxiliary' position powers up the reactor console and turning a key either to the 'automatic' or 'manual' mode allows for a different degree of control by the 'driver', the Licensed Operator. These actions move vertically a cadmium control rod located centrally in the reactor core, resulting in the reactor being started up, operated automatically or manually, and shut down. Aside from its ability to produce radioisotopes, the important features of this reactor are the reproducibility and stability of its neutron flux, its operational reliability due to its simplicity, its capability to be operated by licensed, yet relatively non-nuclear personnel, and its versatility so that these 'operators' remain primarily devoted to their applications and research. As well as being undemanding in its operation, its maintenance requirements consist of a few hours of weekly checks shared by the Licensed Operators and, every few years, an addition of a shim of beryllium reflector to maintain the excess reactivity of the reactor core (as opposed to the more usual method of refueling).

The acceptance and innocuousness of this nuclear research reactor is indicated by the location of the Facility in a corner of a two-floor engineering laboratory at RMC/CMR (Figure 1). From the second floor laboratory rooms of the Facility, radioisotopes can be produced by sending sealed samples pneumatically via 'rabbit' systems to the reactor core below. Although some of this production is used directly as radioisotopes for the users' intent, most radioisotopes are returned to gamma spectroscopy systems for analysis to determine the samples' elemental concentrations, and then appropriately discarded. This analytical technique of Instrumental Neutron Activation Analysis (INAA) was the main application envisioned for this reactor and remains so today. A second and recently established application is neutron radiography whereby objects are exposed to neutrons guided through a beam tube tangential to the Reactor Container (Figure 2). At the end of this beam tube at the top of the Reactor Container, few radioisotopes are produced in the object being inspected or imaging devices and are avoided by careful handling of these parts after each exposure. The desire is to view the attenuation of neutron-scattering or absorbing elements such as hydrogen found in moisture trapped inside the flight control surfaces of the CF-18 aircraft. These images are developed as radiographs or viewed on a monitor by means of a near real-time radioscopy system in a room adjacent to the Reactor Room.

The third direct application of the reactor is the irradiation of polymers in the reactor pool by the available neutrons, gamma radiation, and stripped electrons to produce radiation effects in the polymers. The samples are lowered to this location via an 'elevator' which can place several samples reproducibly near the reactor core. The effects of this radiation combination on materials such as adhesives and nitrocelluloses can then be studied.

In another room adjacent to the Reactor Room, an already-produced radioisotope (0.2 TBq Am-Be neutron source) is used to calibrate several low-level neutron-sensitive detectors which are flown on jet aircraft for measurement of the cosmic radiation exposure to aircrew at altitude (which is about half dose-equivalent neutrons). This room is also used for calibration checks of detectors using their own sources or other small sources located there.

An example of the external support use of the Facility is that for the Canadian Forces' Nuclear Emergency Response Teams (NERTs). Radioisotopes are used for the calibration of gamma spectroscopy systems located here and in laboratories on each coast near the berths for visiting nuclear submarines. Long-lived multi-element radioisotopes that mimic the range of fission products expected in a released plume are commercially obtained in counting geometries such as Marinelli beakers and are used to ensure compliance for these accredited laboratories.

As indicated by these various applications, various radioisotopes are unavoidably produced, acquired and used in this Facility for which there are a wide variety of applications, some extending beyond its walls.

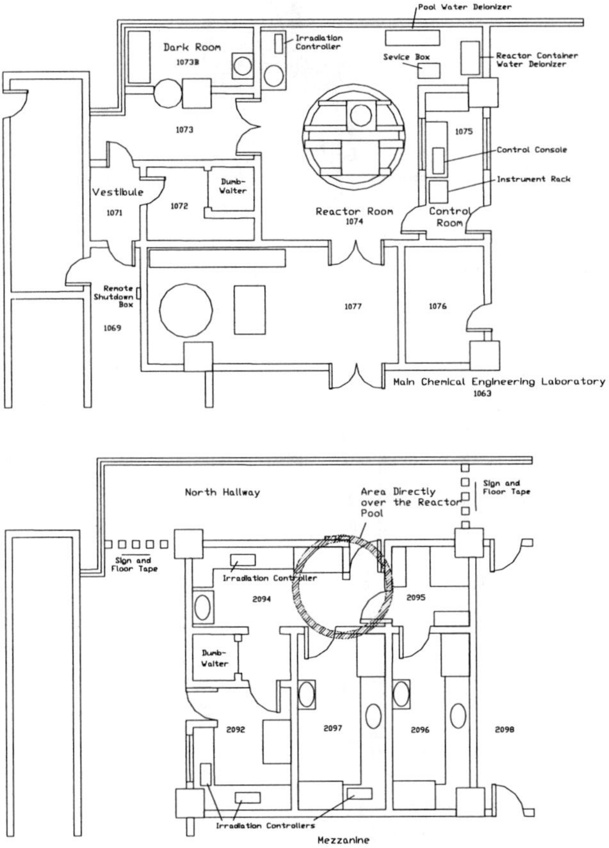

Figure 1: Plan layout of the SLOWPOKE-2 Facility at RMC/CMR

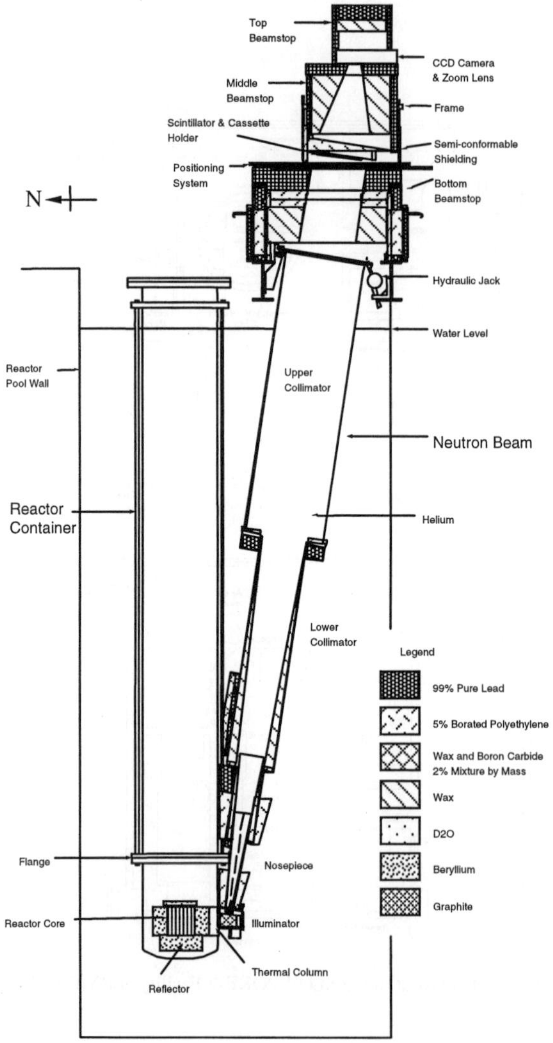

Figure 2: SLOWPOKE-2 research reactor installation at RMC/CMR

RAPID ANALYSIS OF NATURALLY OCCURRING RADIOACTIVE MATERIAL (NORM) USING HIGH RESOLUTION GAMMA-RAY SPECTROMETRY.

M. JOHN M. DUKE

SLOWPOKE *Nuclear Reactor Facility, University of Alberta, Edmonton, Alberta, CANADA.*

Numerous extraction and processing industries, including for example the oil and gas, fertilizer, and uranium milling industries, can concentrate significant quantities of naturally occurring radioactive material (NORM). When present in sufficient amounts NORM may pose a significant health hazard. Of particular concern are the long-lived radionuclides: ^{226}Ra and ^{210}Pb (part of the ^{238}U-decay chain), gaseous ^{222}Rn (also a member of the ^{238}U-decay chain), and ^{228}Ra, ^{228}Th, and ^{224}Ra (from the ^{232}Th-decay chain). Because of health related concerns NORM is subject to strict regulations in many countries. In Canada it has recently been proposed that the release limits for diffuse NORM in solids be lowered from 10 to 0.3 Bq/g for many of the U- and Th-series radionuclides listed above. If the proposal is adopted (which appears probable) it will result in a significant increase in the amount of NORM contaminated material (which would previously have been classified as *de minimus*) which will require special management (*i.e.*, handling and disposal). In addition, adoption of the proposed changes will result in an increased need for the accurate and rapid (hence most likely non-destructive) measurement of these radionuclides in samples.

High resolution γ-ray spectrometry, employing a hyperpure Ge detector, offers a non-destructive alternative to the much more expensive, time consuming and destructive measurement of a number of NORM radionuclides of interest using α-spectrometry, for example.

While ^{226}Ra may be determined directly *via* measurement of its 186.1 keV γ-ray (3.28 % γ-yield) poor sensitivity and a potential spectral interference from ^{235}U (185.7 keV, 53 % γ-yield) limits this approach. A more sensitive and interference-free alternative is the indirect determination of ^{226}Ra *via* its much more intense γ-emitting progeny ^{214}Pb and ^{214}Bi. A major hindrance to a more rapid sample turnaround time for specific activity determinations of NORM samples by γ-ray spectrometry has however, been the time required to ensure secular equilibrium between ^{226}Ra and ^{214}Pb and ^{214}Bi, due to the possible loss of the intermediate gaseous daughter ^{222}Rn ($T_{1/2}$=3.82 days) during sample collection and preparation. Typically, in preparation for analysis samples are sealed in containers and left a minimum of 2 to 3 weeks (~ 3 to 5 half-lives of ^{222}Rn) to ensure the re-establishment of secular equilibrium between ^{226}Ra and its progeny ^{214}Pb and ^{214}Bi. A more rapid alternative to this conventional approach (but which still utilizes the more sensitive ^{214}Pb and ^{214}Bi measurements) has been developed at our laboratory. The approach involves measuring the degree of sample disequilibrium, due to incomplete in-growth of ^{222}Rn in the sample, and correcting for this as necessary. Using this approach sample turnaround time can be reduced to 2 or 3 days if necessary. This approach, together with the NORM analysis protocol developed at our laboratory over the past ten to fifteen years, will be presented.

Email address: *John.Duke@UAlberta.CA*

NUCLEAR ANALYTICAL TECHNIQUES IN THE SUGAR CANE AGROINDUSTRY

ELISABETE A. DE NADAI FERNANDES [1], EPAMINONDAS S. B. FERRAZ [1,2], MÁRCIO A. BACCHI[1], VIRGÍLIO F. NASCIMENTO FILHO [1,2]

[1]Centro de Energia Nuclear na Agricultura, Universidade de São Paulo, CP 96, 13400-970, Piracicaba, SP, Brazil, E-mail: lis@cena.usp.br; [2]Escola Superior de Agricultura Luiz de Queiroz, Universidade de São Paulo,CP 9, 13400-970 Piracicaba, SP, Brazil

Brazil is the world's largest producer of ethanol (13 billion liters per year) and sugar (12 million tons per year) from the milling of 240 million tons of sugar cane. It also has the potential to provide 240,000 tons of dry yeast per year, derived from ethanol production, for use as a protein supplement to animals. Mechanization of field operations has changed the quality of the cane supply due to significant amounts of soil entering the mill. Drawbacks of the soil on factory performance and costs have prompted an investigation to develop a reliable method for assessing its content in the cane supply. Nuclear methods, involving the determination of elements present in much higher concentrations in soil than in cane, such as instrumental neutron activation analysis (Al, Fe, Hf, Sc, Th), energy dispersive X ray fluorescence analysis (Al, Fe, Si, Ti, Zr) and natural gamma-ray emitter radionuclides (^{212}Pb) were considered. INAA has also been applied in evaluation studies of the efficiency of the cane washing system, wearing rates of industrial machinery and the consequent introduction of metallic elements in the cane flow, elemental dynamics during the fermentation process, and elemental composition of brown and white sugars and yeasts.

1 Introduction

Motivated by the oil crisis in the seventies and the ability to supply cleaner renewable fuel - plenty arable land, adequate climate, hydric resources, enormous labor contingency and well-known technology, Brazil has become the largest producer of ethanol from sugar cane juice fermentation by yeasts. The country also ranks first in sugar production. To provide the sugar cane needed by the agroindustry, mechanization of field operations was introduced deteriorating the quality of raw material, with a major impact on factory performance [1,2]. Metallic components were continuously incorporated in the industrial cane flow as a consequence of the machinery wear [3]. Much effort for reducing soil in harvested cane has been given by millers to improve loading and washing systems [4]. Additional costs involved in cane washing, water treatment, sugar losses and equipment maintenance directed investigation toward reliable methods for assessing soil content in cane supply using non-conventional methods of analysis [1,2,5-7].

Evaluation of mineral elements in the fermentation process of ethanol production has received much attention lately [8], since most distilleries are recovering surplus yeast to produce dry yeast of high nutritional quality for use as protein supplement in animal food formulations [9]. Nuclear and related techniques are very useful for such purposes as demonstrated in this review.

2 Materials and Methods

Instrumental neutron activation analysis (INAA) [1-4,6,8,9], energy dispersive X-ray fluorescence analysis (ED-XRF) [5] and natural gamma-ray emitters spectrometry [7] were used for determination of chemical elements and radionuclides that could be used as tracers in the industrial processing of cane for sugar and ethanol production. Samples of soil, sugar cane stalks, shredded cane, bagasse, cane juice, yeast, must, cream, wine and mill metallic parts were oven-dried (solids) and freeze-dried (liquids) and ground to small particles. Sample size varied from 200-500 mg (INAA, XRF) to 2 kg (natural radioactivity). Details of each procedure are given in the related papers.

3 Results and Discussion

In principle, elements predominating in soil over cane matrix can be used as tracers for soil entering the industries with sugar cane stalks. Scandium was the element of choice since it is present in levels which are at least a thousand times higher in soils than in cane. It ranged from 4 to 66 ppm in 10 different soils from the sugar cane cultivated area of the São Paulo state, while remained below 3 ppb for all sugar cane plants (variety SP71-1406) grown on such soils. Other soil elements could be used with restrictions dictated by the relatively higher levels in plants or unfavorable INAA analytical condition [1]. The INAA method for Sc seems potentially better than the conventional ash method used in mills, achieving more accurate and precise results, mainly in presence of minimum amounts of soil [2].

Results from 9 field trials evidenced that the amount of soil removed from cane stalks during harvesting and washing is dependent on the soil characteristics. The most significant factor observed in this study is that errors in estimation of soil in cane materials using the INAA method can arise in cases of fractionation of soil in samples because elemental segregation occurs between size particles [6].

Al, Fe, Si, Ti and Zr were determined by ED-XRF [5] in the 10 soil samples analyzed previously by INAA [1]. The elemental ratios Fe/Si and Fe/(Si + Al), established to identify soils, proved to be useful for measurements of soil in cane.

^{212}Pb was chosen as the best tracer among the gamma-ray emitting radionuclides of natural occurrence in soils, providing consistent and reliable results when compared to the traditional ash method [7].

Wear intensity is large for the components, like shredder hammers and mill rollers, in continuous contact with bagasse and juice in motion. Iron, Cr, Zn and Co, elements taking part in composition of such metallic materials, were significantly increased as the industrial cane flux proceeded across the extraction system [3].

Scandium, Fe, Th and Hf showed much higher concentrations in soil than in cane so that efficiency of washing system could be evaluated simply on the basis of differences in concentrations between industrial cane before and after washing, demonstrating the practicability of the method in routine applications [4].

Sugar cane characteristic elements (Br, K, Rb, Zn) were similarly distributed in the 5 compartments of the fermentation unit, namely must, yeast suspension, wine, non-yeast wine and yeast cream. An accumulation trend was observed for terrigenous elements (Ce, Co, Cs, Eu, Fe, Hf, La, Na, Sc, Sm, Th) in the yeast cream, which may create problems in centrifugation process such as minimizing cell recuperation and damaging equipment because of increased wear rates [8].

Elemental composition of yeast biomass taken at different conditions of ethanol production indicated the presence of many trace elements, e.g. As, Br, Ca, Ce, Co, Cr, Cs, Eu, Fe, Hf, K, La, Na, Rb, Sc, Sm, Th and Zn in varied concentrations. Thus, yeast suitability needs to be assessed to comply with maximum tolerable levels of such elements in feed formulations for domestic animals [9].

4 Acknowledgements

Authors are thankful to FAPESP, CNPq and PRONEX for financial support and COPERSUCAR and Usina Santa Bárbara S.A. for technical assistance.

References

1. Fernandes, E.A.N. Scandium as tracer in the sugar and alcohol agroindustry. *J. Radioanal. Nucl. Chem.*, Articles. **168**-1 (1993) pp 41-6.
2. Fernandes, E.A.N. and Bacchi, M.A. Nuclear and traditional methods for soil determination in sugar cane industry - Validity of sampling procedure. *Biological Trace Element Research.* **43/45** (1994) pp.643-8.
3. Fernandes, E.A.N., Bacchi, M.A. and Ferraz, E.S.B. Sugar cane mill wear investigation using a nuclear analytical technique (INAA). *Int. Sugar Journal.* **95-**1129 (1993) pp 33-6.
4. Bacchi, M.A. and Fernandes, E.A.N. Evaluation of the washing system efficiency in sugar cane mills by neutron activation analysis. *Biological Trace Element Research.* **43/45** (1994) pp. 663-9.
5. Fernandes, E.A.N., Bacchi, M.A., Nascimento Filho, V.F. and Simabuco, S.M. X-Ray fluorescence for soil characterization in shredded cane. *International Sugar Journal* **1154-**97 (1995) pp.92-5.
6. Fernandes, E.A.N. Dirt in cane removal influenced by soil characteristics. *J. Radioanal. Nucl. Chem.* **216-**2 (1997) pp.285-8.
7. Bacchi, M.A. and Fernandes, E.A.N. Natural radionuclides as dirt tracers in sugar cane consignments. *J. Radioanal. Nucl. Chem.* **235/**1 (1998) pp. 179-83.
8. Nepomuceno, N., Fernandes, E.A.N. and Bacchi, M. A. Dynamics of chemical elements in the fermentation process of ethanol production. *J. Radioanal. Nucl. Chem.* **216**-2 (1997) pp.289-92.
9. Fernandes, E.A.N., Nepomuceno, N., Trevizam, A. B. and Amorim, H.V. From potential to reality: yeasts derived from ethanol production for animal nutrition. *J. Radioanal. Nucl. Chem.* **234/**1-2 (1998) pp. 113-8.

ADVANTAGES OF INAA OVER COMPETING CHEMICAL ANALYSIS TECHNIQUES FOR INDUSTRIAL QUALITY ASSURANCE

J.ST-PIERRE AND G.G.KENNEDY

Ecole Polytechnique, P.O.Box 6079, Downtown, Montreal, Quebec, H3C 3A7, Canada

The recent decrease in the number of research reactors around the world and the development of powerful analytical techniques like ICP-MS have raised questions about the future of Instrumental Neutron Activation Analysis. In this work we will show that INAA is more than ever the method of choice for industrial quality assurance because of the penetrating power of neutrons and gamma rays. Because of the absence of matrix effects, INAA is the only method that gives accurate concentrations in bulk samples with no sample preparation. The elimination of the sample dissolution step makes it possible to avoid contamination and loss of material and reduces the analysis time. The turn-around time for elements producing short half-life radionuclides is about one day. In our laboratory at Ecole Polytechnique, the number of industrial samples analyzed by INAA increased from 2,000 to more than 10,000 per year in the last ten years. Materials that are difficult to dissolve (wood, plastics and metals) represent more than 80% of the samples analyzed. The use of a very stable neutron source (SLOWPOKE reactor) and state-of-the-art gamma ray spectrometers and software make it possible to guarantee accurate and reproducible results over long periods of time, while automation with pneumatic irradiation systems and mechanical sample changers increases the productivity of the laboratory.

1 Introduction

From 1975 to 1996 the number of nuclear research reactors in the world has decreased from 351 to 265 and nothing indicates a change in the trend; indeed 75% of the operational reactors are more than 20 years old and only twelve new reactors are under construction (1). During the same period, powerful analytical techniques like ICP-MS were developed and have gained in popularity. However, it is false to think that instrumental neutron activation analysis is obsolete; in fact, we believe that INAA is the method of choice for industrial quality assurance because it is the only method that gives accurate concentrations in bulk solid samples with no sample preparation. Also, industrial users require results which are accurate and reproducible (even over long periods of time) and the turn-around time must be reasonably short. In this paper we will show how INAA can reach these goals.

2 Advantages of INAA

By far the most important advantage of INAA over the other analytical techniques is the absence of matrix effects, due to the penetrating power of neutrons and gamma-rays, which permits accurate analysis with no sample preparation. Not only does this facilitate the analysis, but it increases the accuracy since it eliminates the possibility of sample loss or contamination. An INAA laboratory should therefore specialize in the analysis of solid samples which are difficult to dissolve. In our laboratory, wood, paper, plastics and metallic samples constitute more than 95% of the industrial samples we analyze on a regular basis. The only pre-treatment needed for some concentrated solid samples is dilution in an inert matrix like sugar.

Since no sample preparation is needed, INAA can be fast, especially for elements determined with short-lived radio-isotopes. Each sample received in our lab is logged, weighed if necessary, and placed in a polyethylene vial and is ready for analysis. The use of an automated pneumatic transfer system and mechanical sample changers increases our productivity. Analysis reports are often faxed or e-mailed the same day the samples are received. This fast turn-around time is essential because INAA results are often necessary to solve urgent production problems or to verify the quality of material before shipment; the constant availability of the analytical service is therefore essential. The SLOWPOKE reactor at Ecole Polytechnique is essentially maintenance free, average one day down-time per year and one month down-time every twenty years for refuelling.

To ensure the consistency of their products, many industries require an analytical technique which is reproducible over long periods of time. Recent studies have shown that the neutron flux at our SLOWPOKE reactor is reproducible to within 2% in normal operation (2). Three digital gamma-ray spectrometers were recently added to our equipment; this new technology eliminates the possibility of amplifier gain shifts and permits the counting of samples with higher activities.

To verify the stability of our system, we analyzed the same iodine standard (KI solution dried on a piece of wood) about 900 times since 1990. All data obtained are presented in Figure 1 where the central line represents the real iodine concentration (218μg) while the upper and lower lines show a variation of plus or minus 2% . The average concentration obtained is 218.7 μg with a standard deviation of 2%.

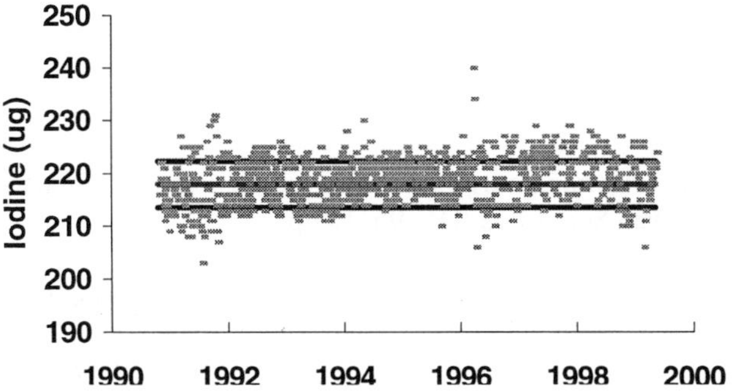

Figure 1 Repeated analysis of an iodine standard

3 INAA at Ecole Polytechnique

The usefulness of INAA is shown on Figure 2, which gives the number of industrial samples irradiated in our reactor each year since 1981. The number of analyses is now limited to about 10,000 per year by the capacity of the laboratory staff, not by the number of potential users.

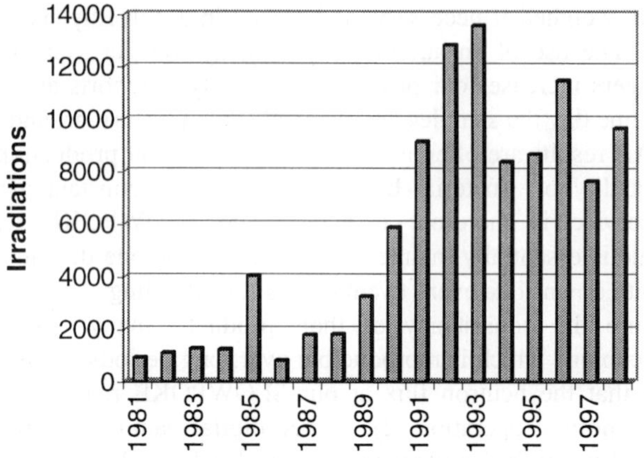

Figure 2 . Number of irradiations of industrial samples at Ecole Polytechnique

4 Conclusion

INAA has potentially a bright future in the surviving nuclear analytical labs but they should concentrate their effort on the analysis of solid samples for which INAA is unquestionably the best method.

References

1. International Atomic Energy Agency, Nuclear Research Reactors in the world, Reference Data Series no3, Dec 96, 127 p

2. St-Pierre J. and Kennedy G., Effects of reactor temperature and sample mass on the activation of biological and geological materials with a SLOWPOKE reactor, Biological Trace Element Research, (in press 1999)

Long Range Transport of Heavy Metals over the Great Lakes as Determined by Neutron Activation Analysis

S. LANDSBERGER[1], S. R. BIEGALSKI[2]AND F. ISKANDER[1]

[1]*University of Texas at Austin, Nuclear Engineering Teaching Lab*
Pickle Research Campus, R-9000, Austin, Texas 78712
E-mail: s.landsberger@mail.utexas.edu

[2]*Center for Monitoring Research*
1300 N 17th St, Suite 1450
Arlington, Virginia, 22209
E-mail: biegalski@cmr.gov

In 1990 Canada and the United Sates signed an agreement to monitor air and water quality of the Great Lakes. Over the past decade we have been involved in the analysis of airborne particles collected at two Great Lakes stations and one rural station to monitor the concentration of heavy metals that may deposit into the waterways. We have employed several non-traditional methods in neutron activation analysis including Compton suppression to achieve very detection limits for heavy metals at these sites. Elements such as antimony, arsenic, cadmium, copper, nickel, selenium, tin, vanadium, zinc and others were routinely determined. Several statistical evaluations were performed including, enrichment factor analysis, elemental ratios and factor analysis. With these results we have been able to identify several important source signatures including those attributed to mining activities. Our findings have also shown that long-distance travel of bromine during depleted ozone events can be seen in Central Ontario, many hundreds of kilometers downwind from the Arctic Circle.

1 Introduction

Over the past decade there has been an abundance of air pollution monitoring work done to elucidate the potential impact of air toxics depositing into the Great Lakes. These substances include a wide variety of organic chemicals, sulphur, and various trace metals such as lead, cadmium and mercury [1]. It is well known that there are many sources that can emit airborne particles that can then re-enter into freshwater potentially contaminating the ecosystem. More recently, there have been concerns of emissions from smelters from various mining activities, a very important economic base for Canada and its various provinces. To accurately quantify the heavy metals contained in airborne particles at low concentrations sensitive analytical techniques are required. We have successfully utilized a variety of neutron activation analysis methods that can be judiciously employed to determine many trace elements at the level of several nanograms. These results are then effectively used in statistical interpretations for factor analysis, trace element ratios, enrichment factors, and other source-receptor models in conjunction with back air trajectories.

2 Methodology

Aerosol Sample Collection

Three sampling sites at Point Petrie (between Toronto and Kingston on Lake Ontario), Burnt Island in Lake Huron and Egbert (inland station 70 km north of Toronto) chosen by the Canadian Atmospheric Environment Service were used to collect aerosols. The samplers were Sierra Anderson high-volume equipped with a 10 μm inlet and operated between 18-19 liters per second every six days for a 24-hour period. Results for Whatman-41 air filters are presented here are for sampling periods between 1992-1994.

Neutron Activation Analysis

Various non-destructive neutron activation analysis (NAA) methods were used to determine thirty-four elements in the filters. Various thermal and epithermal NAA methods in conjunction with Compton suppression techniques were used. The combination of these methods allowed the determination of several key elements usually not reported in routine NAA at very low elemental mass. These included copper, antimony, arsenic, indium, silver, tungsten and tin. These elements are very important in identifying smelter activities in source-receptors models. A complete description of the NAA methodology has been previously described [2].

Quality Assurance and Quality Control

Strict quality assurance and quality control protocols were followed. These included the use of single elemental solutions for calibration, various standard reference materials, careful monitoring of neutron flux variations, use of separate isotopes and irradiation procedures to determine the mass of the same element. Treatment of data below the limit of detection and blank correction of data were carefully evaluated to achieve a high degree of confidence in the final results before being used in the various statistical manipulations.

3 Results and Discussion

In a multiple component variable system, such as airborne particles, it is now well accepted that that certain groups can follow similar trends. Factor analysis can yield results that strongly indicate which groups co-vary together. For Burnt Island we were able to identify seven factors. Factor one contains Al, Ca, Ce, Fe, Mn, Sc, Si, Ti and V; indicating a crustal source. Factor two contains Fe, Mn, Sb, Se and Zn; indicating a possible mixed soil source with an incinerator. The third factor contains Ca, Fe, K, Sc and Zn; indicating to be a second crustal source. The fourth factor contains Br, I and Na; which clearly represents a salt source. The fifth factor contains, Ag, As, and Se; which can represent smelting activity in Sudbury. The sixth factor contains Sn, while the seventh factor contains only In. The last two factors are clearly indicative of smelter operations. Similar factors were seen for Point Petrie and Egbert. While seasonal plots of individual elemental air concentrations may be of some significance, elemental ratios can reduce the background variation substantially so those individual pollution events can be more distinctly seen. Elemental ratios of As/Sb are such a case. Over a three-year period we were able to identify anomalies either in elevated As or Sb concentrations. Enrichment factor analysis for all three sampling stations indicated high concentrations (as compared to the earth's crust) for Sn, In, As, Sb, Ag, Se, Ni, W, and Zn. Crustal elements such as Si, Sc, Ti, Ce, Th, Fe, etc, showed little or no enrichment.

4 Conclusions

We have demonstrated that vanguard NAA techniques can be ideally implemented in a monitoring program that can result in the determination of several unique source signatures. The precision and accuracy of the methods can greatly benefit source receptor modeling in samples collected in rural/remote areas of Ontario. These results can also benefit the calculation of heavy metal deposition into the Great Lakes.

References

1. Sonzogni, W. C. and Swain, W. R. In *Toxic Contaminants in the Great Lakes*, Nriagu, J.O and Simsons, M. S., Eds., John Wiley, 1980, pp1-29.
2. Biegalski, S. R. and Landsberger, S., Improved detection limits for trace elements on aerosol filters using Compton suppression epithermal irradiation techniques, *J. Radioanal. Nucl. Chem.*, **192** (1995) 195-204.

Session C3

Radioisotope Production

CURRENT STATUS OF INER'S CYCLOTRON-PRODUCED RADIOISOTOPE PRODUCTION

WUU-JYH LIN, LIE-HANG SHAN, HENTON HUNG AND GANN TING

Institute of Nuclear Energy Research, P.O. Box 3-27, Lung-Tan, Taiwan

E-mail:bblin@usa.net

It has been six years since the installation of TR30/15 cyclotron (purchased from EBCO) at INER. The Division of Isotope Application is engaged in the production of radioisotopes for medical applications. Five radioisotopes are produced routinely namely ^{123}I, ^{201}Tl, ^{67}Ga, ^{18}F and ^{81}Rb/^{81m}Kr generator. ^{123}I is used for the development of Na^{123}I, ^{123}I-MIBG, ^{123}I-IBZM, ^{123}I-BMIPPA, ^{123}I-IMP, ^{123}I-OIH, ^{123}I-•CIT. ^{201}Tl is made for ^{201}TlCl injection solution and is applied commercially to local hospitals. ^{67}Ga is produced for ^{67}Ga citrate and is also provided commercially to local hospitals. ^{18}F is synthesized as ^{18}FDG for clinical trial in Taiwan University Hospital. ^{81}Rb/^{81m}Kr generator is also at clinical trial stage in Tri-general Hospital. Besides isotopes mentioned above, ^{111}In, ^{57}Co, ^{62}Zn/^{62}Cu and ^{68}Ge/^{68}Ga are also produced less frequently for research purpose. ^{123}I is a turnkey target bought from Nordion. We have developed the chemical processes for the rest radioisotopes. The application of legal approval Na^{123}I oral solution was submitted to the Department of Health (DOH) and has been approved in May 1999. To produce ^{124}I for related iodine radiopharmaceuticals and $^{18}F_2$ for ^{18}F fluoro-Dopa are our next goals. Because commercial supply of isotopes is one of INER's mission, quality assurance and Good Manufacture Practice (GMP) are the requirements. According to Taiwan DOH regulation, all the pharmaceutical manufacturers must be GMP certified. To fellow this trend, Cyclotron and Isotope Production Groups passed the criteria of ISO-9002 qualification and were certified by TÜV in July 1998. Radiopharmaceutical Production Group has been approved as a radiopharmaceutical GMP production unit in November 1998.

1 Production of ^{123}I

^{123}I is a turnkey gas target system bought from Nordion. It is produced via ^{124}Xe(p, 2n) ^{123}Cs $\xrightarrow{\beta}$ ^{123}Xe $\xrightarrow{\beta}$ ^{123}I nuclear reaction. We replaced target windows when the cumulated irradiation time reached 20000μAh. The first replacement looked fine from the surface of the replaced windows. For the second and third replacements, we have found a dark spot upside on both the windows. In May 1998, a window leakage was found during irradiation. ^{124}Xe gas was soon pulled back to storage tank. About 80% of ^{124}Xe was recovered. The original beam diagnosis collimator is a round circle. The real beam profile can not be detected. In order to optimize beam condition and to adjust it easier by operator, a four-slit collimator was mounted in front of diagnosis box. From the beam current on the target and on the four slits, it is very clear to judge if the beam is well spread or not? It has been 16 months since the four-slit collimator was installed. It is working fine so far. Most of the ^{123}I is used for the research on nuclear medicine, such as Na^{123}I, ^{123}I-MIBG, ^{123}I-IBZM, ^{123}I-BMIPPA, ^{123}I-IMP, ^{123}I-OIH, ^{123}I-•CIT. Na^{123}I oral solution was approved by the DOH In May 1999 and will be supplied to local hospitals.

2 Production of ^{201}Tl, ^{67}Ga, ^{57}Co , ^{111}In and ^{18}F

The ^{201}Tl process is modified from Qaim's method •1•. It is a semi-automatic system, partially controlled by switch and partially controlled by manipulator. A series of analysis methods are also developed to analyze nuclides, radiochemical impurities and chemical impurities for ^{201}Tl. Subsequently, the radioisotope was used to manufacture ^{201}Tl-thallous chloride radiopharmaceutical.^{201}Tl-thallous chloride injection solution was approved by the DOH in June 1996 and then routinely supplied to local hospitals.
The chemical processes to produce ^{67}Ga, ^{57}Co and ^{111}In are very similar to the process of ^{201}Tl expect different ion exchange resin, different acid to dissolve metal and different elute solution. ^{67}Ga, ^{57}Co and ^{111}In production apparatuses are also made with semi-automatic systems.^{67}Ga-gallium citrate injection solution was approved by DOH in 1998. It is already on Taiwan market. ^{57}Co and ^{111}In productions are in trace amount. They are only used for research purposes.
^{18}O(p,n)^{18}F nuclear reaction was chosen for the production of ^{18}F. Laser beam weld titanium target chamber with graphite slit beam diagnosis system was constructed. The amount of ^{18}O enriched water in target chamber is about 1.5mL. The average production yields of ^{18}F are 80mCi/μAh, 71mCi/μAh and 65 mCi/μAh, respectively for 30minutes, 60minutes and 90minutes irradiation. This is a semi-automatic system. Water injection and transfer are outside controlled to make the continuous production possible without much radiation dose to the operator. After irradiation ^{18}F with ^{18}O enriched water is delivered to hot cell for further applications.

3 The production of ^{81}Rb/^{81m}Kr, ^{62}Zn/^{62}Cu and ^{68}Ge/^{68}Ga

^{81}Rb/^{81m}Kr Generator can be provided from a medium cyclotron by utilizing nuclear reaction of ^{82}Kr(p,2n)^{81}Rb. Experimentally, the yield of ^{81}Rb is 1.85GBq(50mCi) with 99.9% enriched ^{82}Kr gas target under the irradiation of proton energy at 25-30 MeV and a beam current of 5μA for 30 minutes•2•. The recovery of ^{82}Kr from target chamber at the end of bombardment is greater than 99.99%. ^{81m}Kr can be eluted continuously from the prepared millicuries generator with air flush at 1 lpm. The yields of ^{81m}Kr is 90% with respect to the radioactivity of ^{81}Rb loaded and the radionuclide purity of greater than 99.999% is quite satisfactory to meet the specifications for lung imaging in nuclear medicine applications.
The parent isotope, ^{62}Zn, is usually separated from the nuclea reaction ^{63}Cu(p, 2n) ^{62}Zn with proton bombardment of natural copper foil target. The yield of ^{62}Zn was estimated approximately of 2 mCi/μAh under the irradiation of proton energy at 22MeV. A simple and efficient procedure for the preparation of ^{62}Zn/ ^{62}Cu radionuclide generator was developed with a single anion-exchange column in HCl medium•3•. The results from hot run experiments reveal that above ten-millicurie of ^{62}Zn/^{62}Cu generator can be readily provided with the irradiation of 0.1mm x13mm• Cu foil at proton energy between 22 and 30 MeV and a beam current of 5 μA for several hours. High radio- and chemical purity of ^{62}Cu that eluted

from the generator system with 3ml of 1.0M HCI is suitable for radiopharmaceuticals labeling.

A simple procedure was developed for the preparation of a binary Ga/Ag alloy solid target electrodeposited on silver surface of target holder in a strong alkali solution containing cyanide electrolyte•4•. Experimentally, $^{68}Ge/^{68}Ga$ generator can be provided by using a generic tin dioxide as adsorbent with the target material under the irradiation of proton energy at 19MeV and a beam current of 10-100μA. The yields of ^{68}Ga with ridionuclide purity greater than 99.99% can be obtained with 4ml of 1.0M HCI.

4 ISO-9002 and GMP Certified Manufacturer.

To be a radioisotope supplier, our primary goal is customer satisfaction by preventing nonconformity and to keep the whole facilites in reliable condition. Preventive Maintenance was implemented for cyclotron and isotope production. Furthermore, to achieve international certification of ISO quality system is necessary. Quality assurance document, standard operation procedures for every detailed production operation steps and standard procedures in contract, purchasing, customer's complaint, internal audit etc. are all included. In July 1998, the production of ^{201}Tl, ^{67}Ga isotopes obtained the ISO-9002 certificate from TÜV. We just passed the second year evaluation in July 1999. The properties of radiopharmaceuticals are short-lived with highly specific activity and in small amount and are different from ordinary pharmaceuticals. INER has accomplished more than 160 GMP documents for the production of radiopharmaceuticals. INER's radiopharmaceutical production was approved as a radiopharmaceutical qualified GMP production manufacturer in November 1998.

References

1. Qaim S. M., Weinreich R. and Ollig H., Production of Tl-201 and Pb-203 via Proton Induced Nuclear Reaction on Natural Thallium, *International Journal of Applied Radiation and Isotopes,* **30** (1979) pp. 85-95.
2. Lo A. R. and Cheng W. L., Study on Millicuries $^{81}Rb/^{81m}Kr$ Generator Production with an Automated Controlled System, *Nucl. Sci. J.* **36** (1999) pp.13-26.
3. Cheng W. L., Wang W. M., Cheng T. H. and Tsai Y. M., Research and Development of Copper-62 Radiopharmaceuticals, *Nucl. Sci. J.* **34** (1997) pp.276-290.
4. Cheng W. L., Gao Y., Lee C. S. and Lo A. R., Preparation of $^{68}Ge/^{68}Ga$ Generator with a Binary Ga/Ag Electrodepositions as Solid Target, Tenth International Conference on Modern Trends in Activation Analysis, Maryland, USA, April 19-23, (1999). (To be published).

A TECHNETIUM -99m GENERATOR FOR THE 21st CENTURY

Michael Druce and Peter Anderson
Radiopharmaceuticals Division (Australian Radiopharmaceutical)
Australian Nuclear Science and Technology Organisation
New Illawarra Road, Lucas Heights, NSW, Australia

The Australian Nuclear Science and Technology Organisation (Ansto) has been producing Technetium-99m generators since the late 1970s. The current generator has been on the market since 1982 and has served the Australian nuclear medicine community well.

Recently, however, the leaders in the nuclear medicine community have been asking for improvements in the generator. Principally the requests have been for an eluate, which does not contain nitrate ions, and for a means of easily varying the elution volume. It was decided 18 months ago to develop a generator having these features. The challenge for us was to develop this generator and introduce changes to our plant while still continuing the manufacture of the existing generator in the same plant.

The result of our work is a generator, which is a modification of the existing generator. Externally it looks and operates quite differently, while internally many of the components are the same or similar to the current model. This has allowed a smooth manufacturing transition. Significantly the new model incorporates dry bed technology. The eluate is free of nitrate ions and by simple manipulations the eluate volume can be accurately controlled. This is a generator to meet the needs of nuclear medicine in the 21st century.

Introduction

Since the early 1980s we have been producing generators which utilise a plastic bag of saline as the elution liquid. The generator column stays wet with saline after elution. Ninety parts per million nitrate ions are added to the saline as an oxidant to prevent the reduction of pertechnetate ions on the alumina column. If the pertechnetate ions are reduced there is a substantial reduction in the generator elution efficiency. Recently some major customers have complained that the presence of nitrate in the generator eluant decreases the technetium-99m labelling efficiency when labelling some diagnostic imaging agents.

It therefore became imperative that the nitrate is removed from the saline solution and another method found to maintain satisfactory elution efficiencies.

The Challenge

The present generator was designed with certain features. These were that it was extremely robust, it is easy to use and the components can be easily recycled. The robust features were initially designed so that the overall generator shipping package, because of the wet bed nature, would survive a 9 metre drop test. Over the years this feature became very important as it improved the life of the component parts and helped reduce production costs. Also we have considerable capital investment in our generator production facility and it was seen as desirable to continue to utilise this investment.

The challenge therefore was to design a generator that would have the same or similar features to our existing generator and would also utilise the existing production facility.

The Solution

The method adopted was to use saline in glass vials and to leave the generator column dry after elution. This would enable the same column and lead pot to be used hence enabling us to use the same in cell equipment to be used to manufacture the generator.

It was found that saline in glass vials (USP type 1 glass) did not have an adverse effect on the generator. In addition if the column was left dry after the elution oxygen in the air was sufficient to ensure that the pertechnetate ions remained oxidized and the generator gave good elution efficiencies.

An added advantage of using glass vials was that the volume of elution was accurately known and could be easily adjusted by aseptically removing some of the saline before elution.

The change to glass vials meant the elution system for the generator had to be changed. When using a plastic bag, the bag simply collapses as saline is removed. However when using the glass vials a vent system had to be devised to allow the elution to proceed. Also the evacuated vial at the elution end had to be larger in volume than the saline vial to allow all of the saline to be removed from the column.

Changing the elution system led to radical changes in the external case of the generator. A lot of engineering effort has been put into this design with key customers being consulted as part of the process. Although different in appearance to the current generator the new case maintains the same key features of being robust, easy to use and easy to recycle components.

Testing

A rigorous schedule of testing was undertaken. The testing was split into three: chemical, microbiological and operational.

i. **Chemical**

   Generators were manufactured and the subjected to full pharmacopoeia testing. The results were compared with the existing wet bed generator. No significant differences were detected. Elution efficiency for both types of generators remained high.

Importantly it was found that the dry bed generator had improved labeling efficiency of diagnostic kits. Results for radiochemical purity of two proprietary kits is listed below:

Diagnostic Kit	Wet Bed %Radiochem. Purity	Dry Bed %Radiochem Purity
Cardiolite (Dupont)	95.35	98.08
MAG-3 (Mallinkrodt)	98.82	99.34

ii. Microbiology

The eluate from the generator should be sterile. The change in elution method gave considerable microbiological concern. The saline vial must be changed for every elution and this introduces an opportunity for microbiological contamination. Also the vent into the vial must contain a 0.2 micron filter to stop microbiological contamination of the generator system.

Generators were tested by using TSB media for elution instead of saline and the eluates tested for sterility. Generators were also maintained at higher temperatures (40°C) to provide a worst possible scenario. In all tests the generator was shown to provide a sterile eluate.

iii. Operation

Initially mock-ups of the design were made and sent to key customers for testing. This, along with our own testing of mock-ups provided the basis for the final design. When the plastic moldings were available repeated manufacture and elutions has continued. Final designs have now also been sent to the key customers for confirmatory evaluation.

Conclusion

The result of our work is a generator, which is a modification of the existing generator. Externally it looks and operates quite differently, while internally many of the components are the same or similar to the current model. This has allowed a smooth manufacturing transition. Significantly the new model incorporates dry bed technology. The eluate is free of nitrate ions and by simple manipulations the eluate volume can be accurately controlled. This is a generator to meet the needs of nuclear medicine in the 21st century.

PRODUCTION OF RADIOISOTOPES IN THE BR2 REACTOR

BERNARD PONSARD

Radioisotopes Project Manager
Belgian Nuclear Research Center SCK.CEN
Boeretang 200, B-2400 Mol, Belgium
E-mail: bponsard@sckcen.be

Research reactors continue to play a key role in the production of radioisotopes for various applications in the nuclear medicine (diagnostic, therapy, palliation of metastatic bone pain), industry (radiography of welds, ...), agriculture (radiotracers to monitor the movement of groundwater, ...) and basic research (study of the nuclear properties of heavy elements, ...). The beneficial uses of radioisotopes is still a growing market in which the BR2 reactor is very active by bringing its contribution to the need to maintain a continuous supply of products as ^{99}Mo (^{99m}Tc), ^{131}I, ^{133}Xe, ^{192}Ir, ^{89}Sr, ^{186}Re, ^{153}Sm, ^{90}Y, ^{32}P, ^{60}Co, ^{117m}Sn, ^{169}Yb, ... and the promising ^{188}W (^{188}Re). The BR2 reactor, put into operation in 1963 and refurbished from July 1995 till April 1997, is a 100 MW high-flux Materials Testing Reactor, using 93% ^{235}U enriched uranium as fuel, light water as coolant and beryllium as moderator. The core configuration provides thermal neutron fluxes up to 1.2 10^{15} n/cm².s for the production of radioisotopes characterized by high specific activities within standard irradiation cycles of 3 weeks operation. The productions of radioisotopes are carried out in dedicated in-core devices; some of them allow the loading and unloading of irradiation targets during the operation of the reactor. Hot-cells and storage facilities are available for the loading of the irradiated targets into suitable containers to enable their transport to dedicated processing facilities.

1 Introduction

The BR2 reactor is a high-flux Materials Testing Reactor operational since 1963 and refurbished in the period 1995-1997. It is operated by the Belgian Nuclear Research Center in the framework of many international programs concerning the development of structural materials and nuclear fuels for various types of reactors. The availability of high thermal, epithermal and fast neutron fluxes allows also an important routine production of radioisotopes characterized by high specific activities for medical and industrial applications. A standard irradiation cycle consists of 3 weeks operation at an operating power between 50 and 70 MW_{th}, depending on the core configuration, which is easily adaptable to the experimental load. The present operating regime consists of 5 irradiation cycles per year.

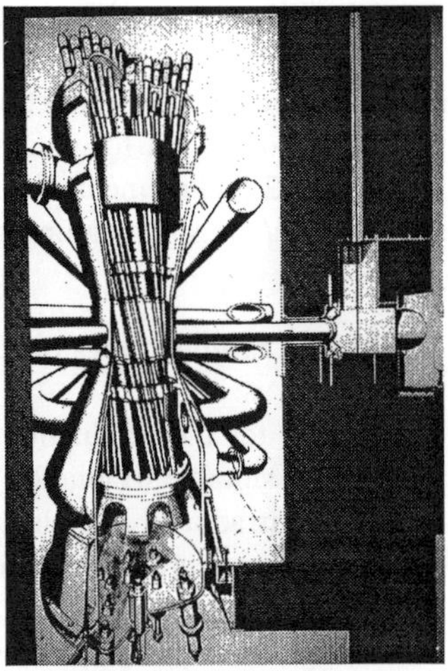

Figure 1. BR2 reactor

2 Core configuration and irradiation positions

The BR2 reactor uses 93% ^{235}U enriched uranium as fuel and is moderated by light water and beryllium. The core is composed of beryllium hexagons with central irradiation channels of 200, 84, 50 or 33 mm diameter. The cooling water is pressurized at 12 bar and has a temperature of 40-45°C. The aluminium pressure vessel is located in a pool (Figure 1) filled with demineralized water. The actual core configuration (Figure 2) is centered on the central H1 channel, characterized by thermal neutron fluxes up to 1.2 10^{15} n/cm².s, in order to optimize the productions of radioisotopes of high specific activities as ^{192}Ir, ^{89}Sr, ^{188}W, ^{32}P, ... Other irradiation positions are provided in peripheral reflector channels (thermal neutron fluxes from 1 to 3.5 10^{14} n/cm².s) for the production of ^{99}Mo, ^{153}Sm, ^{186}Re, ^{90}Y, ...

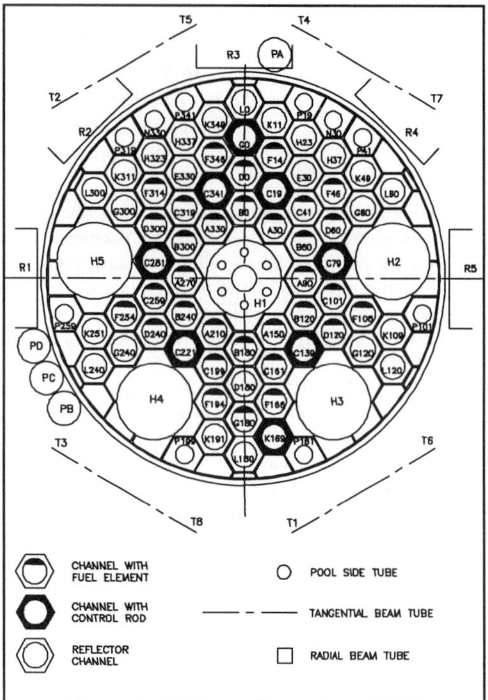

Figure 2. BR2 configuration (20D)

The thermal neutron fluxes of 4.5 10^{14} n/cm².s available at the bottom of the control rods are used for the production of ^{60}Co. If fast neutron fluxes are required, for the production of ^{117m}Sn for example, the targets are loaded into an irradiation basket inside a BR2 fuel element where the flux above an energy of 0.1 MeV is between 1 and 5 10^{14} n/cm².s, depending on the location of the fuel element in the core.

3 Major radioisotopes produced in the BR2 reactor

The targets for the production of radioisotopes are supplied by the customers and loaded into suitable irradiation capsules. After commissioning tests, the capsules are loaded and irradiated in dedicated in-core devices; some of them allow their loading and unloading during the operation of the reactor. After irradiation, temporary storage for cooling and, if required, passage through the BR2 hot-cells for decanning, the targets are loaded into suitable shipping containers in order to be sent back to the customers for processing.

The major isotope produced in the BR2 reactor is **^{99}Mo** ($T_{1/2}$=66 h). It is obtained by irradiation of high enriched uranium targets (93% ^{235}U) of 4 g ^{235}U. Four irradiation devices are routinely loaded in reflector channels as G60 (Figure 2), providing a total irradiation capacity of 24 targets at the same time, i.e. 72 targets per reactor cycle. A typical irradiation of 150 hours in a thermal neutron flux of 2.5 10^{14} n/cm².s yields up to 1200 Ci per target at the end of the irradiation "EOI", i.e. 180 Ci per target at calibration date ("EOI" + 7.5 days). In its processing facilities, the customer recovers the ^{99}Mo activity - in order to manufacture **^{99}Mo/^{99m}Tc** ($T_{1/2}$=6 h) generators - and other fission isotopes as **^{131}I** ($T_{1/2}$=8.02 d) and **^{133}Xe** ($T_{1/2}$=5.24 d).
The production of **^{192}Ir** ($T_{1/2}$=74 d) consists in the irradiation of natural iridium discs in the 6 peripheral channels of the central beryllium plug H1, characterized by thermal neutron fluxes up to 1.2 10^{15} n/cm².s. The maximum irradiation capacity is 36 capsules loaded with 5 g iridium each; in this case, an additional fuel element is loaded in H1/central in order to compensate the large antireactivity effect of the irradiation baskets. The specific activities achieved after one irradiation cycle of 21 days are 500 up to 750 Ci/g "EOI", depending on the size of the discs, the loading of the capsule and its axial position in the basket.
The productions of **^{188}W** ($T_{1/2}$=69.4 d) for the fabrication of **^{188}W/^{188}Re** ($T_{1/2}$=17 h) generators, **^{89}Sr** ($T_{1/2}$=50.5 d) and **^{32}P** ($T_{1/2}$=14.3 d) are, as ^{192}Ir, carried out in the central beryllium plug H1. The quartz ampoules are loaded in standardized cold-welded aluminium "CSF" capsules containing an inner graphite or aluminium cylinder and helium gas. The typical specific activities obtained after one irradiation cycle of 21 days are 1.2 Ci ^{188}W/g of ^{186}W, 0.3 Ci ^{89}Sr/g of ^{88}Sr and 50 Ci ^{32}P/g of ^{31}P.
The productions of **^{186}Re** ($T_{1/2}$=3.8 d), **^{153}Sm** ($T_{1/2}$=46.7 h), **^{90}Y** ($T_{1/2}$=64 h) are carried out in "CSF" capsules loaded in dedicated irradiation devices. Two such devices are routinely loaded in reflector channels as E30 (Figure 2), providing a total irradiation capacity of 18 targets at the same time. The typical specific activities obtained in a thermal neutron flux of 3.5 10^{14} n/cm².s are 815 Ci ^{186}Re/g of ^{185}Re after 7 days irradiation, 4500 Ci ^{153}Sm/g of ^{152}Sm after 4 days irradiation, 60 Ci ^{90}Y/g of ^{89}Y after 9 days irradiation.
The irradiation during 1200 days of the capsules containing 175 g of cobalt at the bottom of the control rods yields **^{60}Co** ($T_{1/2}$=5.26 y) specific activities of 300 Ci/g.

4 Conclusion

The extensive refurbishment program of the BR2 reactor, performed in 1995-1997 after more than 30 years utilization, provided a life extension of more than 15 years. The BR2 reactor will thus continue in the next future to bring its large contribution to the continuous supply of radioisotopes, especially ^{99}Mo in Western Europe, through yearly well coordinated operating schedules of the reactors BR2 (Mol; Belgium), HFR (Petten; The Netherlands) and OSIRIS (Saclay; France).

FEASIBILITY STUDY ON ^{123}I PRODUCED BY PHOTONUCLEAR REACTION ON XENON AT ELECTRON LINAC OF 200 MEV *

Y.J.PEI, Q.L.LIU, Y.X.LI, W.M.CHEN, B.XU, X.F.LUO, B.G.SUN
University of Science & Technology of China, Hefei 230029

Y.X.WANG, SH.G.SHENG, X.H.ZHANG, B.H.CAO, G.D.WANG
Shanghai Institute of Nuclear Research, the Chinese Academy of Sciences, Shanghai 201800

Isotope ^{123}I with good nuclear characteristics ($T_{1/2}$, E_γ=159 keV, etc.) has been used widely in medicine. A facility for producing ^{123}I by photonuclear reaction on natural Xenon at an electron LINAC of 200 MeV of National Synchrotron Radiation Lab, USTC was established and described here. Some research results, such as the ^{123}I yield, γ-ray spectrum of isotopes of Iodine, nuclear purity, chemistry purity and so on, were given in this paper. The ^{123}I yield of 42μci has got, when the electron energy is of 200 MeV, average current of 2.7μA, gas target with natural Xenon of 2.6g/cm^2 was irradiated by bremsstrahlung beam via a γ/e tungsten convertor of 5 mm and irradiation time of 2.4 hrs.

The results from our tests shown that ^{123}I yield of 380 mci is possible if average current to be increased to 12μA, a pure $^{124}X_e$(100%) instead of the natural Xenon and gas target thickness of 5.1g/cm^2.

1 Introduction

^{123}I has been recognized since1962 as most useful of the iodine radioisotopes for in vivo study [1]. Its advantages lie in its decay characteristics, these being the absence of β emission, short half-life and a γ-ray spectrum with photopeak energy of 159 keV which is well suited to diagnosis and cure on medicine.

The production of ^{123}I is based presently on the generation method of obtaining radioiodine from radioxenon:

$$^{123}X_e \xrightarrow{\beta^+,2.1h} {}^{123}I \xrightarrow{EC,13.2h} \tag{1}$$

This method ensures high radionuclear purity of ^{123}I specimens and relatively simple experimental arrangements. $^{123}X_e$ is obtained in following nuclear reactions of $^{127}I(p,5n)$ $^{123}X_e$ on proton accelerator with proton energy of up to 80 MeV[2]; reaction of $^{127}I(d,6n)$

* This work supported by the Ministry of Science and Technology of China

$^{123}X_e$ on synchrotron with deuteron energy of up to 108 MeV[3]; less often reactions of $^{124}X_e(\gamma,n)\,^{123}X_e$ on electron accelerators [4] and so on[5].

An electron linear accelerator of 200 MeV existed is an injector for a storage ring of HLS at NSRL(National Synchrotron Radiation laboratory)[6]. So we can use the electron beam to study the following reaction:

$$^{124}X_e(\gamma,n)\,^{123}X_e \rightarrow {}^{123}I \tag{2}$$

The ^{123}I radiation activity of 42μci has got, when the electron energy is of 200 MeV, average current of 2.7μA, gas target with natural Xenon of 2.6g/cm^2 was irradiated by bremsstrahlung beam via a γ/e tungsten convertor of 5 mm and irradiation time of 2.4 hrs.

The results from our tests shown that product ^{123}I of 380 mci is possible if average current to be increased to 12μA, a pure $^{124}X_e$(100%) instead of the natural Xenon and gas target thickness of 5.1g/cm^2.

2 Calculation of yield of ^{123}I

2.1 Calculating formula

^{123}I produced by nuclear reaction of $^{124}X_e(\gamma,n)\,^{123}X_e \rightarrow {}^{123}I$ is a nuclear reaction and decay process in sequence. Let us label these processes as A→B→C→D.

Where A—express the mother body $^{124}X_e$, and its molecular number is labeled as N_0,

B—show $^{123}X_e$ produced by reaction (γ,n), and molecular number produced is N_1,

C—express ^{123}I produced by decay of $^{123}X_e$, and molecular number is N_2,

D— express ^{123}I produced itself, and molecular number after decay is N_I.

Yield of ^{123}I produced via irradiation of γ-ray, $N_2(t)$ can be got. Then stopping irradiation, $^{123}X_e$ produced by reaction (γ,n) will be stopped, but ^{123}I produced by $^{123}X_e$ decay will decay itself , so yield $N_I(t)$ of ^{123}I after stopping irradiation can be written

$$N_I(t) = \frac{\lambda_0 \lambda_1}{(\lambda_1 - \lambda_2)} N_0(0) \left\{ \frac{e^{-\lambda_2 t}}{(\lambda_2 - \lambda_0)} (e^{-\lambda_0 t_0} - e^{-\lambda_2 t_0}) - \frac{e^{-\lambda_1 t}}{(\lambda_1 - \lambda_0)} (e^{-\lambda_0 t_0} - e^{-\lambda_1 t_0}) \right\} \tag{3}$$

Where λ_0— is $\eta \int \phi_\gamma(\varepsilon) \sigma(\varepsilon) d\varepsilon$, $\phi_\gamma(\varepsilon)$ is flux of γ-ray produced via electron beam hitting the target of tungsten,

$\sigma(\varepsilon)$ is cross section of photonuclear reaction, η is utility factor of the γ-ray.

λ_1— is decay coefficient of $^{123}X_e$.

λ_2— is decay coefficient of ^{123}I. t_0 is irradiation time, t is from stopping irradiation.

Finally, if retrieve the gas Xenon at t_m, then radiation activity A(t) of ^{123}I can be written

$$A(t) = A_0 N_0(0) \left\{ \frac{e^{-\lambda_2 t_m}}{(\lambda_2 - \lambda_0)} (e^{-\lambda_0 t_0} - e^{-\lambda_2 t_0}) - \frac{e^{-\lambda_1 t_m}}{(\lambda_1 - \lambda_0)} (e^{-\lambda_0 t_0} - e^{-\lambda_1 t_0}) \right\} e^{-\lambda_2 t} \tag{4}$$

Where t is from t_m, $A_0 = \frac{\lambda_0 \lambda_1 \lambda_2}{3.7 \times (\lambda_2 - \lambda_0)} \times 10^{-10}$, unit is ci.

2.2 Calculating results

Some beam parameters are electron energy of 200 MeV, pulse current of 50 mA, beam width τ=1μs, repeating frequency of 50 pps. Target of tungsten and its thickness is of 5 mm. Simulation results show that $\lambda_0 = 1.31\times10^{-12}$/s, $N_0 = 1.22\times10^{18}\times p/cm^3$, where p is pressure of natural Xenon in the reaction chamber. Radiation activity of ^{123}I at different irradiation time t_0 and retrieve the Xenon at t_m, can be got via formula (4), and the results are shown in Fig.1.

When the irradiation time t_0 is of 3 hrs, t_m=7 hrs, the radiation activity of ^{123}I is 100 μci. This means that radiation activity of ^{123}I will be reached 104 mci, if a pure $^{124}X_e$ instead of natural Xenon. Obviously, if electron beam of 100 mA, repeating frequency of 100 pps and pressure of 60 Atm instead of 50 mA, 50 pps, 30 Atm respectively, the radiation activity of ^{123}I will be 832 mci.

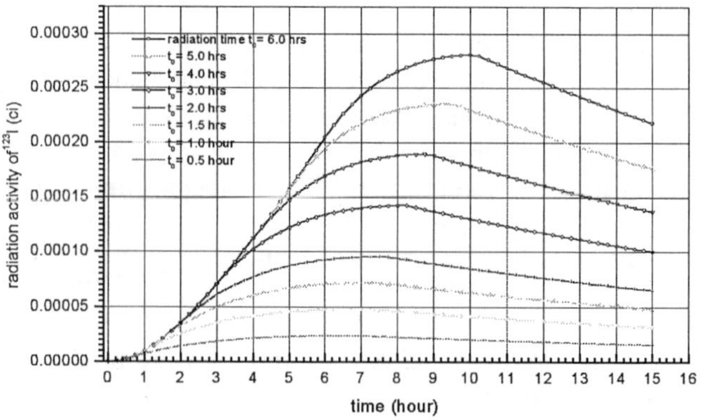

Fig. 1 Radiation activity of ^{123}I yield by photonuclear reaction of X_e

3 Experimental Facility

The experimental facility is located at the end of electron linear accelerator of 200 MeV in tunnel. The sketch of the facility is shown in Fig.2. The facility include mainly convertor of tungsten of 5 mm, reaction chamber of stainless steel which volume is of 50 ml and size is of 20 mm in diameter, 150 mm in length, the chamber is filled by pure natural gas xenon of 30~60 atm.

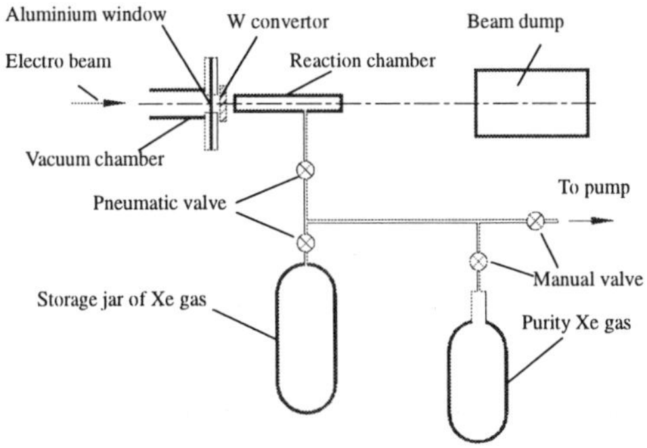

Fig.2 Sketch of the experimental facility

4 Experimental Results

In order to increase the nuclear purity of the products produced by photonuclear reaction, before irradiation the chamber will be exhausted using vacuum pump to reduce the pressure down to 10^{-5} torr. Then cooling the reaction chamber using liquefied nitrogen and opening a pneumatic valve simultaneously to let the natural Xenon to entry the chamber from a storage jar. After that closed the valve and removed the liquefied nitrogen vessel. When the pressure manometer shows the pressure being stable, start to irradiate the reaction chamber using the electron beam of 200 MeV. After irradiation the natural Xenon is retrieved at time t_m by cooling the storage jar and opening the pneumatic valve. Then wash the reaction chamber using pure water and separate the iodide on aluminum column. In our case, irradiation time t_0 is of 2.4 hrs, beam current is of 60 mA(average current of 2.7μA), pressure of the chamber is of 28 Atm. (2.6g/cm^2), radiation activity of the yield ^{123}I is 42 μci. The γ-ray spectrum of the products is shown in Fig 3. Table 1 and table 2 show the radiation activity of the products and their chemistry purity respectively.

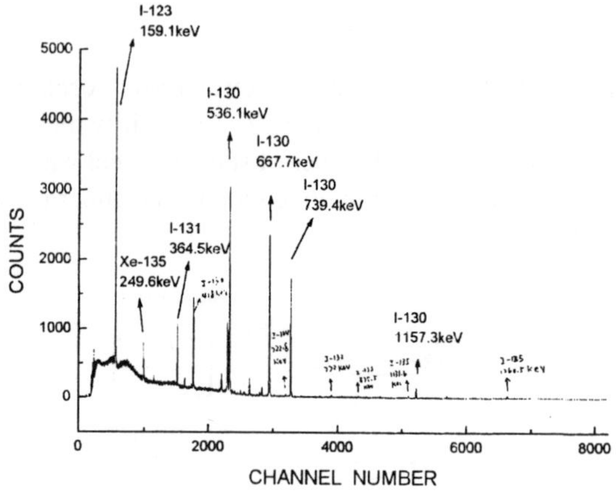

Fig. 3 γ-ray spectrum of iodide produced by reaction(γ,n)

Table 1. Radiation activity of iodide produced by photonuclear reaction

NUCLIDE	ACTIVITY(μci)	CONTENT(%)	Relative cross section	Reaction formulas
I-123	42.02	31.35	100.00%	$^{124}X_e(\gamma,n)\,^{123}X_e \rightarrow {}^{123}I$
I-124	5.038	3.76	11.99%	$^{126}X_e(\gamma,np)\,^{124}I$
I-126	2.161	1.61	2.57%	$^{128}X_e(\gamma,np)\,^{126}I$
I-130	60.75	45.32	6.55%	$^{131}X_e(\gamma,p)\,^{130}I$
I-131	6.281	4.68	0.497%	$^{132}X_e(\gamma,p)\,^{131}I$
I-133	1.454	1.08	0.3%	$^{134}X_e(\gamma,p)\,^{133}I$
I-135	15.87	11.84	4.07%	$^{136}X_e(\gamma,p)\,^{135}I$
Xe-135	0.447	0.33		$^{135}I \rightarrow {}^{135}X_e$

Table 2. Chemistry component (μg/ml) in origin solution and washed solution

element	Cd	Co	Cu	Mn	Pb	Mg	Zn	Ca	P	Cr	Fe
origin	0.01	0.01	0.02	0.07	0.05	0.1	0.02	0.62	0.05	0.02	0.03
washed	0.02	0.01	0.01	0.01	0.40	.05	0.06	0.43	0.92	0.04	0.01

Table 2 con't

element	Ni	Mo	Si
origin	0.30	0.01	0.45
washed	0.01	0.06	0.45

5 Conclusion

The test results as mentioned above show that no other nuclides were found, that means the nuclear purity of the products are very high and their chemistry purity is also good enough, all elements in the product solution are less than 1μg/ml. If a pure gas of $^{124}X_e$ instead of the natural Xenon, and increase the current up to 100 mA, repeating frequency up to 100 pps, pressure in the reaction chamber increase to 60 Atm.(5.1g/cm^2), then the radiation activity of the yield ^{123}I will be 381 mci which can be labeled and provide a pharmaceuticals of ^{123}I for hospital.

Reference

1. W.G.Myers and H.O.Angers, J. Nucl. Med. **3** (1962) pp183
2. N.G.Zaitseva, Z.Hons, V.A. Khalkin, et al., Intl. J.Appl.Radiat. Vol. 36, No. 11, (1985) pp 897
3. Y.Legoux, M.Cieur, R.Goutheraud, et al., Int. J.Appl.Radiat. Vol.36, No.1,(1985) pp 63-67
4. A.A.Arhipov, N.G.Zaitseva, O.E.Kolakun et al., Production of ^{123}I by Photonuclear Reaction on Xenon at Linac LUR-40 of FLNP JINR, Dubna, (1994) pp 6-94-402
5. J.Zheng, Y.X.Wang• Preparation of ^{123}I by Cyclotrons and electron accelerators• Nuclear Techniques Vol.19, No.4, (1996) pp 241-248
6. Y.J.Pei, Phase II Project of NSRL, Proc. of 6th China-Japan-Joint Symposium on Accelerators and their Applications, (1996) pp 32-35

RADIONUCLIDE PRODUCTION USING FAST FLUX REACTOR

YE.A. KARELIN, V.N. EFIMOV, V.T. FILIMONOV, R.A. KUZNETSOV, YU.L.REVYAKIN,O.I. ANDREEV, I.YU. ZHEMKOV, V.G. BUKH, V.M. LEBEDEV, YE.N. SPIRIDONOV

Russia State Scientific Center "Research Institute of Atomic Reactors",Dimitrovgrad-10, 433510, RUSSIA

The production of ^{89}Sr, ^{32,33}P, ^{35}S via (n,p)- reaction and ^{117m}Sn, ^{153}Gd via (n,γ)-reaction using BOR-60 fast flux reactor was experimentally studied. Test samples were irradiated in an active core of BOR-60 reactor with fast neutron flux 1.10^{15} cm^{-2}s^{-1} for 40-100 effective days. Gadolinium-153 was produced in a radial blanket cell, characterized by modified ("softened") neutron spectrum. Data of target materials, procedure of irradiated targets reprocessing, radionuclide yield and specific activity are summarized in the report. The results of the experiment showed large-scale production of radionuclides listed above is possible using fast-flux reactor.

Thermal neutron capture nuclear reactions, characterized by high rate of neutron reactions, are currently used for major radionuclide production. In many cases, using of these reactions' in fast-flux reactors is not effective, or requires «softening» of neutron spectrum in the irradiation position. At the same time, many threshold reactions, of (n,p), (n,α), (n,2n) type, taking place in fast-flux reactors, can be effectively used for radionuclide production. But only minor experience has been gained as fast-flux reactors are seldom used in this field [1].

This presentation reports the results of experimental study of possibility of large-scale production of radionuclides using fast-flux reactor BOR-60, operating at RIAR. Radionuclides ^{32,33}P, ^{89}Sr, ^{153}Gd, ^{117m}Sn, ^{35}S were as an example.

1 Strontium-89

The ^{89}Y(n,p)^{89}Sr reaction (threshold •... = 2,5 MeV) is used for ^{89}Sr production in fast-flux reactor. Its principal advantages (as compared to «routine» ^{88}Sr(n,γ) reaction) are as follows:

- high-specific activity ^{89}Sr (no carrier-added grade) is produced
- production of ^{85}Sr as well as ^{90}Sr impurity has low output
- large irradiation capacities of fast reactor providing large batches (tens of Curies) production.

The optimal target irradiation time necessary for maximum output of ^{89}Sr depends on (n,p)-reaction rate and consequently is determined by absolute flux density of neutron with energy E > 2,5 MeV, i.e. – irradiation position and reactor operation power. The typical campaign of BOR-60 reactor (up to 100 days) provides accumulation of 10-12 Ci of ^{89}Sr (at EOB) in one target assembly, containing ~1 kg of yttrium oxide. At the same time irradiation duration should be limited, because of ^{90}Sr production via $^{89}Sr(n,\gamma)$ reaction or fission of uranium or thorium impurities. Typically irradiation for 60 days gives strontium-90 content approximately (3-4).10^{-5}%.

2 Gadolinium-153

The chain of nuclear reactions $^{151}Eu(n,\gamma)^{152}Eu \rightarrow ^{152}Gd(n,\gamma)^{153}Gd$, based on irradiation of natural europium is used for high-specific ^{153}Gd production. Reactor production of this radionuclide has some peculiarities:

- large resonance integrals of nuclides, preceding ^{153}Gd production, cause expediency of using of neutron spectra with large portion of epithermal neutrons (hard neutron spectrum) for irradiation,
- ^{153}Gd yield and specific activity increase both while neutron spectra hardness increasing and while temperature of neutron Maxwell distribution at thermal energy increasing; the latest effect takes place due to sharp decreasing of ^{153}Gd nuclei burn-out while their production rate decreases much less.

Gadolinium-153 was produced in a radial blanket cell, characterized by modified neutron spectrum, «softened" by ZrH_2 moderators. Due to this spectrum transformation the portion of epithermal neutrons was 14 times increased. The average neutron energy in initial spectrum was 0,359 MeV, after modification – 0,140 MeV.

Irradiated europium targets were reprocessed via previously developed technology [2].

The experimental results has showed the yield of ^{153}Gd is inversely proportional to the europium concentration in the target due to depression of the neutron flux. Irradiation of target material containing 0,2 g/cm3 of europium for 2 reactor campaign (240 days totally, 34 days shut-down) gives ^{153}Gd output 11,2 Ci/g europium, specific activity 141,5 Ci/g (measured by mass-spectrometry). Production of ^{153}Gd of specific activity greater then 150 Ci/g is possible without changing reactor operation schedule.

3 Phosphorous-32,33

Samples of sulfur of natural isotope abundance (for ^{32}P production) or isotopically enriched (>99%) ^{33}S (^{33}P production) were irradiated. Isolation of $^{32,33}P$ were carried out using technology, based on vacuum sulfur distillation, dissolution of phosphorus in the diluted hydrochloric acid, purification of phosphate – ions via ion-exchange. Parameters of irradiated targets are presented in Table 1.

Table 1. ^{32}P and ^{33}P output, irradiation conditions

Radio-nuclide	Irradiation conditions		Radionuclide output at EOB, Ci/g
	Irradiation time, days	Reactor power, MWt	
^{32}P	47	40,9	4,1*
	100	50	5,5**
^{33}P	87	39,3	2,6*
	100	50	3,4**

* experimental data

** prognosis for indicated reactor operation power and irradiation time

The technology described above allowed production of ($^{32,33}P$)-orthophosphoric acid with the following characteristics

	^{32}P	^{33}P
Molar activity	> 9000 Ci/mMol	>4200 Ci/mMol
Portion of orthophosphoric acid	99,9%	99,5%
Phosphorus impurity content	<5% (^{33}P)	<2%(^{32}P)

4 Tin-117m

A metal tin sample of natural isotopic composition was irradiated at 40,9 MWt operation power for 47 days to estimate output of tin-117m. Activity of radionuclides was measured by gamma-spectrometry (Table 2).

At nominal reactor mode of operation (100 days at 50 MWt) ^{117m}Sn output (specific activity) reaches 1,7 Ci/g for natural mixture of tin isotopes. This radionuclide is produced by two nuclear reactions $^{116}Sn(n,\gamma)^{117m}Sn$ and $^{117}Sn(n,n')^{117m}Sn$. Partial outputs of these two reactions calculated on the basis of activation experiment results are 5,5 Ci/g and 12,5 Ci/g. Obviously, ^{117m}Sn specific activity can be increased by using enriched ^{117}Sn as a target.

Table 2. Specific activity of radionuclides in irradiated tin sample

Radionuclide	Specific activity, Ci/g
^{117m}Sn	1.25
^{113}Sn	0.063
^{123}Sn	0.011
^{124}Sb	$0.36.10^{-3}$
^{125}Sb	0.003

5 Sulphur-35

High purity sodium chloride was used as a target material for ^{35}S production. All targets were irradiated for 72 days (64,8 effective days), the integral neutron flux was $1{,}1\cdot10^{22}$ cm^{-2}. Reprocessing procedure included dissolution of KCl in diluted hydrochloric acid, simultaneous adsorption of sulfate- and phosphate-ions on alumina, selective desorption of sulfate-ion with ammonia solution, purification of sulfate-ion from cation impurities using Dowex-50 resin.

Experimentally determined sulphur-35 output was 2,3 Ci/g of KCl, at nominal operation power of BOR-60 reactor it will reach 3,5 Ci/g of KCl.

6 Conclusion

Analysis of accumulation parameters of short-lived radionuclides (^{32}P, ^{33}P, ^{117m}Sn) has showed that output values of these radionuclides are $\sim 25\%$ lower and irradiation time is several time more longer for BOR-60 fast reactor then for high-flux SM reactor, routinely used by RIAR for isotope production. The BOR-60 reactor operation schedule is mainly oriented for power production, so, using of BOR-60 reactor for routine production of short-lived radionuclides was not recognized expedient. This way is effective only as a supporting one of continuous radionuclide production at the period of SM reactor service shut-down.

Production of ^{35}S, ^{89}Sr, ^{153}Gd using fast-flux reactor seems to be prospective as their half-lives are relatively long and production (irradiation) schedule of these radionuclides can be easily superposed with reactor operation schedule. In this case large-scale production of radionuclides with required parameters can be established.

References

1. Wootan D.W., Rawline J.A., Carter L.L., et al. *Nucl. Sci. Eng.* **103** (1989) pp. 150-156.
2. V.A.Tarasov, Yu.G.Toporov, V.T.Filimonov et al. *Radioactivity & Radiochemistry* **8** N2 (1997) pp.27-33.

Session D1

Radiochemical Synthesis — II

PREPARATION AND QUALITY CONTROL OF ^{131}I-MIBG

HUANG YONG LIU YISHU TIAN JIANCHUN SONG DAOHAI
Nuclear Power Institute of China (NPIC)

P.O.Box 291-107, Chengdu, China, 610005

Based on "MW2" solid-state exchange technique, a simplified method for labeling meta-Iodobenzylguanidine (mIBG) with Iodine-131 without purification step is presented. This procedure allows in both high radiochemical yield (>97%) and high radiochemical purity (>98%) at 160• heating within 1h, and leads to highly reproducible results.The method described here is simple and efficient ,very suitable for the routine preparation of ^{131}I-mIBG injection.The variable parameters affecting labeling efficiency, such as mass of ammonium sulfate, reaction temperature and time, and activity of Iodine-131, were investigated in detail and optimum values obtained. Radiochemical purity of the product was determined by ascending paper (Whatman No.1) chromatogram using n-butanol: glacial acetic: water (5:2:1 v/v/v) as developing solvent. To date, the Iodine-131 labeled mIBG prepared by this method has been used in several hospitals for clinical trials and good clinical results achieved.

1 Instruction

The radioiodinated-blocking agent ^{131}I-mIBG (meta-iodobenzylguanidine) has been one of the most successful of present radiopharmaceuticals, and now it is very valuable for diagnosis and treatment of pheochromocytoma and neuroblastoma etc.

Published methods for 131I-mIBG preparation differs greatly. The procedure for the routine production of 131I-mIBG should meet the following criteria: (1) the radiochemical yield should be close to 100%;(2) the labelling procedure should be fast and simple;(3) the method should lead to reproduceble results. The "MW2" method, which exchange occurs in solid-phase under the mildly acidic and oxidizing condition and using ammonium sulfate as a solid state transfer catalyst, is suitable for the high specific exchange radioiodination of mIBG. But the overall radiochemical yield is low and products with high radiochemical purity only after passage through anion-exchange column. Based on "MW2" method [2], simple procedure for routine preparation of high quality 131I-mIBG without further purification step has been established by NPIC.

2 Labeling procedure

The procedure for the routine production of ^{131}I-mIBG as following illustrated (Fig.1) is developed by NPIC.

The unlabeled mIBG as sulfate salt was synthesized by the procedure described by Wieland et al [3]. To verify that the products indeed mIBG, following analysis have been performed: melting point, U.V, IR, NMR, mass spectroscopy and quantitative elemental analysis, the chemical purity of the product above 99.9%. The Na^{131}I solution was produced by dry process of NPIC with high quality and radioactive concentration (up to 1Ci/ml, reductant and carrier-free), other reagents are all analytical reagent.

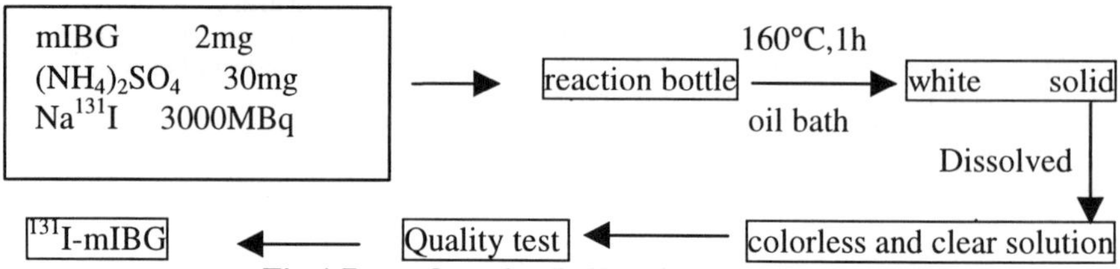

Fig.1 Procedure for Iodine-131 labeling mIBG

To simplified the labeling procedure and obtain optimum labeling parameters, the variable parameters affecting labeling efficiency, such as mass of ammonium sulfate, reaction temperature and time, and activity of Iodine-131, were investigated in detail and optimum values obtained, and established the labeling procedure as above illustrates (Fig. 1). In this procedure, first, 1mL exchange solution (contains 2mg mIBG and 30mg ammonium sulfate) and 3000MBq Na^{131}I solution were added into reaction bottle, and the reaction solution was heated to dryness(the water in the closed reaction bottle was drove out by heated at 120• through a vent and condensed in another bottle), then, reaction temperature was increased to 160• and kept for 1h, finally, the contents in the reaction bottle were dissolved in 20mL acetate buffer (pH: 5.5), and assayed for radiochemical purity. In our experience, a typical run with the procedure illustrated above, which the oxidizing condition is eliminated, reagent mass, reaction time and temperature are in order, high quality products with radiochemical purity of ^{131}I-mIBG of 98% and high radiochemical yield with activity losses about 2~3% will be obtained, and that further purification step is unnecessary.

3. Quality control

The radiochemical purity of the products were checked by ascending paper (Whatman No.1) chromatography developed for 10cm in a jar containing a mixed developing solvent (n-butanol: glacial acetic: H_2O = 5: 2: 1(V/V/V)). After developing, we found the free iodide near the origin (Rf value: 0.2~0.3), ^{131}I-mIBG

migrated with solvent front (Rf: 0.9~1.0). We also have applied comparative analysis with HPLC quality control procedure to verify the radiochemical purity of product; the results of the two methods were found to correlate very well.
The ^{131}I-mIBG solution is very unstable without stabilizers at room temperature due to the radiolysis. If add benzyl alcohol to 1% as radical absorber and bactericidal agent, and ascorbic acid to 0.1 mg/ml as reductant into solution, only of 2~3% Iodine-131decomposed from ^{131}I-mIBG in 5 days when products with radioactive concentration approximately 185MBq/mLstored at room temperature.
To date, ^{131}I-mIBG injection has been prepared by this procedure successfully of NPIC and used for diagnosis and treatment in several hospitals every week, the labeling efficiency has been consistently above 97%. Patients have been investigated with good clinical results, no reported side effect.

4. Conclusion

In conclusion, the labeling procedure without purification step described above, is simple and fast, perfectly fulfills the requirement for the routine preparation of ^{131}I-mIBG.

Reference

1. Luigi Trocone, et al: The diagnostic and therapeutic utility of radioiodinated metaiodobenzylguanidine (MIBG) 5 years experience, Eur. J. Nucl. Med. 16: 325~335,1990.
2. Thomas J. Mangner, Jiann-Long Wu, Donald M.Wieland, Solid-phase exchange radioiodination of aryl Iodides, Facilitation by ammonium sulfate, J. Org. Chem. 47,1484~1488, 1982.
3. Donal M. Wieland Jiann-Long Wu, Thomas J. Mangner et al, Radiolabelled adrenergic neuron blocking agents: Adrenomedullary imaging with [^{131}I] Iodobenzylguanidine, J. Nucl. Med. 21,349~353, 1980

THE PREPARATION OF A THIRD GENERATION REVERSE TRANSCRIPTASE INHIBITOR (RTI), RADIOISOTOPICALLY LABELED WITH CARBON-14 OR TRITIUM, OR STABLE ISOTOPICALLY LABELED WITH CARBON-13 AND NITROGEN-15.

WAYNE T. STOLLE

Drug Metabolism Research
Pharmacia & Upjohn, 301 Henrietta St., Kalamazoo, MI 49007, USA
E-mail: wayne.t.stolle@am.pnu.com

Considerable effort has been focused over the past decade on the development of HIV-1 specific non-nucleoside reverse transcriptase inhibitors (NNRTIs) due to their safety, selectivity, and antiviral profiles for the treatment of HIV-1 infection and AIDS. Such efforts at Pharmacia & Upjohn have led to the identification of ***1***, as a candidate for clinical development. The preparation of radioisotope and stable isotope labeled PNU-142721E was requested for drug absorption, distribution, metabolism, and excretion (ADME) studies.

The initial preparation of this isotopically labeled chiral compound followed a convergent synthetic pathway in which an isotopically labeled pyrimidine moiety was coupled with the furopyridine portion of the molecule. Preparation of 4-amino-6-chloro-2-mercaptopyrimidine, with a carbon-14 radioisotope at the 2-position, or with carbon-13 and nitrogen-15 stable isotopes at the 2-, and 1,3-positions respectively, of the pyrimidine ring, was accomplished starting from isotopically labeled urea. Coupling of the carbon-14 labeled pyrimidine with 5[(1R)-1-chloroethyl]furo[2,3-c]pyridine gave 26.5 mCi (overall 18% yield, 6-steps) of ***1a***, with a specific activity of 58 mCi/mmol (93 atom %), a radiochemical purity of 99% by both HPLC and TLC, and enantiomeric purity of at least 99% by HPLC. Likewise, 298 mg of stable isotope labeled ***1b***, (99 atom % ^{13}C, 98 atom % ^{15}N) having excellent chemical purity, was prepared using the same synthetic pathway.

To obtain a complete metabolic profile of ***1***, isotopic label in both major regions of the molecule was needed. PNU-142721 was thus labeled with a tritium radioisotopic label in the furopyridine portion of the molecule, which also provided material with high specific activity for binding studies. Tritium – Halogen exchange from a prepared analog provided 46.1 mCi of ***1c***, having a specific activity of 12.5 Ci/mmol, radiochemical purity of 98.7% by TLC and HPLC, and enantiomeric purity of 99% by chiral phase HPLC.

PNU - 142721E : A Third Generation NNRTI

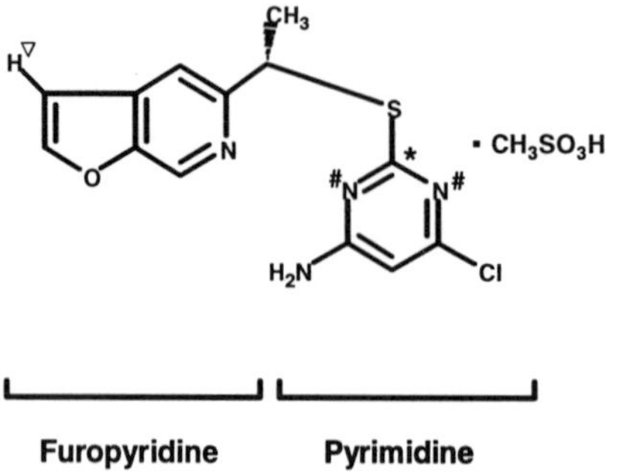

1a. * = ^{14}C, # = ^{14}N, ▽ = ^{1}H

1b. * = ^{13}C, # = ^{15}N, ▽ = ^{1}H

1c. * = ^{12}C, # = ^{14}N, ▽ = ^{3}H

1 Introduction

The use of antiviral agents in combination therapy has brought a new array of tools for the treatment of HIV-1 infection and a new strategy to fight the ever-present problem of drug resistance. This concern with the development of mutant HIV-1 viral forms has led to the development of. PNU-142721E, a Third Generation NNRTI and a member of a new class of antiviral agents known as furopyridine pyrimidine thioethers.

2 Results and Discussion

The preparation of PNU-142721E followed a simple convergent pathway; the coupling of the furopyridine moiety with an appropriately functionalized thiopyrimidine. Our initial attention focused on labeling the thiopyrimidine moiety.

Isotopically Labeled PNU-142721E

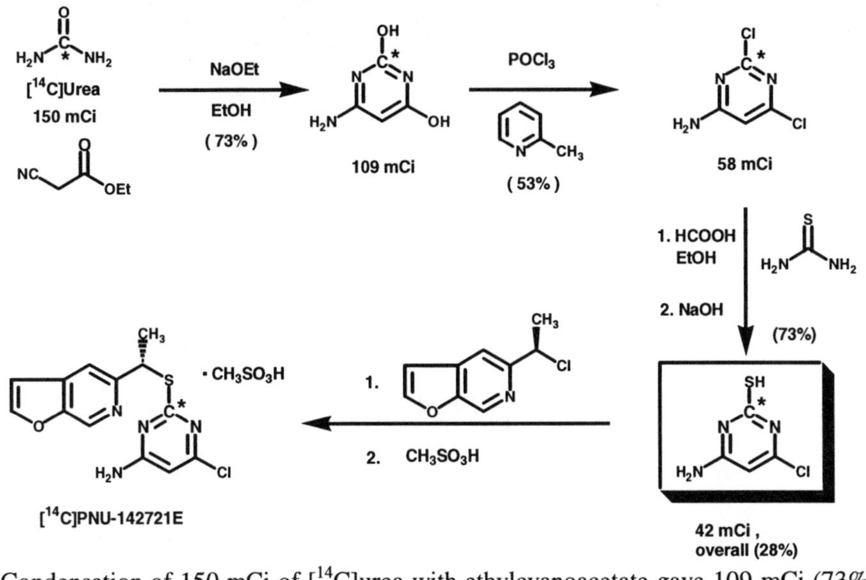

Condensation of 150 mCi of [^{14}C]urea with ethylcyanoacetate gave 109 mCi (73%) of the amino-pyrimidine diol. Chlorination was successful using phosphorous oxychloride with two equivalents of 2-picoline to give 58 mCi of the amino-dichloro-pyrimidine. Introduction of the sulfur functionality was attempted by refluxing the above compound with thiourea in ethanol, but purity and yield were greatly improved by the addition of formic acid as a catalyst for the reaction. The pyrimidine thiol was prepared following hydrolysis, with a purity of 97%, representing an overall yield from [^{14}C]urea of 28%. Regioselective alkylation of the labeled pyrimidine with the furopyridine moiety gave 33.8 mCi of ***1a*** representing an overall yield of 18% from labeled urea. Using the same chemistry, 298 mg of the stable isotope labeled ***1b*** was prepared from [^{13}C, $^{15}N_2$]urea in an overall yield of 13%.

To further explore metabolic events occurring in the left-hand region of the molecule, and to prepare material suitable for binding studies, tritium labeling in the furopyridine moiety was examined. Several schemes were investigated using halogenated analogs or precursors of PNU-142721 in which Tritium-Halogen exchange would give site-specific tritium labeled product.

Preparation of [^{3}H]PNU-142721

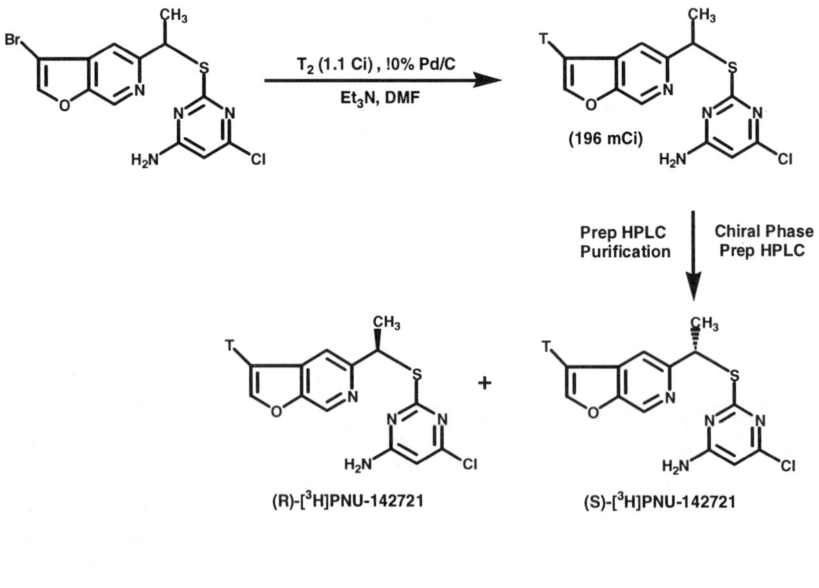

The bromo analog was exposed to 1.1 Ci of tritium gas in the presence of 10% Palladium on carbon in 2% triethylamine/DMF. Purification by HPLC gave 49.5 mCi of the desired (S)-PNU-142721, with a specific activity of 12.5 Ci/mmol, radiochemical purity of 99% by TLC and reverse phase HPLC, and enantiomeric purity of 99.7%. Likewise, 50.1 mCi of the (R)-PNU-142721 was obtained with similar purity.

JUMBO COLUMN' GENERATOR FOR ^{99m}TC : FEASIBILITY FOR UTILISATION OF ^{99}MO OF MEDIUM SPECIFIC ACTIVITY AND CONCENTRATION OF ^{99m}TC ELUATES USING TANDEM ION-EXCHANGER

S. K. SARKAR, G. ARJUN, P. SARASWATHY, N. RAMAMOORTHY

Board of Radiation and Isotope Technology (BRIT), BARC Vashi Complex, Navi Mumbai - 400 705, INDIA

E-mail : nrama@magnum.barc.ernet.in

In the efforts towards user-friendly column generators for ^{99m}Tc using easy to produce, inexpensive $(n,\gamma)^{99}$Mo, there has been a new possibility based on the method of concentration of 'nca' ^{188}Re eluates using a tandem cation-anion exchanger reported by Knapp et al., who also showed the potential for similar concentration of ^{99m}Tc eluates. The latter applicability, for at least a central radiopharmacy, was hence studied by preparing large alumina column and zirconium molybdate gel column generators. The possibility to use commonplace ion-exchangers, instead of specialised ones, has been explored. ^{99}Mo of specific activity 200-500 mCi/g, upto 40 g of acidic alumina columns and generators containing upto 250 mCi were used. The increase in concentration of ^{99m}Tc activity was about 6-6.5 times with commonplace ion-exchangers, with >90% recovery. In other words, initial eluate volumes of about 50ml could be accommodated with this technique. The quality of pertechnetate and compatibility for preparing ^{99m}Tc formulations were satisfactory. Similar results were also obtained in the case of zirconium molybdate-^{99}Mo gel (30g) based generators.

1 Introduction

The most popular source of ^{99m}Tc is the alumina column chromatography generator, which requires the use of ^{99}Mo of very high specific activity. (n,γ) ^{99}Mo of medium specific activity warrants the use of large alumina beds and in turn, higher eluent volumes, leading to low radioactive concentration (RAC). Knapp et al. [1,2] have reported a new concept for extended utilisation of the precious ^{188}W in ^{188}W -^{188}Re generators. They have exploited the no-carrier-added (nca, NCA) nature of rhenium content of these generator eluates and retained the 'nca' perrhenate on a tiny weak anion exchanger column, after pre-treatment of the primary eluate from the generator so as to render the eluate bereft of any macroscopic anions. Triggered by this success, Blower [3] and ORNL group [2] have explored the utility of this concept for the concentration of pertechnetate from ^{99}Mo -^{99m}Tc generators.

At our centre, the development of zirconium molybdate - ^{99}Mo (ZrMo) gel concept is being pursued using $(n,\gamma)^{99}$Mo with promising results [4]. We have investigated the preparation of large size ('Jumbo') alumina column chromatography generators as well as ZrMo gel generators followed by concentration of the large volume $^{99m}TcO_4^-$ eluate.

2 Methods

^{99}Mo was available ex-stock as sodium molybdate in ~2M NaOH, 150mg Mo/ml, 30-60 mCi/ml. Cation exchanger : Dowex 50 W x 8 , H+ form, 100-200 mesh (Fluka). Anion exchanger : DEAE Cellulose, partially quaternized, micro-granular form (Sigma).

^{99m}Tc column chromatography generators prepared by standard method, but on 40g pre-conditioned acidic alumina, were eluted with 0.1/ 0.3 M NH_4OAc solution (pH ~5). Zr^{99}Mo gel generators containing ~30 g gel were prepared but elutions were performed with water. ^{99m}Tc eluate in 20 ml was passed through a series of pre-conditioned cation exchanger columns (2x800 mg) to render an eluate solution of pH ~3. This was then passed through an anion exchanger column (200mg). The column effluent was collected along with a few ml of washings as active waste. $^{99m}TcO_4^-$ was recovered from the anion column by elution with normal saline in 0.5/1 ml fractions. Quality of $^{99m}TcO_4^-$, after concentration, was evaluated by formulating ^{99m}Tc compounds using the more sensitive kits, such as kit for Tc-ECD, Tc-MDP etc.

Table 1 : Concentration of ^{99m}Tc eluates from alumina column and gel generators

Generator Reference	^{99}Mo (mCi)	Primary elutes 20mL		Final Product 3mL		Conc. factor	Y/X (%)
		^{99m}Tc (mCi) (X)	RAC mCi/mL	^{99m}Tc (mCi) (Y)	RAC mCi/mL		
A	152.3	38$^{\phi}$	1.9	36.4	12.13	6.38	95.7
4g	120.3	72	7.2	68	22.66	3.14*	94.4
alumina	57.7	37.6	3.76	34	11.33	3.01*	90.4
B	254	197.4	9.87	185	61.66	6.24	93.7
40g	198	152.4	7.62	140	46.66	6.13	91.9
alumina							
Jumbo	42	26.8	1.34	25.2	8.4	6.26	94
gel$	20	14	0.7	13.1	4.36	6.22	93.6
	12	7.4	0.37	7.2	2.4	6.48	97.3

^{99}Mo Sp. Act. in Gen A : ~ 7 Ci / g ; Gen B : 434 mCi/g ; 750 mg Mo loaded.
ϕ ^{99m}Tc growth time : 4 h * Primary eluate volume : 10 ml; $ 30 g Zr^{99}Mo gel

3 Results and Discussion

The results obtained from 4g and 40g alumina column chromatography generators eluted with 10 ml and 30 ml of 0.3M and 0.1M NH_4OAc and those from $Zr^{99}Mo$ gel generator (~30g gel, 10g Mo) are shown in Table. ^{99m}Tc was eluted with 20 ml of water in the latter case, which was found to be sufficient to elute >90% of theoretical ^{99m}Tc activity on the 30g size gel generator. In both the cases, 3 ml of normal saline was required to elute >90% of ^{99m}Tc activity from the anion exchanger column. Concentration of ^{99m}Tc by about ~6 times was thus obtained by the above process, even when using commonplace ion-exchanger.

The purity of the concentrated pertechnetate eluates was satisfactory and compatibility for formulation of even sensitive products, such as ^{99m}Tc-ECD and ^{99m}Tc-MDP, with both fresh kits as well as post-expiry kits, were found to be quite high, in terms of RC purity, stability and organ uptake in test animals.

The above procedure can be applied for both the types of column generators to increase the RAC of ^{99m}Tc at the end of shelf life of the generator and can be used beneficially wherever higher RAC of ^{99m}Tc is essential. It is felt that up to about 1g of Mo can be absorbed per Jumbo alumina generator (i.e. 300 – 500 mCi ^{99}Mo) and ^{99m}Tc obtained finally in 3ml volume. It may be possible to simplify the method by using a suitable eluent for the generator in order to obviate the need for the cation exchanger bed(s) and concentrate $^{99m}TcO_4^-$ by a single anion exchanger column alone. Based on the above studies it is felt feasible to establish suitable method for making use of $(n,\gamma)^{99}Mo$ as the source for a reasonably user-friendly delivery system for ^{99m}Tc.

References

1. Singh J, Reghebi K, Lazarus CR, Clarke SEM, Callahan AP , Knapp FF Jr, Blower PJ, Studies on the preparation and isomeric composition of ^{186}Re and ^{188}Re - pentavalent rhenium dimercaptosuccinic acid complex, *Nucl. Med. Commun,* **14** (1993) pp.197-203.
2. Knapp FF Jr, Beets AL, Mirzadeh S, and Guhlke S, Use of a new tandem cation / anion exchange system with clinical -scale generators provides high specific volume solutions of technetium-99m and rhenium -188, *IAEA-TECDOC-1029* (1998) pp.419-425.
3. Blower PJ, Extending the life of a ^{99m}Tc generator: a simple and convenient method for concentrating generator eluate for clinical use, *Nucl. Med. Commun.* **14** (1993) pp. 995-997.

CARRIER-FREE RADIOISOTOPES ISOLATED BY THE GAS CHEMISTRY METHODS

V.P. DOMANOV
Joint Institute for Nuclear Research, Flerov Laboratory of Nuclear Reactions, Dubna, Moscow reg., 141980, Russia

E-mail: domanov@nrsun.jinr.dubna.su

The selective isolation of carrier-free radioisotopes of some transition elements from the gas phase has been performed. The developed methods are based on the difference between volatilities of compounds of separated radioelements - products of nuclear reactions and between the chemical and physical-chemical behaviour of micro- and macro-components. ^{99m}Tc, ^{188}Re, ^{101m}Rh and ^{188}Ir were isolated in the form of volatile oxides, ^{100}Pd and $^{188,191}Pt$ - in the form of sulphides and $^{105,106}Ag$ and ^{196}Au in the form of pnictides.

A short review of the developed gas-chemical procedures of isolation of some carrier-free radioisotopes is presented. The gas chemistry methods of isolation are practically free of the radiolysis effect on the separation process, they are fast and require, as a rule, simple experimental set-ups. Here the thermochromatographic (TC) technique [1] was used in the majority of the cases. For isolation of the desired products, they were converted into volatile oxides, chalkogenides or pnictides, transported from the starting zone in a stream of a carrier gas along the open-tubular quartz TC column and adsorbed on its walls at certain temperatures.

Novel procedures of isolating of the ^{99m}Tc and ^{188}Re radioisotopes from the generator systems ^{99}Mo - ^{99m}Tc and ^{188}W - ^{188}Re are proposed. For production of the mother activity MoO_3 was irradiated by thermal neutrons to induce the reaction $^{98}Mo(n,\gamma)^{99m}Tc$ and for inducing the reaction $^{186}W(2n,2\gamma)^{188}W$ tungsten trioxide was irradiated by high-intensity resonance neutrons. To isolate ^{99m}Tc, the MoO_3 target was placed into a quartz container and heated in the atmosphere of air at 690-700°C. Volatile $^{99m}TcO_3$ was collected at < 450°C. Each subsequent isolation of ^{99m}Tc was performed under a higher heating of the target than a previous one. The ^{188}Re radioisotope was also produced as a volatile trioxide. The WO_3 target was placed into a quartz TC column and on the first stage it was heated in the H_2 stream at 600-650°C. To isolate ^{188}Re, the resulting WO_2 was heated at 750-800°C in a stream of a dry O_2-containing gas. These redox cycles were repeated as the daughter radioisotope was accumulated.

The discovery of differences in the physical-chemical behaviour of micro and macro- components in the gas phase, particularly for transition elements [2-3], is of obvious theoretical and practical interest. Based on this phenomenon convenient methods of TC isolation of radioisotopes of platinum metals and some other transition elements are developed.The ^{101m}Rh radioisotope, produced by the $^{99}Tc(4He,2n)^{101m}Rh$ reaction, was purified of the target matter at moderate heating in an air stream. The residue, having been placed into a TC column, was annealed in an O_2 stream at 1100°C. The volatile oxide ^{101m}RhO was adsorbed at ~775°C. Separation of the ^{188}Pt-^{188}Ir pair was also carried out in the stream of a dry O_2-containing gas. Being with the carrier (ponderable form), ^{188}Pt remained in the starting zone while carrier-free ^{188}Ir in the form of volatile IrO_3 was adsorbed at ~300°C [4].

Selective isolation of ^{100}Pd from a mixture of different radionuclides, which were adsorbed on the surface of SiO_2 powder, was carried out at 900°C in a He stream containing S vapour [5]. An analogous procedure was proposed for isolation of $^{188,191}Pt$. The centres of the adsorption zones of PdS_2 and PtS_2 were at 510°C and 550°C respectively. Methods for selective isolation of $^{105,106}Ag$ radioisotopes in the presence of Sb vapour and of ^{196}Au in the presence of P or As are also presented. The formations are perceived to be AgSb and AuP (or AuAs). They were adsorbed at 770°C (AgSb) and 660°C (AuP). The procedure for isolation ^{196}Au in the form of a volatile arsenide is preferred.

Acknowledgements

Prof. B.Eichler took an active part in the development of the iridium-188 isolation procedure and Prof. I.Zvara favoured the development of the method for rhodium-99m isolation. The author is grateful to them for their contribution.

REFERENCES

1. Merinis J.H. and Boussieres G. Anal. Chim. Acta, **25** (1961) pp.498-504.
2. Domanov V.P., Eicher B., Zvara I. Radiokhimiya, **26** (1984) pp.66-75 (in Russian).
3. Domanov V.P., Zvara I. Radiokhimiya, **26** (1984) pp.770-777 (in Russian).
4. Domanov V.P. and Eichler B. Invitation Certificate USSR N 580000, Report N 42, 1977 (in Russian).
5. Domanov V.P. Proc. of 4th Int. Conf. on Nuclear and Radiochemistry, St. Malo, France, 8-13 Sept., 1996, v.1, B-P9.

Session D2

Nuclear Analytical Applications of Radioisotopes — II

URANIUM ANALYSIS OF ORES AND TAILINGS: A COMPARISON OF DELAYED NEUTRON COUNTING, NATURAL GAMMA-RAY SPECTROMETRY, INAA AND XRF.

M. JOHN M. DUKE[1], CLAIRE A. J. TAYLOR[2] AND ANGELO FERNANDO[3].

[1] SLOWPOKE *Nuclear Reactor Facility, University of Alberta, Edmonton, Alberta, CANADA. T6G 2E3.* [2] *Oceanography Laboratories, Department of Earth Science, University of Liverpool, Liverpool, ENGLAND. L69 3BX, and* [3] *Key Lake Operation, Cameco Corporation, Saskatoon, Saskatchewan, CANADA.*

In a comparative study of instrumental analytical techniques for uranium analysis almost two hundred uranium ore and mill tailings samples, ranging from over 50 % to 0.017 % U, were analysed using a number of nuclear analytical techniques following their initial analysis by X-ray fluorescence. Using the same sample aliquot throughout (thus avoiding sample heterogeneity) samples were sequentially analysed using natural gamma-ray spectrometry (employing a high resolution Ge detector), delayed neutron counting (DNC), and instrumental neutron activation analysis (INAA).

The major drawback of uranium quantification by natural gamma-ray spectrometry is that it is an indirect approach which assumes secular equilibrium between the parent ^{238}U and its gamma-ray emitting progeny, ^{226}Ra, ^{214}Pb and ^{214}Bi. Consequently, the use of natural gamma-ray spectrometry is inappropriate for determining the uranium content of samples which are in a severe state of disequilibrium, *e.g.*, uranium tailings samples from which the uranium has been extracted. However, the method does have value in assessing the radiological health hazard posed by such deposits.

DNC continues to offer the advantages of being an extremely rapid, sensitive and essentially interference-free method for uranium quantification. The method is versatile and can be employed to analyze samples with uranium concentrations ranging from the sub-µg/g to the weight percent level.

In this study uranium was determined by INAA using the uranium activation product ^{239}Np produced *via* the reaction $^{238}U(n,\gamma)^{239}U \rightarrow {}^{239}Np$ (T½ = 2.355 d). In the course of the study it was found that a second neutron irradiation (in addition to that employed for the DNC analysis) was unnecessary for the INAA determination of uranium. Because of the high uranium content of the majority of the samples encountered in this work the neutron irradiation employed for the DNC analysis (20 s) was ample to activate the sample producing sufficient ^{239}Np for quantification purposes. Furthermore, it was found that it was possible to measure the natural gamma emitters ^{214}Pb and ^{214}Bi (ranging in energy 242 keV to 1764 keV) while simultaneously measuring the 228 keV and 277 keV gamma-emissions of ^{239}Np - this eliminates the need for a separate natural gamma-ray measurement.

For ore samples with a high uranium content the gamma-emissions associated with the decay of ^{239}Np may suffer significant self-attenuation leading to an underestimation of the sample uranium content. An empirical correction for such self-absorption can be determined using the variable attenuation of the numerous gamma-emissions of ^{214}Bi and ^{214}Pb.

In summary the analysis of samples by DNC, followed by the combined INAA/natural gamma-ray measurement, can provide accurate and independent determinations of the uranium content of uranium ore and tailings samples together with information on the state of equilibrium in the ^{238}U-decay chain plus the specific activity of ^{226}Ra in samples. This information is valuable for assaying uranium ore, for monitoring mill feed and recovery, and for the characterization and environmental assessment of uranium mill tailings.

Email address: *John.Duke@UAlberta.CA*

ASSAY OF Hg IN LARGE ELECTROCHEMICAL CELLS BY ISOTOPE DILUTION

HENRY C. GRIFFIN
Dept. of Chemistry, Univ. of Michigan, Ann Arbor MI 48109-1055, USA
E-mail: hcg@umich.edu

JAMES E. MARTIN AND CHUL LEE
School of Public Health, Univ. of Michigan, Ann Arbor MI 48109 USA
E-mail: jemartin@umich.edu

Mercury (Hg) electrodes can be and are used in the "chlor-alkali" process for production of elemental chlorine (Cl_2) and 50% sodium hydroxide solution. A given cell might contain 2000-6000 kg of Hg in a complex system of flowing electrode, reaction vessel, pumps, and collection tanks; some may be in the form of a semisolid "butter". Prudent practice requires that an inventory of the Hg be maintained. We have applied isotope dilution with ^{203}Hg to measure the Hg in each cell without interrupting the operation of the cell. Spikes and standards are prepared in a radiochemical laboratory and are shipped to the chlor-alkali plant. Samples taken from the spiked cells are counted on-site or back in the radiochemical laboratory. This process gives overall uncertainties of <0.5% (1σ) per cell and approximately 0.1% for a collection of cells.

1 Introduction

In the chlor-alkali process a salt solution (NaCl or KCl) is electrolyzed to produce chlorine and sodium hydroxide [1]. The negative electrode is a thin "sheet" of liquid mercury flowing in a slightly inclined tray about a meter wide and 10-20 meters long. The salt solution flows over the mercury, and a fixed positive electrode (often made of titanium) completes the circuit. Chlorine gas is produced at the upper electrode, from which it bubbles into a collection system. At the negative electrode Na^+ ions are reduced to Na metal, which dissolves in the Hg and is carried to a decomposition tank. In the decomposer Na reacts with pure water to produce NaOH solution and H_2 gas.

A brute force method to measure the Hg in a cell is to stop the flow and collect all of the Hg in one place. This is neither reliable (errors can be several percent if Hg is trapped in the system) nor convenient. The Na-Hg amalgam is quite fluid, but impurities can collect in the Hg and produce a semisolid "butter". Isotope dilution involves thoroughly mixing a known amount of tracer into the pool of mercury in a cell, measuring the amount of tracer in a sample taken from the mixed pool, and inferring the amount of mercury from the dilution factor.

2 Methods and Results

Neutron irradiation of mercury is the most convenient way to produce radioactive Hg tracer. When the irradiated material has a high concentration of mercury (e.g., Hg metal or a pure compound), the yield will depend on geometry because ^{199}Hg (σ_a = 2200 b) is a significant neutron sink. The two main radioactive products are ^{197}Hg and ^{203}Hg. Relevant properties of these isotopes are given in Table 1. Although 64-h ^{197}Hg is produced in much higher yield than ^{203}Hg in a short irradiation, its half-life is inconveniently short and its radiations tend to be absorbed by bulk Hg in a cell sample. Even for ^{203}Hg, self-shielding is significant. The half-thickness in Hg for 279-keV photons is about 1.6 g cm^{-2}, which corresponds to 1.2 mm in Hg.

Table 1. Radioisotopes of Hg suitable as tracers.

Isotope	Half-life	Yield[a]	Main γ ray
Hg-197	64.1 h	32300	K x rays ~90%
Hg-203	46.60 d	656	279.2 keV 81.5%

[a]Disintegrations s^{-1} μg^{-1} for Hg irradiated for one day at 10^{13} n cm^{-2} s^{-1} [2]

The self-shielding effect leads to very low radiation levels in the vicinity of a spiked cell. The ^{203}Hg γ rays escape the surface at a rate of 5-10 photons per cm^2 per second, which corresponds to <5 μR/h as compared to about 20 μR/h for typical background levels.

High specific activity is not needed. In fact, it is more convenient to start with low specific activity material because the bulk mercury is effective in shielding. In any event, a batch of spike material must be very well mixed. We confirm uniformity by taking 7 aliquots of each batch, diluting each aliquot to a specific activity typical of the final cell samples, and confirming that the count rates agree with dilution to better than 0.5% per sample.

Reproducible detection efficiency for the cell samples is essential. We use 18-mm (od) polyethylene vials to place 80.0 g of Hg in the 19-mm (id) wells of our NaI(Tl) detectors with for counting. Typical count rates are 100-200 s^{-1}, and 10-min counts give statistical uncertainties (1 σ) of <0.4 %.

The background for counting 80-g samples of Hg metal must be obtained with 80 g of Hg in the counting position. The mercury contributes to the shielding of the detector. In practice we need to know both the detector background and the "cell background", the residual activity from tracer previously put into the cell. It is not sufficient to use previous count rates and apply decay factors, because mercury may have been transferred between cells or fresh mercury may have been added in the course of normal plant operation. Therefore we take a sample from each cell just prior to spiking and use the increase in count rate of a sample from the spiked cell to compute the dilution factor. This process corrects for detector background and cell "background".

Small variations in the shape of vials can change counting efficiency by ~0.1%. We have found that selecting vials that have the same mass (within about 0.02 g out of ~3.8 g) makes these variations insignificant.

We have done many tests of the apparent approach to equilibrium of the specific activity of cell samples, but fewer direct tests (that require taking a cell off-line) of the activity of butter in a cell. However, those tests show (see Table 2) that mixing is quite effective. It has been our practice to allow the cell to operate for 48 hours before taking samples on which the inventory is based. The data in the table show that, even after 24 hours, the activity of the butter is not statistically different from the free Hg in the cell.

Table 2. Analysis of butter samples vs. time of sampling. The rates are in counts per minute, and the statistical uncertainties in the count rates are ±4.0 cpm

	Free Hg	Butter			
	48 h	24 h	48 h	72 h	96 h
Count rate	770.8	769.9	762.7	769.0	763.4
Deviation	+3.7	+2.8	-4.4	+1.9	-3.7

3 Acknowledgments

We acknowledge the support of Richard A. Copeland and Rad Chem Services, Inc.

References

1. J. Kroschwitz, exec. ed., *Encyclopedia of Chemical Technology*, vol. 1, 957-960, Wiley (1991).
2. G. Erdtmann, *Neutron Activation Tables*, Kernchemie Einzeldarstellungen, vol. 6, Verlag Chemie (1976).

A STUDY OF AIR LUMINESCENCE FOR ACCURATE LIQUID SCINTILLATION COUNTING OF LOW-LEVEL RADIONUCLIDES AND THE USE OF AIR LUMINESCENCE FOR CALIBRATING DETECTORS TO MEASURE ^{222}RN

YUKO MORITA-MURASE AND YOSHIO HOMMA

Laboratory for Radiopharmaceutical Chemistry, Kyoritsu College of Pharmacy 1-5-30, Shibakoenn, Minato-ku, Tokyo 105, JAPAN

The spectral response (300-520 nm) of current bialkali photomultiplier tube, PMT, matches the air luminescence spectrum which are produced by the interactions of α–particles from ^{222}Rn and its α–emitting daughters with nitrogen molecules in air. Therefore, the presence of ^{222}Rn and its α-emitting daughters in the vicinity of the PMTs of liquid scintillation spectrometers or NaI(Tl) scintillation counters produce additional background counts, which vary based on a number of factors. We discuss methods for reducing the background counts due to the air luminescence. We also describe the effect of air luminescence counts on determining ^{222}Rn by liquid scintillation counting: ^{222}Rn and its α-emitting daughters in the air space above the liquid scintillator do produce air luminescence counts which lead to errors in quantitative determination of ^{222}Rn unless necessary corrections for the counts are made. On the other hand, we standardize ^{222}Rn by measuring air luminescence counts of ^{222}Rn samples and use them to calibrate detectors for determining ^{222}Rn. A significant advantage of this calibration method is that the time required for the preparation of known amounts of ^{222}Rn for calibration is only 3.5 h. In contrast, using conventional methods to prepare known amounts of ^{222}Rn usually requires much more time (30 days).

1 Introduction

The passage of α-particle in air results in the ionization and excitation of nitrogen and oxygen [1]. The emission from excited nitrogen is in the sensitive spectral region (300 - 520 nm) of the current bialkali photocathode. Therefore, the air luminescence spectrum due to ^{222}Rn and its α-emitting daughters in air can be measured with commercially available liquid scintillation spectrometers showing peak at 2.4 keV, and endpoint at 18 keV [2,3]. What we have done is to examine in some detail effects of the air luminescence counts on the background level of the instrument equipped with the multiplier phototubes, PMTs, and to evaluate the factor influencing low-level liquid scintillation counting of ^{3}H, ^{14}C and ^{222}Rn etc.[3,4,5].

On the other hand we have extended the air luminescence counts due to ^{222}Rn to the calibration of detectors measuring ^{222}Rn: We standardized ^{222}Rn in the counting vial by measuring air luminescence counts with a commercially available liquid scintillation spectrometer [5,6]. We used these ^{222}Rn samples to calibrate detectors measuring ^{222}Rn. The results obtained by this calibration method are in good agreement with those obtained by other independent methods.

2 Changes in the background level due to the air luminescence count

Changes of radon (^{222}Rn and ^{220}Rn) concentration in air can be very detrimental for low-level liquid scintillation counting, because it will lead to changes in the background level of the liquid scintillation spectrometer : By turning off the switch of the ventilation system of the room where the instruments are located, ^{222}Rn concentration in the room, and hence, in the sample chamber of the liquid scintillation spectrometer, gradually increases. The background count due to the air luminescence produced by ^{222}Rn and its α-emitting daughters consequently increases. The fact that the variation of the background is based on the ^{222}Rn concentration in air is clearly shown by subtracting the pulse-height spectrum (B) from that of (A) in Fig .1: Residual spectrum (C) represents the typical air luminescence spectrum.

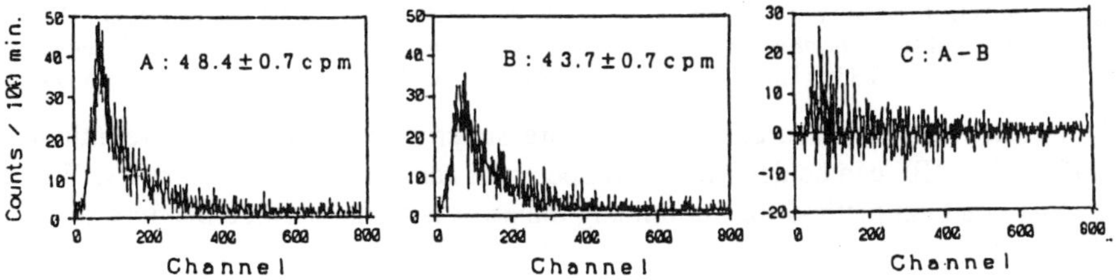

Fig. 1 Effects of ventilation on the changes in the background counts of a liquid scintillation spectrometer. (A), the ventilation systems are off; (B), the ventilation systems are on ;(C), (A)- (B).

A factor, which is of prime importance in determination of low-level radioactivities, is the background level of the detector. In addition to the general background due to the presence of traces of uranium, thorium, potassium, etc., in every laboratory, we have to take into account the background due to the air luminescence counts produced by ^{220}Rn, ^{222}Rn and their α–emitting daughters, which are hitherto neglected.

A method to obtain low and constant background is allowing ^{222}Rn - free gas, e.g., N_2 gas, to flow through the sample chamber where sample vial and PMTs are located.

3 Effects of the air luminescence counts in determination of ^{222}Rn by liquid scintillation counting

The fact that ^{222}Rn is appreciably soluble in liquid scintillator and the use of the integral counting technique make liquid scintillation counting of ^{222}Rn practical. However, ^{222}Rn and its α-emitting daughters in the gaseous space above the liquid scintillator in the counting vial produce the air luminescence counts which lead to errors in quantitative determination of ^{222}Rn by liquid scintillation counting. The error, which varies depending on the volume of liquid scintillator (LS) used for ^{222}Rn measurement, is estimated to be : 15.7% for 4.0 mL-LS, 6.3% for 8.0 mL-LS, 3.0% for 12.0 mL-LS and 1.4% for 16 mL-LS [5].

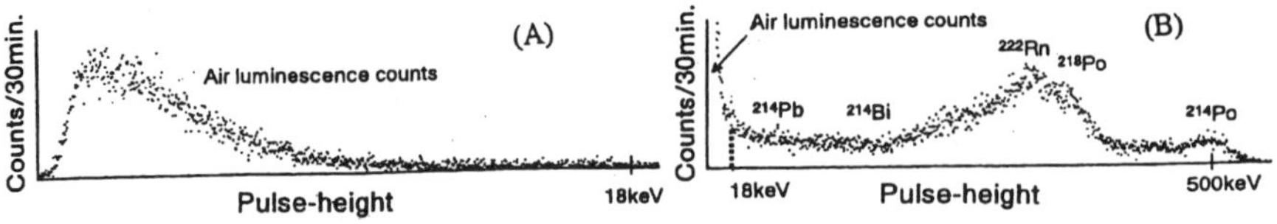

Fig.2 Pulse-height spectrum of ^{222}Rn and its daughters measured with (A) 1024-analyzer of an Aloka LSC-3500 liquid scintillation spectrometer which was set to measure from 1 to ca. 20 keV; (B) another 1024 channel analyser of LCS-3500 liquid scintillation spectrometer which was set to measure ca. 1 to 500 keV.• @

The air luminescence counts, which are superimposed upon the • -continuum of 214Pb and 214Bi, can be determined by spectrum analysis of 222Rn samples. However, the most practical method for minimizing the error is to measure 222Rn in a scintillator volume of no less than about 16 mL. The results obtained now seem to be very important since most of earlier measurements of 222Rn with liquid scintillation spectrometers appear to have been more or less subject to uncertainties due to the air luminescence.

4 Rapid calibration using the air luminescence method

We have extended the air luminescence due to ^{222}Rn to calibrate the detector measuring ^{222}Rn. Until now, detectors measuring ^{222}Rn have been commonly calibrated with known amounts of ^{222}Rn from a standard ^{226}Ra source. Such calibrations, however, require usually a rather long waiting period (30 days), which makes the calibration time-consuming and tedious.

A significant advantage of this calibration method is that time required for preparing known amounts of ^{222}Rn necessary for calibration is only 3.5 h. Moreover, the ^{226}Ra solution, from which ^{222}Rn is collected, is not necessary to be standard solution.

4.1 Standardization of ^{222}Rn

^{222}Rn was collected from 0.1 M HCl solution of $^{226}RaCl_2$ with a syringe and added to a standard counting vial which was full of air and closed with a silicon rubber stopper and was allowed to remain for 3.5 h before measurement. During this time, the daughters of ^{222}Rn came to transient equilibrium with ^{222}Rn. The air luminescence counts of ^{222}Rn samples were measured with an Aloka liquid scintillation spectrometer, Model LSC-3500 (Aloka Co. Ltd. Tokyo, Japan). Counting efficiency of the air luminescence for the liquid scintillation spectrometer was determined to be 74.4% by comparing the air luminescence count rate of a standardized ^{210}Po source with its standardized dpm [4,6].

4.2 Calibration of detectors measuring ^{222}Rn

Detectors measuring ^{222}Rn calibrated in this study by the air luminescence method are the PICO-RAD ^{222}Rn detector cartridge (Packard Instrument Co.,Inc. IL USA)and a cellulose nitrate track detector (Kodak LR 115 Film).

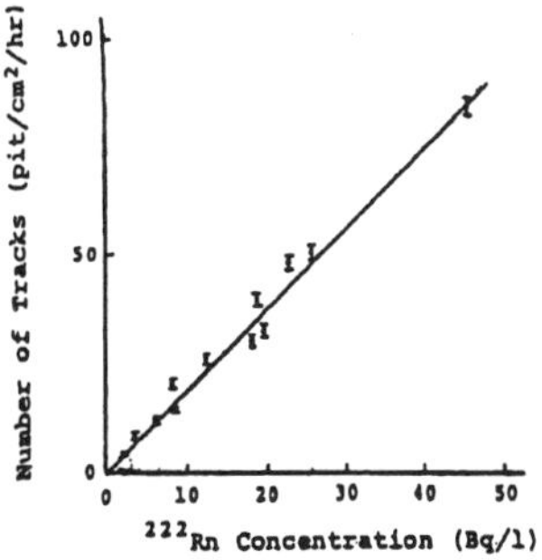

Fig. 3 The relation of number of tracks and ^{222}Rn concentration in air

The cartridge was exposed to standardized ^{222}Rn of different concentrations in the plastic vessel for 24 and 48 h with temperature of 24 – 28• • and relative humidity ranging from 30 - 40 %. After the exposure, the cartridge was added 10 mL of Insta-Fluor scintillation cocktail and kept at room temperature for an additional 17 – 18 h and measured with a Packard Tri-Carb liquid scintillation analyzer, Model 2550 TR/LL (Packard Instr. Co. IL, USA) using the PICO-RAD radon analysis software.

Agreement between the average ^{222}Rn concentrations determined by the air luminescence method and those determined by the PICO-RAD is satisfactory.

The relation between the net number of tracks of the track detector produced by α particles from ^{222}Rn and its α-emitting daughters and the average ^{222}Rn concentration of standardized ^{222}Rn is linear over the entire ^{222}Rn concentration studied. The reproducibility has proven to be within 10 %, and the slope corresponds to 100 tracks/cm^2 for an hour exposure of 54 Bq/L of ^{222}Rn.

Advantages of this calibration method are:

1. the time required for the preparation of known amounts of ^{222}Rn necessary for calibration is much shorter (3.5 h) than that of conventional methods (30 days),
2. ^{226}Ra solution, from which ^{222}Rn is collected, is not necessary to be standardized one, and
3. .^{222}Rn can easily be standardized using a commercially available liquid scintillation spectrometer.

References

1. Dondes S., Harteck P. and Kunz C., A spectroscopic study of alpha-ray-induced luminescence in gases. *Radiat. Res.* **27** (1966) pp. 174-210.
2. Takiue M. and Ishikawa H., Alpha-ray measurement due to air luminescence employing a liquid scintillation spectrometer. *Nucl. Instr.meth.* **159** (1979) pp.139-143.
3. Murase Y., Homma Y., Takiue M. and Aburai T., Determination of air luminescence spectra for alpha-emitters with liquid scintillation spectrometers . *Appl Radiat. Isot.* **40** (1978) pp. 291-.294.
4. Homma• @Y., Murase Y. and Takiue M., Determination of ^{222}Rn by air luminescence method. *J. Radioanal. Nucl. Chem.* **119** (1987) pp. 457-465.
5. Murase Y., Homma Y. and Takiue M.,Effect of air luminescence counts on determination of ^{222}Rn by liquid scintillation counting. *Appl Radiat. Isot.* **40** (1989) pp. 295-298.
6. Homma Y., Murase Y., Handa K., Koyama S., Suzuki N. and Horiuchi K., Rapid calibration of detectors for determining ^{222}Rn using air luminescence counting. In *proceedings of the International Conference on Advances in Liquid Scintillation Spectrometry 1994* ed. by Cook G.T., Harkness D.D., MacKenzie A.B., Miller B.F. and Scott E.M. Radiocarbon (1996) pp. 111-116.

PHOTON ACTIVATION ANALYSIS OF TRACE ELEMENTS IN LIVERS, KIDNEYS, AND SPLEENS OF PIG FETUSES

M. FUKUSHIMA, K. KAJIWARA, AND M. OIZUMI

Ishinomaki Senshu University, Ishinomaki, Miyagi, 986-8580, Japan

E-mail: fukusima@isenshu-u.ac.jp

The levels of trace elements in livers, kidneys, and spleens of pig fetuses were analyzed by two methods: photon activation analysis (PAA) and flame atomic absorption spectrometry (FAAS) after solvent extraction. Fetuses were removed from the mother pigs of different pregnancy periods. Approximately half of 30 fetuses were used for PAA and the rest were for FAAS. Twelve elements (Br, Ca, Cu, Fe, I, Mg, Mn, Na, Rb, Sn, Sr, and Zn) by means of PAA, and additional three elements (Ni, Pb, and K) by FAAS could be determined.

1 Introduction

The purpose of this study was to determine the normal elemental levels of pig fetuses at different stages of pregnancy. From the levels of trace elements in organs, we can obtain information about the relationship between of elemental accumulation and growth stage of organs. Such basic data about mammal fetuses, especially about domestic animals, are few. Honda *et. al.* [1] studied the elemental levels found in Japanese Serow (*Capricornis crispus*), and they determined total concentrations of 9 heavy metals in whole body of the fetus. It is well known that the elemental concentrations in same organs of animals can differ from one habitat to another. Most domestic animals reside under almost the same habitat. Therefore we selected several organs of pig for this purpose. PAA is a suitable method for multi-element analysis of biological samples because interfering activities induced from matrix elements such as Na and P are low compared to neutron activation analysis. Moreover, PAA characteristics permit the convenient determination of I and Mg.

2 Experimental

2.1 Samples

Fetuses were removed from healthy mother pigs of different pregnancy periods after slaughtering the animals. Livers, kidneys, and spleens were taken from fetuses and were preserved in formaldehyde within a week. Body lengths of fetuses were 12–35 cm, and the pregnancy periods of mother pigs were about 60–120 days [2]. Wet weights of livers, kidneys, and spleens were 4–22 g, 1–7 g (for both halves of the kidneys), 0.06–0.9 g, respectively.

2.2 PAA

Livers and kidneys were first freeze-dried then pulverized and prepared for irradiation as described in a previous paper [3]. Spleens were freeze-dried and used as such for further analysis. NIST SRM-1566a Oyster Tissue and SRM-1577b Bovine Liver were used as elemental comparator standards. Samples and several standards were sealed in a silica tube, and irradiated for 3 h at 30 MeV bremsstrahlung (120 μA beam current) at the linear accelerator facility of Tohoku University in Sendai, Japan. Under this condition, (γ,n) reaction is the main nuclear reaction for middle-to-high Z elements. The gamma-ray spectra were recorded for 10 min, 1 h, 3 h, and 6 h after cooling times of 3 h, 1 d, 7 d, and 1 month, respectively, using hyper-pure Ge detectors (ORTEC, GEM-40200P and GEM-30185) with multichannel pulse-height analyzers. The target nuclides, nuclear reactions, product nuclides, half-lives, gamma-ray energies (keV) of the elements of interest are as follows; Br-79(γ,2n)Br-77 (2.3765d, 657.8); Ca-44(γ,p)K-43 (22.3h, 617.8); Cu-63(γ,2n)Cu-61 (3.41h, 282.9); Fe-58(γ,pn)Mn-56 (2.579h, 846.8); I-127(γ,n)I-126 (13.0d, 388,6, 666.3); Mg-25(γ,p)Na-24 (15.02h, 1368.6); Mn-55(γ,n)Mn-54 (312.2d, 834.8); Na-23(γ,n)Na-22 (2.602a, 1274.6); Rb-85(γ,n)Rb-84 (32.9d, 881.6); Sn-118(γ,n)Sn-117m (14.0d, 158.6); Sr-88(γ,n)Sr-87m (2.8h, 388.4); and Zn-66(γ,n)Zn-65 (244.1d, 1115.5).

2.3 FAAS

Organs were cut into pieces, and digested with nitric and hydrochloric acids. The mixture obtained was transferred into a 50-mL volumetric flask, filled up to the mark with 0.1M nitric acid, and then filtered using Millipore cellulose nitrate membranes. To 15 mL of this solution, 10 mL of 10% diethyldithiocarbamic acid (DDTC) solution and 5 mL of a buffer solution (pH 6.0) were added. After 5 min, 10 mL of 4-methyl-2-pentanone (MIBK) were added, and the mixture was shaken for 5 min. Organic and aqueous phases were separated and then sprayed into a flame. The concentrations of elements were calculated from calibration curves obtained using mono-element comparator standards.

3 Results and Discussions

3.1 The growth of organs

The relationships between body length (cm) of fetuses and the wet weight (g) of each organ are shown in Fig. 1. It is evident that the weights of three organs increase with increasing body length, and that the curves appear to reach a plateau near 20-cm. This observation is related to body length and pregnancy period [2].

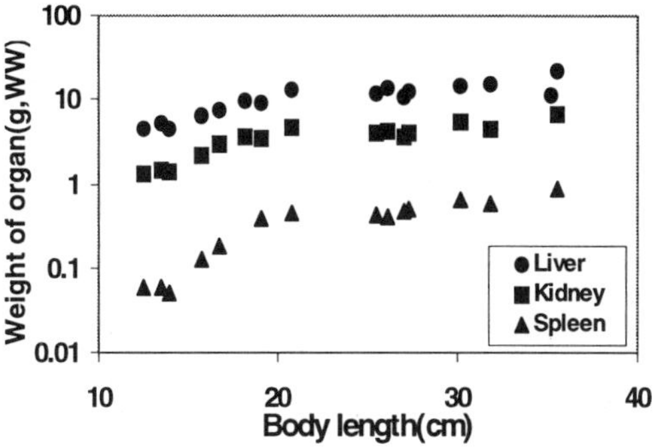

Figure 1. Weight of organs and body length

3.2 Wet weight and dry weight of organs

The ratios of dry-weight to wet-weight were 0.201±0.019 (n=16), 0.136±0.014 (n=15), and 0.206±0.041 (n=14) for livers, kidneys and spleens, respectively. A slight decrease in the ratio for spleens according to body length was observed.

3.3 Variation of elements in organs

The ranges of concentrations for several elements in various organs are shown in Table 1. The elements are divided three groups: (i) high concentrations at the early stage of pregnancy: Br, Mn, Ni, and Pb in spleen, Sn in kidney, and Zn in all three organs; (ii) high concentrations towards the end of pregnancy: Na, K and Rb in all three organs; and (iii) constant concentration over the pregnancy period investigated: Mg, Ca, Fe, and Cu again in all three organs. A comparison of the levels of elements in fetuses and adult pigs shows that Ca in livers and spleens are higher in fetuses and that Fe in spleens are lower in fetuses than those of adults.

4 Acknowledgment

We would like to thank the staff and Dr. T. Otsuki of the Laboratory of Nuclear Science for their assistance in irradiation and measurements. We thank Prof. T. Sakata and the staff of the Senpoku Meat Inspection Center for providing the organ samples of pig fetuses. This work was financially supported in part by a grant from The Mitsubishi Foundation.

Table 1. Elemental concentrations obtained for pig fetuses.(unit; μg/g, dry-weight)

	Liver		Kidney	Spleen	
	Fetus (n=15)	Adult [4] (n=5)	Fetus (n=15)	Fetus (n=14)	Adult [4] (n=5)
Br	6.2 ~ 17.3	20.2 ± 3.7	10.0 ~ 78.5	11.9 ~ 168	29.3 ± 10.8
Ca	4230 ~ 5620	192 ± 42	3650 ~ 5190	4080 ~ 5520	274 ± 47
Cu	202 ~ 462		~83.8	~85.7	
Fe	408 ~ 1087	665 ± 226	133 ~ 291	117 ~ 848	1440 ± 660
I	0.1 ~ 1.3		~3.8	~1.8	
K*	162 ~ 2560		573 ~ 3340	395 ~ 3230	
Mg	913 ~ 1120	674 ± 23	925 ~ 1140	918 ~ 1410	939 ± 54.9
Mn	4.6 ~ 12.0	17.4 ± 2.1	0.2 ~ 10.8	~26.4	19.4 ± 6.7
Na	2020 ~ 6220	2210 ± 180	2950 ~ 9430	2320 ~ 7540	3050 ± 310
Ni*	~1.1		0.1 ~ 2.7	2.2 ~ 22.9	
Pb*	~17.3		~8.3		
Rb	4.4 ~ 9.3	26.2 ± 2.2	5.3 ~ 13.0	3.1 ~ 11.8	46.0 ± 2.7
Sn	6.5 ~ 11.6		4.0 ~ 65.1	0.3 ~ 15.1	
Sr	0.2 ~ 3.2		0.2 ~ 1.6	0.2 ~ 1.4	
Zn	115 ~ 537	321 ± 73	15.1 ~ 278	19.9 ~ 825	122 ± 19

* determined by FAAS, unit; μg/g, wet-weight

References

1. Honda K., Ichihashi H., and Tatsukawa R., Arch. Environ. Contam. Toxicol.16 (1987) pp.551-561.
2. Niwa T.J. of Zootech. Sci., 251 (1954) pp. 5.
3. Fukushima M., Tamate H., Sato S., Terui S., and Mitsugashira T., J. Radioanal. Nucl. Chem., 239,3 (1999) pp.595-599.
4. Fukushima M., J. Radioanal. Nucl. Chem., 195 (1995) pp.145-153.

Session D3

Separation and Applications of Stable Isotopes

PRODUCTION TECHNIQUES OF STABLE METAL ISOTOPES: CURRENT STATUS AND FUTURE TRENDS

DARREN BROWN

Trace Sciences International. 15 Wertheim Court, Suite 404. Richmond Hill. Ontario, Canada. L4B 3H7

Tel: (905) 707-7000 E-mail: Darren@isotopetrace.com

DR. STEVEN HARRISON

Trace Sciences International. 15 Wertheim Court, Suite 404. Richmond Hill. Ontario, Canada. L4B 3H7

TEL: (905) 707-7000 E-mail: steve@trace.on.ca

The February 1st, 1998 shut down of the ORNL calutrons signaled a loss in the world production capacity for *Stable Metal Isotopes*. This occurred at a critical time, as the major market for these products, namely the radiopharmaceutical industry, had undergone a shift from diagnostic agents to those which were therapeutic in nature. Thus, the demand for new agents and hence their precursors, was increasing globally. Demand from industrial sectors has also witnessed substantial growth in the last five years. Through advances in technology and with the support of several isotope facilities in Russia and Siberia, little or no impact is anticipated on the world market with the ORNL closing. A collaborative effort, co-ordinated by Trace Sciences International, has enabled a unique "production structure" to form. Trace has obtained marketing rights from Electrokhimprebor (EKP), Urenco, The Kurchatov Institute and the Siberian Group of Chemical Enterprises (TOMSK). This has enabled the world capacity to not only expand but to do so in a structured and directed manner. EKP and Kurchatov act as primary and secondary manufacturers for Calutron produced Isotopes where as Urenco and Tomsk perform this function for centrifuge based elements. This structure has greatly supported the radiopharmaceutical industry as it moves from diagnostic to therapeutic agents. Advances in centrifuge technology have also decreased the burden on traditional calutron production techniques with the ability to manufacture isotopes of Zn, Cd, Cr, Ni, Sn, S, Si, Ir, W, and Te. This has further allowed the use of calutrons to become more focused on the production of Sr, Sm and Pd. Thus, ORNL's shutdown has been demonstrated to have an overall positive impact in co-ordinating the world production capacity of *Stable Metal Isotopes*.

1 Historical Perspective

In conjunction with the discovery of the elements of the periodic table, scientists also recognized the existence of naturally occurring isotopes. With this discovery the need naturally arose to study these unique species within the greater subset, and hence the need for isotope separation. The earliest isotope studies focused on radionucleid production and reaction, which in turn expanded the periodic table. Further scientific curiosity lead to the discovery of the "isotope effect" and a new wave of commercialization occurred. It was not long before science learned to then capitalize on the effects of purified and separated isotopes. Prior to World War II most of the separation methods focused on large scale production for the War effort. Most notably, that which occurred in Ryukan Valley in Norway, where the Norwegians established the first heavy water production facility (for the purposes of this discussion, only separation methods of heavy stable metal isotopes will be reviewed). During the same time period, the development of a device anakronymed a "CAL - U - TRON" was envisioned at the University of California by adapting the configuration of a cyclotron. This device specialized in the separation of elements from Magnesium (ATM 24.305) to Uranium (ATM 236.2). As part of the War effort the predecessor to the DOE (Department of Energy), the then AEC (Atomic Energy Commission) established a full-scale production facility to be constructed in the rural farming community of Oak Ridge Tennessee. At the height of construction over 100,000 people were based in this community, working towards the common goal of the separation of Uranium. Curiously enough, the calutrons at the Y-12 facility did not produce production quantities of Uranium; rather the K-25 facility was established utilizing thermal diffusion techniques for isotope separation. From a post War effort the Calutrons were utilized as a research tool to provide science and emerging medical industries with a much-needed source of stable isotopes.

2 Calutron Methodology

The Calutron separation method essentially "rips" apart the element magnetically and centrifically into its isotopic constituents while in an excited state. The element is ionized and injected into a magnetic field where it is thereby defracted, or bent according to its mass. Heavier isotopic species are "bent" the least, and the lightest are subject to a greater field effect and thus have a smaller arc. The trajectory of the masses can be accurately calculated to ascertain the best angle of deflection and thus collection. The collectors or "pockets" are then extracted from the calutron, at which time the isotope target is removed and chemically purified in the lab. While considered a scientific dinosaur due to its sheer mass, electrical consumption, manpower requirements, and low throughput, the calutron has an advantage over other more current production methods in that all the isotopes of an element can be separated simultaneously. Although costly, more elements can be separated using this methodology than other techniques.

3 Current Techniques

Given the growing demand for stable metal isotopes, and the heightened purification issues, new and more cost effective separation methods have come into commercial existence in recent years. As scientists learn to exploit and hone their needs for materials, so goes the requirement for large-scale separation. In this regard the nuclear industry has developed a demand for a "depleted isotope". ***DZO*** or *Depleted Zinc Oxide* is now being added to over 50 reactors world wide to assist in the reduction of corrosion, stress cracking and REM count. In essence naturally abundant Zn^{64} (48.6%) is depleted to less than 1%. It is estimated that the potential demand for this product is upwards of 2000 kilograms per year. This is well beyond the milligram-gram output of the ancient calutrons. In other sectors, kilograms of material are required for the laser industry, as well as for nuclear medical applications. In specific, the largest application for stable metal isotopes has been that of a precursor for nuclear diagnostic agents. In the last 10 years however this industry has been expanding into therapeutic applications as opposed to diagnostic. This has a potential of pulling demand for certain isotopes into the multi-kilogram range. Once again, this is well beyond the capabilities of the calutron.

While the calutron is currently in use, the tool of choice to meet today's isotope needs is undoubtedly the centrifuge. The principal of centrifuge separation is based on the induction of a gas phase, thermally stable compound isotope that contains the target element within its matrix. Typical carrier gases have included diethyl zinc and dimethyl cadmium. In production mode the gas is introduced in low volumes into a centrifuge that has a high RPM rotor. The torodal spinning action causes a centrifical effect whereby the heavier isotope is sent toward the outer diameter of the centrifuge. This gas is then collected and introduced into another series of centrifuges; known as a cascade. Depending on the quantity and isotopic enrichment desired, hundreds of individual centrifuges could be utilized to meet requirements. Unlike the calutron, this method of production is quite inexpensive, and is very energy and labour efficient. The are only two disadvantages to this mode of production: 1) only one isotope can be separated at a time, and 2) there are a limited number of elements suitable for separation. With recent advances in centrifuge technology however, the list of available candidates for production has expanded rapidly. This is in large part owing to the fact that fluorine based metal compounds are not used as carriers.

Despite advances in both Calutron and Centrifuge applications however, some elements "fall between the cracks" as candidates for separation. Most notably of which is Mercury (Hg). Mercury is fast becoming of interest as a bio-marker for environmental studies and thus the demand is growing. Historically, Hg was separated at the Oak Ridge Calutron, with a limited annual quantity produced. Since that time, the "experiment" was not successfully duplicated, and the stocks of Hg isotopes available have been rapidly decreasing. However, scientists at the Kurchatov Institute in Moscow, having a modicum of success with their own calutron for Hg separation, ventured along a path exploiting the effect better known as "resonance trapping." In effect this photochemical technique captures specific Hg isotopes in a cover gas of 1,3 butadiene. The isotopes are placed in an excited state through the capture of an isotopically specific photon that is generated from a pure Hg isotopic source lamp. The entire apparatus is bench scale yet can turn out gram quantities of isotope through successive batch runs. Despite the beauty and simplicity of this device, no other elements have been found as candidates for isotope separation utilizing photochemical excitation.

4 Future Prospects

In keeping with the "excitation" principle of separation, a great deal of work has been devoted to Laser Isotope Separation. As early as 1974 the principles of laser excitation of an atom have been dealt with in the scientific realm. Most notable the AVLIS (Advanced Laser Isotope Separation) program under the auspices of the DOE has focused on the separation of heavy metal isotopes. In principle a high power YAG laser is tuned to the exact vibration of the isotope in question. The target isotope once "hit" by the coherent beam is placed in an excited state, whereby it can be scavenged by a cover gas. In the academic realm this is very elegant approach to isotope separation, however it carries a host of engineering difficulties. Of the many difficulties in this process the precise tuning of the laser is critical. It has only been in the last 5 years that high-powered YAG lasers have been developed with tuning as precise as necessary to make this process feasible. However the power requirements are too low for commercialization: heavier metals cannot be separated using the YAG laser, and therefore it is not useful for isotopes with a Z value greater than ~50. In higher-powered metal vapor lasers, such as a 50W copper vapor laser, precise tuning is an issue, coupled with the issue of the bandwidth of the laser. Moreover, once the target is in an excited state recovery in a "flow" system is a difficult engineering feat, yet this is what is required for economic cost recovery of the isotope. To date an economical method of laser separation has yet to be developed. Based on this principle only one isotope can be separated at a time, which in turn increases costs.

Other advances in separation technology are in keeping with the centrifuge principle of separation. The PSP or Plasma Separation Process was originally developed by TRW in the early 70's. TRW successfully separated gram quantities of Pd^{102} to an enrichment of 70%. This process places the target element in a plasma whereby it is "spun" via a high intensity RF signal. The high speed spinning action creates a centrifugal effect whereby the heavier species is pushed to the outer rim of the plasma. The key difficulties with this process are collection, and perhaps more importantly what has been dubbed "the minority species phenomenon". In this instance the co-efficient of separation is inversely proportional to abundance. In other words low enrichment targets increase in abundance quickly but as the target becomes the majority species: a "de-enrichment" occurs. To solve this problem the

PSP is looked at as being able to provide better starting material to the CALUTRON. In coming full circle with technology, despite its pitfalls, and considering all factors, such as cost, production, availability, and flexibility the calutron is arguably the best technology for isotope separation.

5 ORNL SHUT DOWN

On February 1st, 1998 the calutrons located at ORNL were placed in "cold stand-by". This signaled a loss of the world's production capacity for stable metal isotopes. This occurred at a critical time, as we have seen as the major market for these products, namely the radiopharmaceutical industry, has undergone a shift from diagnostic agents to those therapeutic in nature. Moreover demand from industrial sectors has also grown rapidly in the last 5 years, as has the demand for specialty research and development items. Through advances in technology and with the support of several isotope facilities in Russia and Siberia little or no impact will be felt on the world market with the ORNL closing. A collaborative effort, co-ordinated by Trace Sciences International, has enabled a unique "production structure" to form. Trace has obtained marketing rights from Electrokhimprebor (EKP), Urenco, The Kurchatov Institute and the Siberian Group of Chemical Enterprises (TOMSK). This has enabled the world capacity to not only expand but to do so in a structured manner. EKP and Kurchatov act as primary and secondary manufacturers for Calutron produced Isotopes where as Urenco and Tomsk perform this function for centrifuge based elements. This structure has given great support to the changes that have occurred in the market. In several instances demand for an item has been beyond the productive capacity of one manufacturer. Through the relationship established by Trace, and the "pooling" of resources, output demand can be met.

With other aspects of ORNL's facility remaining questionable in the future, Trace has expanded its laboratory facilities. Currently ORNL operates the finest stable isotope lab in the world today making a wide variety of compounds and foils at a "micro-level". This specialized knowledge is only gained with experience as typical CAS recipes do not apply to micro scale chemistry. While Trace has been performing routine reductions and oxidation's for almost 10 years we were hampered in the manufacture of foils as we were limited in equipment and thus our milling capabilities were limited to ductile materials. With the addition of our new mill we have been able to expand the list of elements available and also reduce the aerial density of this existing list of materials. Currently we are working on U foils for the test reactor market.

Advances in centrifuge technology have also decreased the burden on traditional calutron production techniques with the ability to manufacture isotopes of Zn, Cd Cr, Ni, Sn, S, Si, Ir, W, Te. This has allowed the calutrons to become more focused on the production of Sr, Sm, Gd, Pd, etc. Thus, the overall production system is coordinated and market driven, and this has improved overall efficiency.

This is turn has lead to cost reductions, helping to expand those isotopes that were price elastic.

6 Conclusions

While the isotope market has expanded and evolved since its inception most of the production capacity is still based on wartime technology. However with advances in alternative production methods the shutdown of the only production facility North America has had an overall positive impact in co-ordinating the world's remaining production capacity. Trace Sciences has stepped in to aid in the coordination effort, and has combined the production capacities of Electrokhimprebor (EKP), Urenco, The Kurchatov Institute and the Siberian Group of Chemical Enterprises (TOMSK), along with milling and laboratory expertise to help fill this gap. The use of Calutron and centrifugal production facilities, along with the upcoming technologies of laser and plasma production, have moved Trace into the forefront as isotope providers for the growing radio-pharmaceutical and research industries.

THE FUTURE OF STABLE ISOTOPES AT THE U.S. DEPARTMENTOF ENERGY

JOHN S.CARTY

U.S. Department of Energy
19901 Germantown Road
Germantown, MD 20874
E-mail: John.Carty@hq.doe.gov

1.Introduction

Historically, The U. S. Department of Energy has been the principal stable isotope supplier for researchers. The Department is committed to continue that role. Reluctantly, the Department has placed the calutrons, which produce stable isotopes, in cold standby since more economical suppliers have emerged.

Currently, the Department has at least five years' supply of most isotopes in its inventory. It will continue to supply researchers from its inventory in the short term. For the long term, the Department is studying alternative production technologies, which will produce research quantities of stable isotopes to replace calutrons. Concurrently, the Department is pursuing an initiative to select a private entity to distribute its stable isotope inventory and operate its stable isotope laboratory. Operation of the selected replacement technology may be contracted to this entity.

Isotope Programs defined its stable isotope supply goals and established criteria to measure alternative separation technologies using the listed qualitative and quantitative factors.

1.1 Qualitative Factors

- Separation Factor (High, Moderate, Low)
- Single Pass Enrichment
- Multiple Isotopes (Produce more than one isotope in a single pass)
- Throughput
- Operating Experience
- Operating Versatility/Flexibility Change from one element to another).
- Number of Isotopes Separated
- Required Isotope R&D for a new element

1.2 Quantitative Factors

- Production Rate (mg/hr)

Unfortunately, comparable data on actual production rates are not easily obtained since single pass product assays vary significantly among the technologies. However, economic parameters for a consistent production rate of specific isotopes have been estimated.

2.Technologies under Consideration

- Electromagnetic –
- AVLIS
- Plasma Separator
- Gas Centrifuge

2.1 Electromagnetic Separation

Electromagnetic separation of isotopes is the most versatile and widely used of all separation techniques and has by far the greatest experience base.

2.1.1 Calutrons

Calutrons have a high separation factor, the capacity for single-pass separation, the ability to separate all isotopes of a given element, the capability for moderate throughput, very high experience and versatility, and low required R&D. Calutrons have also been used to separate most of the elements of the periodic table.

2.1.2 Small Laboratory Isotope Separator (LANL)

A laboratory isotope separator, with factor of ten lower throughput than a calutron, is presently available at LANL. It meets some of the Program's objectives, but by itself, would not be able to satisfy the Program's future needs.

2.2 AVLIS

A small AVLIS located at Lawrence Livermore National Laboratory (LLNL) could be made available to Isotope Programs. AVLIS selectively ionizes and excites an isotope by a laser and collects only the excited isotope on an electrode.

The small AVLIS has a moderate-to-high separation factor, single pass separation for moderate assays, cannot produce multiple isotopes simultaneously, has high throughput, low experience, moderate versatility, and high requirements for R&D.

2.3 Plasma Separator

The Plasma Enrichments Process (PEP) uses Ion Cyclotron Resonance Heating (ICRH) to selectively energize only one isotope. Collection is accomplished by discriminating high-energy particles from low-energy particles. A small UCLA machine, which can be operated in a mirror mode to reflect the product isotope back towards the source after one traverse of the chamber and collect it on an annular plate around the source, is expected to greatly increase both the collection efficiency and separation factor.

PEP has a moderate-to-high separation factor, single pass separation for moderate assays, cannot produce multiple isotopes simultaneously, has high throughput, low experience, limited versatility, and few isotopes separated. Its requirements for R&D are moderate to high.

2.4 Gas Centrifuge

The Gas Centrifuge process is characterized by low separation factors and requires the use of many separation stages combined in cascades to achieve the desired separations

3. Economics

An economic comparison analysis was performed among the major alternative technologies using assumed capital cost, operating cost estimates, separation factors, throughput and an isotope production unit cost calculated in \$/mg. The comparison was performed for assumed isotopes with natural abundance of 0.1%, 1%, and 10%. The calculation results show that PSP and AVLIS produce a lower unit cost but a higher total cost than electromagnetic separators due to the large throughput. In addition, lower single-pass enrichments are achieved using PSP and AVLIS than with electromagnetic technologies

4. Conclusions

AVLIS and PEP offer a very high production rate at low unit costs and moderate assays. When combined with the small electromagnetic separator that can deliver high assays at low production rates, they have the potential to exceed the production capabilities of the calutrons at substantially reduced costs. This combination of technologies will allow the maintenance of an independent U.S. capability for production of research isotopes and transition of advanced technologies to commercial production.

5. Path Forward

There is high probability that separation of some useful stable isotopes can be demonstrated within six months to one year at a reasonable cost. The availability of small electromagnetic plasma separation, and AVLIS separators and staffs to operate them in the near-term offers the Isotope Program an opportunity to evaluate these technologies with minimum initial costs. During 2000, the Department of Energy intends to perform demonstration isotope production runs on selected isotopes on PSP, AVLIS, and the laboratory electromagnetic separator. Performance data will be collected to obtain better operating cost data and a higher confidence in alternate technologies both from an operating and performance standpoint and enable the Department to select the best combination of technologies for the 21st century.

URENCO, ENSURING A FUTURE SECURE STABLE ISOTOPE SUPPLY

CHRIS CHATER AND HUBERT RAKHORST

Urenco Nederland B.V., P.O. Box 158, 7600 AD Almelo, The Netherlands
E-mail: isotopes@urenco.nl

In the early 1990s Urenco entered the Stable Isotope market with a strategy to best exploit its world leading advanced gas centrifuge technology. Urenco established itself as a long term, secure, cost effective western supply of stable isotopes, and consequently has seen and will continue to see considerable success in this market. The commitment to expansion is not only in our existing markets but also by developing new markets as can be seen by the recent addition of enriched Iridium to our portfolio. Each of the many technologies currently being used to produce stable isotopes, has its limitations either technically or commercially, and Urenco's technology is no exception. To engineer an element in a centrifuge there must exist a volatile and thermally stable chemical compound, for 31 elements these compounds already exist. Urenco recognises the importance of being able to supply a broad range of isotopes and therefore is currently researching methods whereby the centrifuges limitations are resolved, and furthermore exploring the possible use of laser. In this way Urenco hope to greatly expand its range of commercially available Stable Isotopes in the future, ensuring both the availability of critically needed isotopes and providing a readily available supply for new applications of isotopes. Urenco believe that this is a critical factor in the long term future of the isotope dependent markets.

1 Introduction

In the early 1970s an enrichment plant in Almelo began supplying the world wide civil nuclear programmes with enriched Uranium Hexafluoride. Since that time Almelo has produced, installed and operated 5 generations of centrifuges helping the Urenco Group expand towards 15% market share of the Uranium enrichment market. The Urenco group is not wholly state owned or subsidized, therefore its success can be only attributed to the exploitation of the worlds most advanced gas centrifuge and Urenco's committed customer focus. Urenco continually seeks improvements, and the next generation centrifuge is nearing qualification putting the Urenco group still further ahead of its competition. It is on the basis of this solid and reliable platform that Urenco Nederland B.V can exploit the resources used for Uranium enrichment to provide a secure, long term, cost effective, Western supplier of stable isotopes. The stable isotopes business unit (SI) was formed in the early 1990s and has mirrored the impressive growth shown by the core business.

2 Production and Capabilities

The centrifuges employed by SI consist of fourth and fifth generation machines adjusted where necessary to best enrich and withstand the various feed compounds. The centrifuge consists basically of a cylindrical rotor spinning at high speed in a vacuum on an almost friction free bearing. The centrifugal forces exerted on the feed gas as it enters the machine forces the heavier fraction to concentrate closer to the rotor wall, leaving the lighter fraction more concentrated closer to the centre of the machine, and hence separation occurs. More than one pass through a machine is required, to achieve the desired separation level.

The most effective manner to achieve the correct separation is by the processing of the gas in a cascade whereby a number of machines are connected both in parallel and in series. The sizes and abilities of Urenco's cascades differ widely, allowing Urenco to produce in a highly automated way, a number of different isotopes, in differing quantities to different isotopic concentrations. To complement the cascades, Urenco operates a number of centrifuges singularly whereby process gas efficiency can be tested in a low investment risk environment and ultimately low scale production can be performed. Thereby although Urenco's technology is best suited to large scale enrichment we have found a way to produce low volume isotopes to suit the relevant market demands. That is Urenco can supply from 100'smg of product to over 1000Kg, (and if necessary as with Uranium enrichment 100t) per annum.

3 The Products and Services

The products provided are seldom simply enriched or depleted in one isotope, most customers in fact request an engineered isotopic distribution. These profiles show two products recently produced by Urenco; ^{112}Cd the precursor material for ^{111}In, and ^{129}Xe the NMRI gas.

The enriched ^{112}Cd precursor for ^{111}In has also been depleted in, the most naturally abundant isotope, ^{114}Cd, to prevent ^{m114}In being produced. Similarly the ^{129}Xe product has been stripped of the interfering ^{131}Xe isotope to produce clear NMR images.

Table 1. Cadmium 112 and Xenon 129 Products.

Cadmium Isotope	Natural abundance	112Cd Product	Xenon Isotope	Natural Abundance	129Xe Product
106	1.2%	0.0%	124	0.1%	0.03%
108	0.9%	0.0%	126	0.1%	0.64%
110	12.4%	0.002%	128	1.9%	9.10%
111	12.8%	0.931%	129	26.4%	87.82%
112	24.0%	98.187%	130	4.1%	2.19%
113	12.3%	0.867%	131	21.2%	0.12%
114	28.8%	0.013%	132	26.9%	0.06%
116	7.6%	0.0%	134	10.4%	0.02%
DESIRED			136	8.4%	0.02%
UNDESIRED					

The modern day enricher also needs to be able to convert process gas to the desired chemical format and form. To achieve this a number of in-house state-of-the-art processes for conversion have been. Where our in-house technology is not sufficient Urenco has sought the to provide the most cost effective solution. For example Urenco produces Iridium Black powder which due to its inherently high melting point of over 2400°C could not be pelletised by Urenco. Urenco have worked with expert companies such as Heraeus GmbH (a world leading precious metal company) to offer the market finished pellets available for irradiation. The iridium itself allows equivalent sized sources to be proportionally more radioactive, lasting longer or alternatively gives a mechanism whereby smaller sources can be manufactured.

4 Isotope Portfolio

Each separation method has its own intrinsic limitations, advantages and disadvantages, Urenco's technology is no different. Each process gas must satisfy four basic criteria; a sufficiently high molecular mass, with suitable volatility to allow a mass flow, thermally stable in the gaseous phase and of course to be compatible with the centrifuge materials. Although these criteria initially appear extremely limiting, by working with either the fluoride or the organometal a wide range of isotopes can be processed. Currently isotopes of a range of elements including Xe, Zn, Cd, Ir, Se, Ge, S have been produced. The remaining isotopes fall into two interesting categories, from Urenco's perspective:

1. those isotopes which could possibly be enriched but where large scale markets either do not exist or have as yet not been developed such as Si, Ti, V, Fe, Ni, Ga, Zr, Mo, Ru, Hf, W, Re and Os.

2. finally the isotopes which have interesting markets but as yet there exists no suitable cost effective production method such as for Gadolinium , Palladium.

The first category fall somewhat outside the sphere of influence of an enricher, the second category are the more interesting. Therefore Urenco is at present developing two technologies.

5 The future

Urenco had for some considerable time researched the possible Laser separation of Uranium, the accumulated knowledge and expertise have now been redirected to the enrichment of Gadolinum isotopes. Enriched Gadolinium 157 being a desired poison for LWR fuel assemblies to enable higher burn-ups. Other interesting markets for the Gadolinium isotopes are also developing Gd155 for magneto-calorific effect, and high temperature superconductivity and Gd152 as a precursor for Gd153. Lasers enrich selectively by the produced light being finely tuned to excite a defined atomic or molecular bond, therefore in principle each isotope has a different spectroscopy which usually falls within the capability of the Laser. However each element requires both new spectroscopy and usually complicated ancillary systems resulting in a considerable investment of effort and long lead-times to market. Whereas for the centrifuge enriching an alternative isotope is limited only on the availability of a suitable gaseous compound.

Urenco is convinced of the advantages of centrifuge enrichment for heavier, as well as the lighter elements. It is therefore with this reasoning that Urenco is constantly performing development work to expand the range of centrifugable elements, in parallel to the Laser enrichment. If successful Urenco, with a combination of our cost effective centrifuge technique and the new Laser technology, would have the capability to enrich a large number of the isotopes currently of interest, elements such as the lanthanides and palladium as precursors.

6 Commercial Strategy and Conclusion

Thus in short Urenco has equipped itself to continue making stable isotopes into the 21st century on a scale to match the market requirements. With the backing of one of the most successful nuclear companies in the world to date, we are also undertaking large investments in new techniques to continually expand our production portfolio. We have a vision to be the stable isotope supplier of an extensive range of elements for the 21st century, providing a secure, stable, cost effective and western supply.

ENRICHMENT OF ^{15}N BY COUPLING THREE SYSTEMS OF ION-EXCHANGE CHROMATOGRAPHY COLUMNS

E. MAXIMO; J. A. BENDASSOLLI AND P. C. O. TRIVELIN
Stable Isotopes Laboratory, Center for Nuclear Energy in Agriculture
University of São Paulo, CP 96 CEP 13400-970, Piracicaba-SP
Brazile E-mail: evmaximo@cena.usp.br

Nitrogen tracers are an important tool for N balance in agricultural and biological research. A new methodology of ^{15}N enrichment by ion exchange chromatography method coupling three systems of resin columns is proposed. Each system has seven columns filled with cationic resin Dowex 50W X-8 up to 170 cm high. The nitrogen isotope fractionation takes place in the displacement band chromatography (DBC) of ammonium aqueous with solution of 0.8 mol L^{-1} NaOH. The resin (R) was previously regenerated to NH_4^+-R with solution of 0.75 mol L^{-1} $(NH_4)_2SO_4$. Equilibrium between NH_3 in solution and NH_4^+ bound to actives sites of resin was established during the DBC, enriching the rear of the band with ^{15}N. In the first system an enrichment level of 5 atom % ^{15}N was reached and after connecting with the second system up to 50 atom % was obtained. With the coupling of the second and the third systems 90 atom % ^{15}N was reached. With ammonium enriched in ^{15}N can be produced ammonium sulfate, anhydrous ammonia, aquammonia and urea witch are used as labeled fertilizers. The highest level of ^{15}N enrichment is essential to carry out research with high isotope dilution. Over 100 grams month^{-1} of ammonium sulfate with 90 atom % ^{15}N can be produced. The production cost is 30% lower than FOB price.

1 Introduction

Labeled compounds enriched in the stable isotope ^{15}N are important tools for carrying out studies on nitrogen cycling in agricultural and biological studies. Nitrogen has a natural abundance of 99.634% ^{14}N and 0.366% ^{15}N. Compounds highly enriched with ^{15}N are necessary in studies of systems with high isotope dilution. A limitation for using these compounds is its high cost, which varies proportionately with the level of enrichment. The goal of this study is to present a new methodology using cationic exchange chromatography column method, to obtain a high enrichment level of ^{15}N with cost compatible with the FOB price.

2 Material and Methods

For the separation of ^{15}N, three systems constituted of 7 transparent acrylic columns (systems 1, 2 and 3) 200 cm high, filled with cationic exchange resin Dowex 50W-X8 mesh 100-200 up to 170 cm (resin in the form Na^+-R equilibrated in water) were used. The internal diameter of columns systems 1, 2 and 3 were 14,5 cm; 9,6 cm e 5,2 cm, respectively.

A solution of 0.75 mol L^{-1} hydroxide sodium was introduced with a flow rate of 2 mL cm^{-2} min^{-1} to a regenerated resin column in the NH_4^+ form ($NH_4^+{}_{(R)}$) with 0.75 mo L^{-1} $(NH_4)_2SO_4$. The replacement of the NH_4^+ by Na^+ in the active sites of the resin, results in the formation of ammonia aqua (NH_{3aq}) and a chromatographic band NH_4^+. Equilibrium between NH_3 in solution and the NH_4^+ bound to the resin active sites is established in the band, which is moved down the column with a consequent separation of nitrogen isotopes by the isotope exchange reactions [1] represented in Figure 1.

With the NH_4^+ Displacement Band Chromatography (DBC) in the resin beds of the column of each system, the band becomes richer in ^{15}N at its rear while poorer in ^{15}N at its front. The continuity of the DBC from one column to another is assured by the connection of the columns of each system with nylon tubing.

Coupling of systems 1-2 and 2-3 was carried out by connecting their columns, after the planed ^{15}N-enrichment level was reached. In this coupling, the enriched rear part of the band from systems 1 and 2 was introduced in the band of systems 2 and 3, respectively. The DBC of the new band formed in the system 2 or 3 was carried out and its ^{15}N-enrichment level verified by isotope ratio mass spectrometry analyses.

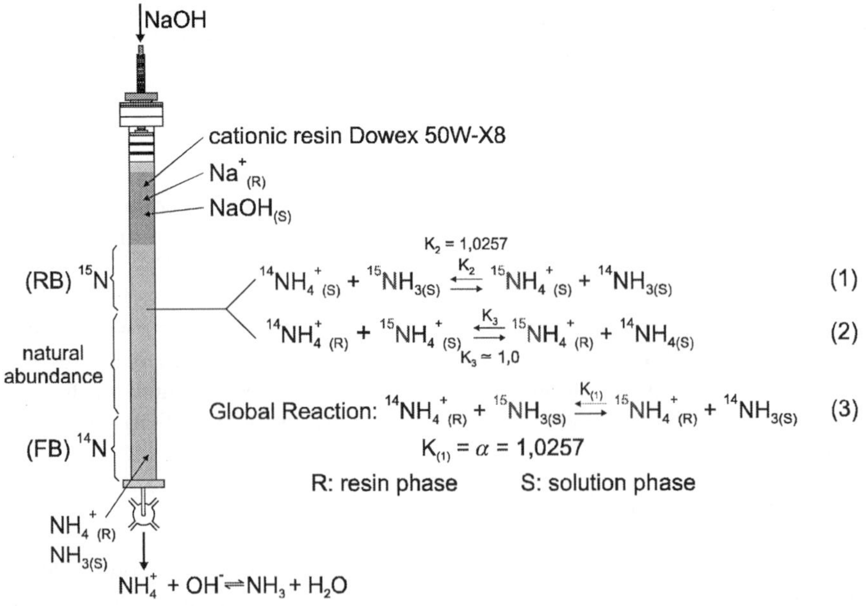

Figure 1. $^{15}NH_4^+$ Displacement Band Chromatography in a cationic resin column.

3 Results and Discussion

A level of 4.499 atom % ^{15}N was reached in the last linear centimeter of the band of system 1 after DBC for 30 m from the ^{15}N natural abundance. After the transference of 10 cm of the rear of the band from systems 1 to 2 (coupling system 1-2), the same enrichment level was obtained with the DBC of system 1 for approximately 17 m.

In the coupling between systems 1 and 2 approximately 2.74 g of ^{15}N was transferred from system 1 to 2. During DBC, the average of ^{15}N-accumulation in the 10 cm rear of the band of system 2 was 49.81 % of the mass from system 1.

After four couplings, the band of system 2 had an enrichment of 64.206 atom % ^{15}N in its last linear centimeter, accumulating 9.230 g of ^{15}N in the last 8 cm (Table 1). That mass was transferred to system 3 where the enrichment level was 30.487 atom % ^{15}N in its last linear centimeter (after DBC of system 3 individually). With the coupling between systems 2-3 followed by DBC for 7 m, a level of 90.447 atom % ^{15}N was obtained in the last linear centimeter of the band. Other nitrogen compounds labeled in the ^{15}N can be obtained with $^{15}NH_4^+$.

Table 1. Data about bands after coupling of the systems 1-2, 2-3 and DBC by 10 m.

Coupling	Band of the system 2			Band of the system 3		
	Mass (g)[A]		at. % ^{15}N[C]	Mass (g)[B]		at. % ^{15}N[C]
	Before	After	Enrich.	Before	After	Enrich.
1° Sist.1-2	1.29	2.64	19.460			
2° Sist.1-2	2.64	4.30	38.000			
3° Sist.1-2	4.30	5.80	44.145			
4° Sist.1-2	5.80	9.77	64.206			
1° Sist.2-3				1.92	5.62	90.447

A. 10 cm band rear edge of system 2 after receives 2.74 g ^{15}N from system 1, by each coupling.
B. 8 cm band rear edge of system 3 after receives 9.23 g ^{15}N from system 2. C. last linear cm of the band.

4 Conclusion

With the coupling of three column systems followed by displacement it was possible to produce 100 g of $(^{15}NH_4)_2SO_4$ 90 atom % ^{15}N a month with a 30% lower cost than the FOB price, that is about US$80.00 per gram.

Reference

1. Spedding, F. H. A laboratory method for separating nitrogen isotopes by ion exchange. *Journal American Chemical Society* **77** (1955) pp.6125-6132.

STABLE ISOTOPES: APPLICATIONS IN COMBINATORIAL CHEMISTRY

D. JOHN ABERHART

Cambridge Isotope Laboratories, 50 Frontage Road, Andover, Massachusetts 01545, U.S.A.
E-mail: aberhart@isotope.com

Applications of stable isotopes in developing methods for solid phase synthesis of combinatorial libraries, and in the encoding of combinatorial libraries will be outlined.

1 Introduction

Combinatorial Chemistry has grown rapidly to become one of the hottest areas of synthetic chemistry. Virtually all pharmaceutical companies and many academic laboratories are now using Combinatorial Chemistry as a tool to supplement or replace more traditional methods of drug discovery.

In Combinatorial Chemistry, large arrays ("libraries") of structurally related compounds are synthesized by sequentially coupling collections of related components. Two basic methods are used: parallel synthesis and mix-and-split synthesis. In parallel synthesis, test compounds are synthesized simultaneously, step-by step, in separate containers, either in solution or on a solid support. In contrast, mix-and-split synthesis is performed only by using solid-phase techniques analogous to those used in modern solid-phase peptide synthesis. Stable isotopes have been used in both approaches, as outlined below.

2 Application of Stable Isotopes in Solid-Phase Parallel Synthesis

Reactions carried out on solid-phase supports are driven normally to completion by using excess reagents. This works very well in the well-established methods for the syntheses of peptides and oligonucleotides. However there is often a need to follow reactions in the development of a combinatorial synthesis strategy. Several approaches to this end using stable isotope labeling have been recently published.

An Affymax group [1] has reported using ^{13}C-enriched compounds to moniter solid-phase preparations using a technique called "Fast ^{13}C NMR". This technique involves monitering reactions on resin samples, placed in NMR inserts by using ^{13}C NMR. In one example, formation of 4-thiazolidinone was followed *on resin* using resin-bound glycine (2-^{13}C). The signal enhancement from using ^{13}C-enriched substrates gave ^{13}C NMR spectra with excellent S/N with very few transients. Similarly, a stepwise reductive amination was followed by a triple-labeling scheme. Sarkar and coworkers [2] at SmithKline Beecham have used ^{13}C labeling of precursor molecules for monitering the progress of solid-phase syntheses.

Russell and coworkers [3] at Zenica have used IR spectroscopy with deuterium labeling to moniter solid-phase reactions. This group took advantage of the nearly unique spectral region (2200-2300 cm^{-1}) where carbon-deuterium stretching frequencies appear, where unlabeled molecules usually have no absorbance. The progress of synthesis of a short peptide was monitered by IR, using as chain extender L-lysine, Nα-Fmoc, Nε-t-BOC labeled in the t-BOC protecting group with t-BOC (D_9), following the increase in intensity of the C-D absorbance.

3 Encoding of Combinatorial Libraries

In the mix-and-split methods of combinatorial synthesis, various types of monomeric starting units, (A_1,A_2,A_3...) are connected to beads *via* an appropriate linker. Each bead is connected to only one type of monomeric starting material *via* the linker. As the synthesis proceeds, beads from preceding steps are *combined* and then *divided* for reaction with the next group of monomeric units, i.e. B_1,B_2,B_3... or C_1,C_2,C_3...etc. At the conclusion of this scheme, the result is a combinatorial library consisting of unique, polymer-bound trimers, ABC.

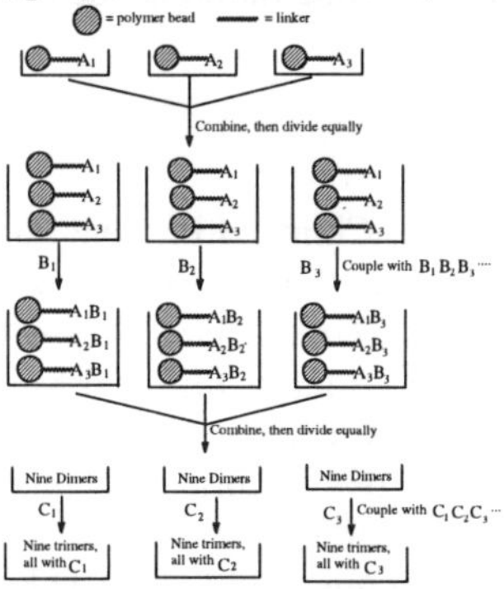

Beads are then separated, and screening performed. As a result of the combine/divide strategy of mix-and-split combinatorial synthesis, the structures of test compounds cannot be deduced from their location in a synthesis apparatus or microtitre plate, and some means of identification by a coding scheme must be used to assign structures. A Glaxo group [4,5] has used stable isotope labeling to encode structures of attached compounds in mix-and-split (also called "one bead-one compound") combinatorial chemistry. A stable isotope-labeled coding element is incorporated into the linker. The code is read by mass spectrometric techniques after cleavage of the linker from the bead. The approach does not require adding a coding element at each stage of the synthesis, as long as the synthesis involves only two addition steps beyond connection of the starting unit, A, to the linker.

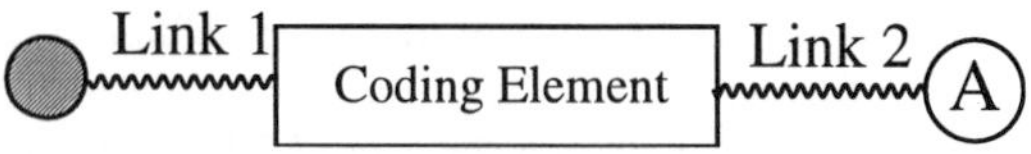

In the technique "Single Peak Positional Encoding", an encoding dipeptide comprised of various stable isotope-labeled and unlabeled glycines and/or alanines is used to encode a unique monomer A_1, A_2... After a mix-and-split synthesis to form compound ABC is completed, the identity of the third monomer C_1, C_2... is defined by the pool from which a particular bead is obtained. After cleavage of link 2 and activity testing of the library, link 1 of promising leads is cleaved, followed by MS analysis of the dipeptide. This allows identification of monomer A. The total mass of the hit compound $A_xB_xC_x$, also determined by MS, allows assignment of the identity of monomer B, provided that all forms of B used in the combinatorial synthesis have different masses. By using dipeptides consisting only of singly or doubly labeled glycines and/or alanines (plus the unlabeled amino acids), 15 unique dipeptides can be created, encoding a library of 3375 products (15^3).

Variations on the above theme involve the use of mixed isotopically distinct monomers ("bar coding") or use of varying ratios of labeled compounds ("ratio encoding"). The Glaxo group, as well as a group at SmithKline Beecham [6] have also used NMR techniques for decoding ratio-encoded combinatorial libraries.

References

1. Look, G.C., Holmes, C.P., Chinn, J.P., and Gallop, M.A., Methods for Combinatorial Organic Synthesis: The Use of Fast ^{13}C NMR Analysis for Gel Phase Reaction Monitering, *J. Org. Chem.* **59** (1994) 7588-7590
2. Sarkar, S.K., Garigipati, R.S., Adams, J.L., and Keifer, P.A., An NMR Method To Identify Nondestructively Chemical Compounds Bound to a Single Solid-Phase-Synthesis Bead for Combinatorial Chemistry Applications, *J. Amer. Chem. Soc.* **118** (1996) 2305-2306
3. Russell, K., Cole, D.C., McLaren, F.M., and D.E. Pivonka, D.E., Analytical Techniques for Combinatorial Chemistry: Quantitative Infrared Spectroscopic Measurements of Deuterium-Labeled Compounds, *J. Amer. Chem. Soc.* **118** (1996) 7941-7945
4. Geysen, H.M., Wagner, C.D., Bodnar, W.M., Markworth, C.J., Parke, G.J., Schoenen, F.J.,Wagner, D.S., and Kinder, D.S., Isotope or Mass Encoding of Combinatorial Libraries, *Chemistry and Biology* **3** (1996) 679-688
5. Wagner, D.S., Markworth, C.J., Wagner, C.D., Schoenen, F.J., Rewerts, C.E., Kay, B.K., and Geysen, H.M., Ratio Encoding Combinatorial Libraries with Stable Isotopes and their Utility in Pharmaceutical Research, *Combinatorial Chem. High Throughput Screening* **1** (1998) 143-153
6. Garigipati, R.S. and Sarkar, S.K., A Binary Coding Method For Use in Combinatorial Chemistry, *PCT Int. Appl.* (1997) WO 9714814

Session E

Industrial Applications and Radiation Safety

RADIOLOGICAL SAFETY AND OPERATIONAL HEALTH PHYSICS OF MEDICAL CYCLOTRONS

BHASKAR MUKHERJEE

Safety Division, Australian Nuclear Science and Technology Organisation (ANSTO), PMB 1, Menai, NSW 2234, AUSTRALIA E-mail: mukherjee@ieee.org

Compact Medical Cyclotrons are now primarily used to produce large activities of neutron deficient radioisotopes for diagnostic nuclear medicine. In recent years, the rapid advancement of accelerator technology and related hardware became the key factors to achieving intense charged particle beams (~ 1000 μA), thus, it became feasible to generate a much higher radioisotope yield at a lower cost. This paper highlights the important operational aspects of room temperature and Superconducting Medical Cyclotrons from the standpoint of a Health Physicist. It also provides valuable information based on scientific analysis of cyclotron operational data and could be used for the planning of new Medical Cyclotron installations. Specific reference will be made to emphasise the advances made by ANSTO Safety Division during the last 9 years of operation of the Australian National Medical Cyclotron.

1 Introduction

The bombardment of cyclotron targets with an energetic charged particle beam results in the production of an intense field of secondary neutrons and gamma rays causing a myriad of radiological hazards. Unlike high energy research cyclotrons operated by large National Laboratories or Universities, the Medical Cyclotrons are usually located in urban hospitals and routinely operated by medical technologists to produce large activities of short lived medical radioisotopes. Medical Cyclotrons handle very high beam currents (p, d) producing a copious number of neutrons and photons enhancing the risk of potential radiological hazard. Therefore, radiological safety and operational health physics play vital roles in the safe and economic operation of Medical Cyclotron facilities. There are now 120 Medical Cyclotrons operating worldwide [1]. Since July 1991 three Medical Cyclotrons are operational in Australia, these are a) CYCLONE 30 (H^- ion, 30 MeV) operated by National Medical Cyclotron (NMC) in Sydney, b) CYCLONE 10/5 (H^-/d^- ion, 10/5 MeV) operated by Austin and Repatriation Medical Centre in Melbourne and c) OSCAR 11 (H^- ion, 11 MeV, superconducting) operated by Peter MacCullum Cancer Institute in Melbourne. In this paper the important aspects of radiological safety operational health physics of commercial Medical Cyclotrons are highlighted.

2 Radiation Hazard pathways of a Medical Cyclotron

During the routine operation of a Medical Cyclotron a large number of high energy photons (gamma rays) and fast neutrons are produced in the target as a result of proton induced (p,xn) nuclear reactions with the target material [2]. These by-product radiation fields may leak out via inadequate cyclotron shielding causing significant personal radiation exposure to cyclotron workers and members of the public. The activation of cyclotron parts [3], production of gaseous radioactive effluent [4] and radiation induced irreversible damage of electronic component of the cyclotron instrumentation and control systems are the major problems associated with the adverse effects of the by-product radiation fields (Table 1).

Radiation Field	**Hazard Pathway**	**Nature of Hazard**
Prompt Neutron	Component activation	Contamination, External exposure
	Vault air activation	External exposure, Inhalation exposure
	Roof / Shield leakage	Skyshine (External) exposure)
Prompt Gamma	Roof / Shield leakage	Skyshine (External) exposure)

Table 1. Showing the type of the radation fields and the associated radiation hazards generated during the operation of a Medical Cyclotron.

3 Radition Safety Devices of Medical Cyclotrons

Sound health physics supervision and a meticulous admistartive access control of the critical work areas by implementing the radiation protection ragulations are the most common radiological safety procedures for cyclotron facilities. However, in cases of modern hospital based Medical Cyclotrons such practices may found to be impractical to apply. Medical Cyclotrons are usually located in hospitals to produce large activities of various short-lived diagnostic radioisotopes for in-situ clinical applications. Hence, high reliability and built-in (engineered) radiological safety devices are vitally important for the safe operation of Medical Cyclotrons and are highlighted in the following sections

3.1 Realtime Radiation Serveillance System

A realtime radiation serveillance system (watchdog) is essential for monitoring the radiation dose levels and other important physical parameters of a Medical Cyclotron in realtime [2,5]. In Figure 1 the locations of the Cyclotron (CYC), PET (P1, P2) and SPECT (T1, T2, T3) targets in the concrete cyclotron vault and target cave, the Gamma (GD1,...GD11) and Neutron (ND1, ND2) monitors, the positron (PD1) and Iodine (ID1) effluent monitors [4,6], the beam current Fraday cups (FC1, FC2) and the delay tank level detectors (LD1, LD2, LD3) are shown. The signal outputs (dose rates, liquid levels, stack releases etc.) are sampled every minute by the datalogger, displayed on computer terminal and permanently stored in a CD ROM. For data analysis. The outputs of the gamma detectors (GD5 and GD6) located at the cyclotron and target vault doors (D) are modified by the datalogger and connected to the safety interlock PLC (Programmable Logic Controller) installed in cyclotron control room regulating the access to the vaults.

3.2 Radiological Shielding

Adequate radiological shielding is the most important requirement for a Medical Cyclotron operating in a hospital environment. Larger high energy (E_p~30 MeV) Medical Cyclotrons like the CYCLONE 30 are usually housed in concrtete vaults with multi-legged entarnce mazes (Figure 2). The phenomenological shielding calculation method [7] using the experimentally evaluated neutron and gamma source terms [2] was adopted to estimate the thickness of the vault wall and mazes [8]. The low energy (E_p~12 MeV) compact self shielded Medical Cyclotrons like TR13 (EBCO), CYCLONE 10/5 (IBA) and HM12 (SUMITOMO) utilise a highly efficient boron based composite material for neutron shielding. The Superheated neutron bubble dosimeters and miniature electronic gamma detectors were used to evaluate the neutron and gamma transmission factors to confirm the shielding efficacy [9].

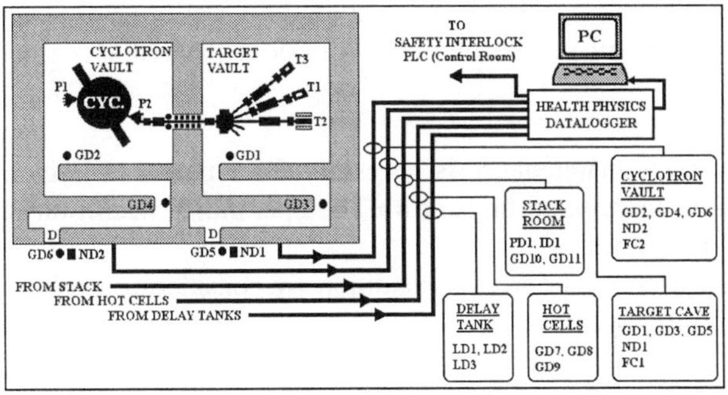

Figure 1. Schematic diagram of the realtime Health Physics monitoring system (Watchdog) developed at Australian National Medical Cyclotron, Sydney.

3.3 Stack Effluent Monitoring

Gaseous radioactive effluent in the form of the compounds of ^{18}F ($T_{1/2}$ = 110 min) and ^{123}I ($T_{1/2}$ = 13.2 h) are released from the radiochemical processing hot cells of Medical Cyclotrons. Relevant national regulatory authorities often require a mandatory hard copy of the stack release data. In our case a real time stack monitoring system was developed at NMC and has been described elsewhere [4,6].

3.4 Personnel Safety Interlocks

In low energy self shielded Medical Cyclotrons a safety interlock system is incorporated within the cyclotron. For a high energy multi-target cyclotron (CYCLONE30 of the NMC) with seperate cyclotron and target cave entries the chance of an accidental radiation exposure is much higher. Therefore the installation of safety interlock systems is recommended [10].

4 Operational Health Physics of Medical Cyclotrons

Operational Health Physics plays an important role in the reduction of personnel radiation exposure and trouble free operation of the Medical Cyclotron facility.

4.1 Routine Operation

A daily check up of radioisotope production areas for radioactive contamination levels and inspection of the radiation exposure (accumulated dose) to radiochemists and medical technologists.

4.2 Trouble shooting and Emeregency Repair

During trouble shooting and emeregency repair of Medical Cyclotrons the technicians may accidentally receive a high radiation dose originating from the activated cyclotron parts. Therefore, a detailed radiation dose survey prior to undertaking the repair work togather with continuous health physics supervision is essential.

4.3 Maintenance

Regular preventative maintenance is necessary for a fault free cyclotron operation and avoidance of unexpected breakdown.The operational health physics procedure is summarised as follows:

a) Determination of the access condition by considering cyclotron cool down time [11], b) Radiation dose survey at critical locations, c) Contamination and activation check of cyclotron components, d) Proper administrative control to prevent contamination and radiation exposure, e) Waste disposal

5 Summary and Conclusion

Use of hospital based Medical Cyclotrons is expanding on world-wide basis. Medical Cyclotrons produce large activities of short lived isotopes and may cause significant radiation hazard. This paper has discussed important key aspects of health physics and radiation safety of Medical Cyclotrons and highlighted, in particular, the relevance of health physics studies carried out over 9 years of operation of the Australian NMC

References:

1. Directory of cyclotrons used for radionuclide production in Member States, IAEA-TECDOC-1007, 1998.

2. B Mukherjee. In *Cyclotrons and their Applications* ed. by E Baron (IOP Publishing, 1999) pp 264-267.
3. B Mukherjee. *Appl Radiat Isot.* **48** (1997) pp 735-738.
4. B Mukherjee. In *Cyclotrons and their Applications* ed. by J C Cornell (World Scientific, 1996) pp 330-333.
5. E. Mishani et al. *Nucl Instr Meth A*. 425 (1999) pp 332-342.
6. B Mukherjee.. *In Proceeding of the 2nd International Conference on Isotopes*, Sydney, Australia, October 1997.
7. B Mukherjee. In *Cyclotrons and their Applications* ed. by J C Cornell (World Scientific, 1996) pp 334-337.
8. B Mukherjee et al. *Appl Radiat Isot.* **48**(1997) pp 453-457.
9. H J Tochon-Danguy et al. In *Cyclotrons and their Applications* ed. by E Baron (IOP Publishing, 1999) pp 105-107.
10. Radiation alarms and Access Control Systems, NCRP Report No. 88, 1986.
11. B Mukherjee. In *Proceeding of the 9th Pacific Basin Nuclear Conference* ed. by C Hardy (Australian Nuclear Association, 1994) pp 645-649.

ISOTOPE RADIOGRAPHY FOR CONTAINER INSPECTION

AN JIGANG, WU ZHIFANG, ZHOU LIYE, LIU YISI
XIANG XINCHENG, WANG LIQIANG, WU HAIFENG

INET, Tsinghua University, Beijing 100084, P.R.China

A ^{60}Co container scanner (TC-SCAN), which uses digital isotope radiography technology to inspect container, had been researched and constructed by INET of Tsinghua University in July 1997. Its demonstration station has been built and begun to work in Fuzhou City (southern China) now. The "Contrast Indicator (CI)" and "Image Quality Indicator (IQI)" for 100 mm steel of this scanner are equal to 0.7% and 2.5% respectively. The "Steel Penetration (SP)" is about 240 mm, the "throughput" is 20 standard containers per hour, and the maximum dose absorbed by the object per scanning is less than 10 μSv. The excellent reliability and stability have been proved by the work experience in the period of 24 months.

1 Introduction

Container inspection equipment is urgently required to control shipment of illegal drugs, arms and contraband. The inspection characteristics of the accelerator type scanners are good enough. But many drawbacks (exp. high price, strong radiation intensity and large amount of space etc.) restrict the wide use of these scanners. The X-ray machine type container scanners are cheap, but their penetration ability is too poor to fit the requirement of container inspection.

A full scale sample system of ^{60}Co scanner (type TC-SCAN), which uses digital isotope radiography technology to inspect container, had been researched and constructed by INET (Institute of Nuclear Energy Technology) in July 1997. It has been working perfectly and was named "TC-SCAN". The demonstration station of TC-SCAN had been built in the Fuzhou City (southern China) and began to work in July 1999. A mobile type of ^{60}Co scanner has also been researched and constructed now.

By using the special sensitive array detector and advance image processing technology, the main difficulty of weak radiation intensity of the 1-11 TBq ^{60}Co source was overcome and the inspection characteristics of the ^{60}Co container scanner are as good as those of accelerator type scanners. Because of the excellent merits of ^{60}Co container scanner on price, safety, reliability and the amount of space, the application prospect of the TC-SCAN systems is very nice not only in China but also in other countries.

2 Principle and Structure

The system composition of TC-SCAN is shown in Fig.1. It consists of the following main components: ① ^{60}Co projector (1-11TBq); ②⑤ front and back collimator; ④ conveyer system; ⑥ array detector; ⑦ signal processing subsystem.

To run the TC-SCAN at Fouzhou City, the ^{60}Co source and the array detector remain fixed while the container is pulled by the conveyer system through the slice type radiation zone for inspection. The array detector invented by INET[1,2] is composed of many independent but orderly arrayed ion-chamber units. The output signal of each ion-chamber unit is proportional to the γ-ray intensity received at its position. By collecting the signals of each ion-chamber unit, arranging them in order and displaying them, a scanning line of the image is seen. As the container advances forward, the scanning lines of the object's image are displayed sequentially. Thus, a 2-D γ-ray projection image, which reflects the material arrangement inside the container, is formed.

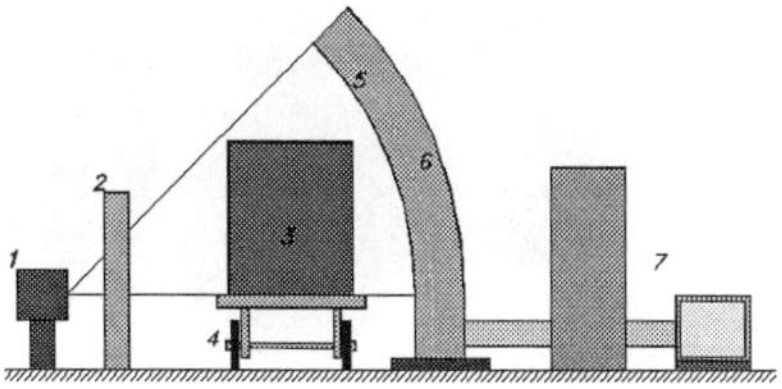

Fig.1 The system composition of TC-SCAN

For the mobile type of TC-SCAN system, the container truck remains fixed while the ^{60}Co source and the array detector make the scanning movement.

3 Inspection Characteristics

The inspection characteristics of TC-SCAN, which are measured according to international standards, are compared with typical accelerator type container scanners in table 1.

Table 1. Comparison of inspection characteristics

indicator / scanner	IQI (100mm steel)	CI (100mm steel)	SP (steel)	maximum dose absorbed per scanning	throughput (per hour)
static accelerator scanner (Paris)	2.5%	0.4%	240mm	0.4mGy	≈60 10-foot containers
static accelerator scanner (England-France Channel)	3%	0.75%	280mm	1mGy	≈20 trucks
Cobalt-60 scanner	2.5%	0.7%	240mm	≤0.01mGy	≈20 40-foot containers

It is obvious according to Table 1 that, the main inspection characteristics of TC-SCAN are similar to those of accelerator type scanners. Especially, the "maximum dose absorbed per scanning" of TC-SCAN is much less. This is excellent for the radiation safety.

The inspection image of a limousine is shown in Fig.2.

Fig.2 Sample image of a limousine

4 Special Features and Application Prospects

TC-SCAN possesses the following special features.

(1) Reliability and stability

The ^{60}Co• photons' energies never change and their intensity can be considered constant during scanning. There is no need for a high voltage microwave or a large electric power supply. Therefore, TC-SCAN is very reliable and stable.

(2) Low radiation and high safety

(3) Small architectural area

The inspection tunnel length is just 6-8 m. The distance from the radiation source to the detector is 5-7 m. The shielding area of TC-SCAN is less than 100 m^2.

(4) Low price

(5) Simple operation and convenient maintenance

(6) Double inspection tunnels

TC-SCAN could be used at customs of boundary, seaports, airports and railway stations, etc.

The mobile type of ^{60}Co container scanner, which is called TCM-SCAN, has been researched and constructed too. The inspection characteristics of mobile systems are equal to those of the fixed type, but its maneuverability is much better.

References

1. An Jigang, Wu Haifeng. Gas Ionization Array Detectors for Radiography. United States Patent, 5 796 110, August 18, 1998
2. An Jigang, Wu Haifeng. Gas Ionization Array Detectors for Radiography. European Patent, EP 0 732 730 B1, November 25, 1998

STAYING AHEAD OF THE GAMMA PROCESSING MOMENTUM

GORDON ASHFIELD, DIRECTOR, MARKETING, SALES & SERVICE
DEBBIE MERRITHEW, MARKETING ASSOCIATE
Industrial Irradiation, MDS Nordion
447 March Road, Kanata, Ontario, K2K 1X8

The last decade has been a period of tremendous opportunity for the gamma processing industry. However, the opportunities have not come without challenge.

This paper will provide an overview of the development of the gamma processing industry and look at some of the key trends in the medical device market, which have influenced the recent momentum. For example, the economic attraction of industry outsourcing; rationalization and industry consolidation; regulatory change and emerging markets have not gone unnoticed.

Waiting to burst out of the starting gate is food irradiation - a highly emotional technology that has great potential for the health of our nations. The paper will touch on the impact of unsafe foods in the public domain, and the status of the FDA and USDA approvals. As well, the presentation will include a brief overview of our newest irradiator design, the *Centurion Irradiator,* an innovative solution for processing food products.

INTRODUCTION

It may not be readily apparent but your child's schoolyard and the International Space Station are linked. The connection is gamma irradiation. Allow me to explain.

Along with some comforting words, a child's scrapped knee is often handled by placing a bandage on the cut. In many cases, that bandage has been sterilized by gamma irradiation. Meanwhile in space, astronauts can chose from a menu of foods, including irradiated beef.

MDS Nordion is the major global supplier of cobalt-60 – the necessary isotope for sterilizing an amazing range of consumer products and foods, from syringes, shampoo and surgical supplies to "pasteurizing" spices, fruits and meats.

This morning, I would like to share with you the history, applications and prospects for what has to be one of the more versatile and important isotopes.

1 The Discovery of Sterilization

Pick up a newspaper and chances are that you will read about another exciting medical breakthrough. We now have needle-less injections, non-invasive surgery and stainless steel joints. However, in the flurry of discovery one criteria has remained constant – protecting the patient from infection. This is where gamma irradiation has its roots.

More than 100 years ago, Sir Joseph Lister identified airborne germs as a source of infection in the operating room. Prior to his discovery, the postoperative mortality rate was as high as 90 percent in some hospitals! Surgeons could not bring themselves to believe *they* were contaminating their own patients by operating without gloves and with unsterile instruments.

Johnson & Johnson took Lister's research seriously, and before the turn of the century, they were treating cotton and gauze dressings by dry heat in an attempt to produce not only an antiseptic product - but a sterile one.[1] However, it was only since the Second World War that the medical profession could turn to more effective sterilizing solutions, including gamma irradiation.

2 Gamma Processing Commercialized

Commercialized in the early 1960s, gamma processing offered large medical device manufacturers a highly reliable and effective sterilization method. The process was introduced in North America in 1964 with the commissioning of 3 MDS Nordion Irradiators for Johnson & Johnson. Gamma processing was – and remains fast and cost effective, particularly for processing significant volumes of product. There is no question that gamma sterilization has been embraced.

To date, MDS Nordion has installed over 110 large industrial gamma irradiation facilities. This represents about 80% of gamma irradiation capacity worldwide. Put another way, we have shipped in excess of 500 million curies of cobalt-60 worldwide – a solid indication of the importance of cobalt as a preferred choice for sterilization for many companies. In fact, gamma is used to sterilize about 40% of the world disposable medical supplies.

However, no marketplace is static. Globalization is reshaping our business. I'll highlight four key trends.

[1] Johnson & Johnson History www.jnj.com

3 In-House vs. Contract Sterilization

Trend #1: The desire to ensure a sterile product – a fairly critical requirement for anyone in the medical device business – has required manufacturers to come to grips with some fundamental business decisions: should sterilization be done in-house or contracted out?

"Contract radiation processing" has grown substantially since it was first used in the 1970s. Today, there are over 80 gamma facilities offering contract services worldwide.
The upshot: it is estimated that 80 percent of medical device manufacturers outsource their sterilization needs, although some both outsource and perform in-house sterilization.[2] Cost effectiveness often drives the decision to outsource or not but there are other factors, too.

Manufacturers have to take a hard look at the economics of contract sterilization. For instance, there are savings for "small-volume" manufacturers in terms of capital, labour, risk, and inventory-warehousing. This must be considered in terms of time and expense associated with transporting product to a contract facility for processing.

Just-in-time inventory systems is causing many medical device manufacturers to contract out. Gamma contractors typically process devices in five days or less, compared with the typical 6 to 10 days offered by most Ethylene Oxide (EtO) contractors.[3]

In short, competition has benefited the medical devices industry: in the 1990s, there has been a proliferation of contract facilities, increased competition among contractors, and customers have the choice of competing technologies. We are pleased to see that contract gamma services has been the fastest growing sector of the medical product sterilization market for the last 10 years. Gamma sterilization is cost–effective and the technology is readily available. We expect this trend to continue into the next millennium.[4]

4 Rationalization and Industry Consolidation

---

[2] www.devicelink.com/pmpn/archive/99/04/003.html. "When In-House Sterilization Makes Sense" Apr.1999
[3] devicelink.com/mddi/archive/97/09/018.html. "Tracking Trends in Industrial Sterilization", Scholla & Wells, DuPont Co. Sep.1997
[4] devicelink.com/pmpn/archive/99/04/003.html. "When In-House Sterilization Makes Sense" Apr.1999

Trend #2: For many industries, the hallmark of the 1990s has to be mergers. One only has to think of financial services or pharmaceutical companies. This is also true for the medical device sector. Over the past five years many, if not all, of MDS Nordion's device customers have been affected by the industry's transformation to some degree. Tyco bought U.S. Surgical and Sherwood-Davis & Geck; Cardinal Health merged with Allegiance; and, Johnson & Johnson acquired Cordis.

The contract sterilization landscape has also changed. The two largest gamma contract sterilization companies have been acquired. Isomedix and SteriGenics were acquired by STERIS and Ion Beam Associates (IBA) respectively.

Interesting is that only five years ago both Steris and IBA were relatively small players in the sterilization industry. Today, based on their David buys Goliath acquisition strategies; we are witnessing unprecedented consolidation in the marketplace.

In the midst of all this, the demand for MDS Nordion's cobalt-60 has reached record highs to meet the demand for medical device sterilization.

5 Regulatory Issues

Trend #3: Harmonization of sterilization standards are assisting with the globalization of the single use medical device industry. Global acceptance of sterility assurance levels provides for distribution and acceptance of products worldwide.

Today, the International Standard ISO 11137 provides the sterilization industry with harmonized requirements for validation, process control and routine monitoring. This standard has been adopted by the American National Standard Institute (ANSI) and the Association for the Advancement of Medical Instrumentation (AAMI) as an American standard. It has also the basis of the Canadian standard, under CSA – The Canadian Standards Association designation.

For devices destined for sale in the European Union, sterilization must be in compliance with EN 552, the present European standard. Beginning this fall ISO and CEN will be working together under ISO leadership with a goal to draft a universally accepted standard for all sterilization processes.

In addition the global adoption and enforcement of environmental and safety regulations will encourage the closure of outdated EtO facilities presenting the opportunity for new state of the art gamma irradiation facilities.

6 Emerging Markets

Trend #4: My fourth and final trend in the globalization of the business is the prospect of major growth in emerging markets. Consider two of the world's largest countries, China and India. According to the Health Industry Manufacturers Association, the Chinese market for medical technology products is growing at an annual rate of 23 percent. The Indian market has a growth rate of 24 percent. In each country, both local and multinational companies are rushing to expand capacity to service this growing demand, and our industry is going to be a part of it.[5]

In this hemisphere, take a look at Brazil, the fifth largest country in the world. Its medical equipment and supplies market is growing at an estimated 10 per cent per year. Three (3) MDS Nordion gamma irradiators have been operating in Brazil for many years. In 1999 alone four additional sites are being constructed, three of the four are being supplied by MDS Nordion.

It is clear that in the midst of significant changes within the global marketplace, gamma sterilization continues to be part of the solution. The battle against bacteria is constant and gamma sterilization is a front line defense.

I will now share with you another important application of cobalt-60.

7 Food Irradiation

When was the last time you ate a rare hamburger? If you had one, did you eat it without a haunting feeling that perhaps, just perhaps, you could be consuming E-coli (O157:H7) bacteria from contaminated meat?

Well, *Time* and *Newsweek* magazines decided that the issue was important enough to devote cover stories to the issue lately. What's seized the media's attention?

The incidence of food-borne illness is staggering. It is a global problem. Consider the U.S., alone. Physicians in the United States deal with 8.2 million cases annually. And it is estimated that as many as 9000 of these people will die from contaminated foods.

Well-known and respected companies are not immune. A Sara Lee subsidiary recalled 35 million pounds of contaminated luncheon meats only several months

---

[5] Competition & Cooperation: Developing New Markets for Radiation Processing, John Morrison, MDS Nordion at IMRP 1997.

ago. The outbreak has been tied to 15 deaths and six miscarriages or stillbirths in 22 states.

In the U.S. this year, there has been at least 36 meat or poultry product recalls initiated by the Food Safety Inspection Service. The consequence can lead to plant closures, bankruptcies and job losses. The health and economic impact of food-borne illness is devastating.

Fortunately, there is an effective solution.

We've been developing a safe process for food pasteurization, using gamma, since 1964. Around the world, some 40 different types of foods are allowed to be irradiated. The World Health Organization, the American Medical Association and the Mayo Clinic, among many others, have declared that food irradiation is safe.

We are aware that consumer acceptance of food irradiation has not kept apace. But let me put this into context. Milk pasteurization took years before becoming widely accepted. Consumers have the right to question the technology. Still, I remain optimistic. One survey found that "approximately 85% of [American] consumers think irradiation could lower the risk of food poisoning from restaurants…". This is a fairly significant finding when you consider that 80% of food-borne bacteria outbreaks come from food prepared outside the home. [footnote: *Consumers' Views on Food Irradiation*, Survey, Food Marketing Institute & Grocery Manufacturers of America, 1998]

We've invested a great deal in working with and educating international health organizations, industry and regulators. The global network of contract sterilizers has also joined in the effort by lobbying regulators and government, informing industry and creating public awareness of the benefits of food irradiation.

There are persuasive reasons to support this technology.

A number of studies released by the federal government's National Technical Information Service concluded that irradiation of meat products not only helps control pathogens, but over the next 20 years may also lower medical costs of food-borne illness by as much as an estimated $171.8 billion.[6]

Once some regulatory hurdles are overcome, I predict that food irradiation - - or cold pasteurization - - will become increasingly widespread to eliminate pathogens in foods.

---

[6] Irradiation May Save Billions in Health-Care Costs, Government Claims. www.mtgplace.com, news article March 24, 1999.

The USDA published in the February 1999 Federal Register its long-awaited draft regulations covering the irradiation of raw meat. The proposal will permit ionizing radiation to be used to treat refrigerated or frozen uncooked meats and certain raw meat-food products to reduce levels of food-borne pathogens. Final regulations are expected at year-end. Poultry irradiation has been legal for several years but has yet to be sold in an irradiated form.

Still, the National Food Processors Association among many others are frustrated by the nearly five years that it took for the government's approval of beef irradiation. On August 23 a broad industry coalition submitted a petition to the FDA to approve the irradiation of ready-to-eat foods such as deli meats, hot dogs and alfalfa sprouts.[7]

8 We're Ready

In the meantime, MDS Nordion and the contract sterilization companies have been actively involved in preparing the network required when the meat regulations are finally approved. We have met with the food industry. We have submitted petitions to government. We have provided information to the media and the public.

MDS Nordion is poised to help preserve the safety of the world's food supply with the *Centurion Irradiator,* an innovative and revolutionary irradiator developed specifically for low-dose processing both fresh and frozen meat and poultry products in a refrigerated environment.

This machine can process up to 200 million pounds of product per year. The food processor would package product ready for distribution. Prior to shipment, the product would be routed through *the Centurion Irradiator*. The process is an excellent final assurance of HACCP practices that the food processors have now been regulated to implement. The food is already in its final packaging, and there is no risk of product re-contamination. As with medical devices the product never comes in contact with the gamma source and certainly does not become radioactive. The dose distribution achieves the highest possible destruction of pathogens, without significantly changing the product's sensory properties such as colour and flavour.

We are making headway. Indeed, people's health depends on it. The next time you order your hamburger at a restaurant, I suspect you'll check whether it is well done or not, and for good reason.

---

[7] FSIS Urged to Finalize 'Overdue" Irradiation Rule, www.mtgplace.com, news article May 7, 1999.

9 Conclusion

With a new millennium at our doorstep, who knows what health risks and remedies, await. But one thing is for certain: the radioisotope cobalt-60 will continue to save lives by eliminating or minimizing harmful bacteria.

Kids will continue to get scrapes and cuts, we'll continue to order our favorite foods from fast food restaurants and, who knows, maybe we will board commercial airlines for space travel. Whatever the situation, gamma sterilization is poised to be part of our lives in the years ahead.

Session F1

Radiopharmaceutical Applications and Medical Imaging — I

SYNTHESIS AND EVALUATION OF A RADIOIODINATED PHOSPHOLIPID ETHER ANALOG (NM-404) FOR DIAGNOSTIC IMAGING OF PROSTATE CANCER

R.E. COUNSELL

Department of Pharmacology, The University of Michigan Medical School, Ann Arbor, MI, 48109 U.S.A.

E-mail: counsell@umich.edu

M.A. LONGINO, A.N. PINCHUK, R.W.S. SKINNER, S.J. FISHER, M.E. VAN DORT, K.J. PIENTA AND R.L. WAHL

Departments of Internal Medicine, Pharmacology, Radiology and Surgery, The University of Michigan Medical School, Ann Arbor MI, 48109 U.S.A.

Imaging procedures play a major role in the current management of patients with prostate cancer. Despite advances in many of these procedures, improvements are still needed, especially in the area of Nuclear Medicine. The radioiodinated phospholipid ether analog (PLE) described here represents a new class of radiopharmaceutical, which has provided excellent images of prostate tumors in animal models and is now undergoing preclinical human pharmacokinetic evaluation.

Design and synthesis of radioiodinated PLE was based on the fact that various animal and human tumors contain higher concentrations of ether lipids than surrounding normal tissues. A number of radioiodinated PLE were synthesized and evaluated by γ-camera imaging using rat tumor models as well as nude and SCID mice bearing human tumor xenografts. Of the several agents that displayed promising results, one candidate (NM-324) was selected for further preclinical evaluation and subsequently studied in cancer patients in an effort to ascertain its ability to be retained in human tumors. These studies revealed that NM-324 was capable of imaging tumors in patients, but the high first pass clearance by the liver severely compromised its clinical utility as a diagnostic radiopharmaceutical. Conversely, this study demonstrated that our animal models were appropriate for the identification of clinical candidates.

In an effort to obtain a more suitable clinical candidate, the present study undertook the synthesis and evaluation of additional radioiodinated PLE with a focus on those displaying good tumor avidity and a prolonged plasma half-life relative to the prototype. Biodistribution analysis and γ-camera imaging of Copenhagen rats bearing Dunning R3327 prostate tumors and SCID mice bearing human prostate cancer (PC-3) revealed NM-404 to display a longer plasma half-life, better tumor/liver and tumor/kidney ratios, and significantly superior imaging properties than the initial prototype, NM-324. (Supported by the U. S. Department of Defense grant DAMD17-98-1-8528 and the SPORE in Prostate Cancer grant P50 CA 65968)

We previously described the remarkable capacity of certain radioiodinated phospholipid ether (PLE) analogs to be selectively retained by a variety of rodent and human tumor cell lines [1]. Moreover, this property made it possible to obtain images of these tumors in rabbits, rats and mice using γ-camera scintigraphy. Based on these and other preliminary results, one of these radioiodinated analogs, 12-(*m*-iodophenyl)dodecyl phosphocholine (NM-324, Figure 1), was approved for pharmacokinetic evaluation in human cancer patients in order to determine whether the results in animals could be confirmed in humans. Although high first pass clearance by the liver compromised the imaging capabilities of NM-324, imaging of the tumors was successful in several patients, and thereby confirmed the potential of radioiodinated PLE analogs for tumor imaging in patients [2].

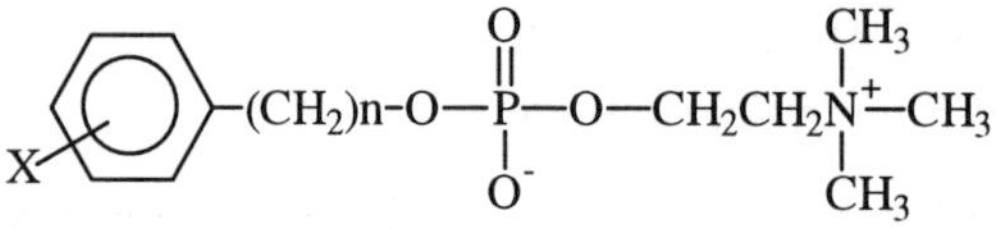

	X	n
NM-324	*m*-^{125}I	12
NM-404	*p*-^{125}I	18

Figure 1. Structures for NM-324 & 404.

In an effort to obtain a more suitable clinical candidate, the present study undertook the synthesis and evaluation of analogs of NM-324 with the aim of improving the tumor retention *vis a vis* the liver and kidneys. Placing the radioiodine in the *para* position and increasing the aliphatic chain length led to NM-404 (Figure 1) which not only increased lipophilicity but also led to the desired properties.

Scheme 1. Synthesis of NM 404

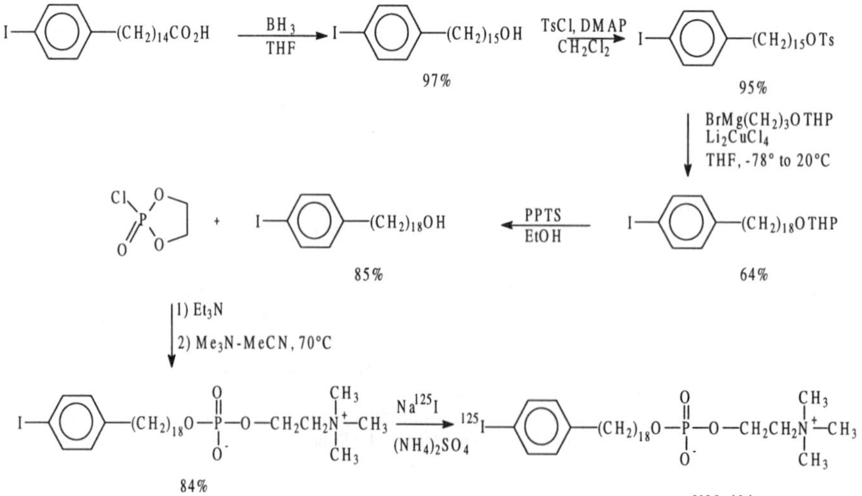

NM-404

Phosphocholination was performed according to Chandrakumar and Hajdu [3], and radioiodination followed the procedure of Mangner, et al. [4].

Biodistribution analysis (Table 1) and γ-camera imaging was performed in Copenhagen rats bearing Dunning R3327 prostate tumors and in SCID mice bearing human prostate cancer (PC-3). Comparison of NM-324 and 404 over several days revealed that tumor visualization was possible in both instances, but radioactivity was only seen to clear from abdominal organs following administration of NM-404.

Based on these results, NM-404 was selected for further preclinical analysis. The Toxicology Research Center at the University Buffalo found an isotonic solution of stable NM-404 to have no physiologic or pathologic effects in rats or rabbits at a dose 200 times the anticipated human dose. Moreover, the above tissue distribution studies along with those in normal Sprague-Dawley rats predicted that ^{131}I labeled NM-404 could be safely injected in humans with thyroid blocking at a dose of 2 mCi.[5]. Phase I studies in humans are planned.

Table 1. Biodistribution of ^{125}I-NM-404 in male SCID mice bearing PC-3 human prostate cancer xenografts, expressed as Dose/gm ± SEM and Target/Non-target Ratio, (n=4).

Tissue	**1 DAY** % Dose/gm (Tumor/Tissue)	**3 DAY** % Dose/gm (Tumor/Tissue)	**5 DAY** % Dose/gm (Tumor/Tissue)	**8 DAY** % Dose/gm (Tumor/Tissue)
Blood	5.74±0.20 (1.59)	3.10±0.13 (4.24)	3.08±0.09 (5.87)	2.17±0.07 (6.91)
Kidney	4.22±0.14 (2.17)	2.14±0.11 (6.13)	2.28±0.09 (7.92)	1.46±0.04 (10.26)
Liver	3.69±0.21 (2.48)	1.93±0.10 (6.81)	1.63±0.06 (11.07)	1.02±0.06 (14.69)
Lung	5.36±0.33 (1.71)	2.60±0.20 (5.06)	2.27±0.09 (7.97)	1.54±0.06 (9.70)
Muscle	0.79±0.03 (11.50)	0.57±0.04 (22.98)	0.49±0.03 (36.95)	0.40±0.03 (37.33)
Prostate	2.60±0.15 (3.51)	1.40±0.27 (9.40)	1.96±0.25 (9.20)	1.41±0.06 (10.64)
Tumor	9.14±0.69 (1.00)	13.14±0.40 (1.00)	18.06±0.80 (1.00)	14.96±0.63 (1.00)

References

1. Counsell R., Tumor visualization with radioiodinated phospholipid ethers, *In Synthesis and Applications of Isotopically Labelled Compounds*, ed. By J. Allen and R. Voges (John Wiley & Sons, New York, 1995) pp. 561-572.

2. Wahl R. L., Fig L. M., Shapiro B., Zasadny K. R., Gross M. D., Weichert J. P. and Counsell R. E., Initial clinical evaluation of I-131 NM-324, a tumor-avid phospholipid ether, in humans with cancer, *Radiology* **205 (T)** (1997) p. 478.

3. Chandrakumar N. S. and Hajdu J., Stereospecific synthesis of ether phospholipids. Preparation of 1-alkyl 2-(acylamino)-2-deoxy-glycerophosphocholine, *J. Org. Chem.* **48** (1983) pp. 1197-1202.

4. Mangner T. J., Wu J. and Weiland D. M., Solid-phase exchange radioiodination of aryl iodides. Facilitation by ammonium sulfate, *J. Org. Chem.* **47** (1983) pp. 1484.

5. Zasadny K. R., Longino M. A., Fisher S. J., Counsell R. E. and Wahl R. L., Predicted dosimetry for I-131-NM-404, a phospholipid ether agent for tumor imaging and possible therapy, *J. Nucl. Med.* **40** (1999) p. 39P.

EVALUATION OF SUCCINYLATED STREPTAVIDIN AS THE RADIONUCLIDE CARRIER IN ANTIBODY PRETARGETING OF PROSTATE CANCER

D.S. WILBUR, D.K. HAMLIN, P.M. PATHARE, K.R. BUHLER AND R.L. VESSELLA

Departments of Radiation Oncology and Urology
University of Washington, Seattle, WA 98195
USA
dswilbur@u.washington.edu

With the exception of kidney, streptavidin does not localize in normal tissue. Importantly, the kidney localization can be blocked by succinylation. This lack of tissue localization of succinylated streptavidin led to an investigation of its use with biotinylated monoclonal antibodies in a two-step pretargeting protocol. Tissue distributions obtained with a PSMA binding antibody, 107-1A4 in an athymic mouse prostate carcinoma xenograft, LNCaP, showed very good tumor-to-blood and tumor-to-muscle ratios with radiolabeled succinylated streptavidin, but RES tissues had elevated concentrations. Further investigation will be required to determine if the elevation in RES concentrations is due to the xenograft model or reagents used.

1. Introduction

Our research group is involved in the development of new methods for treatment of metastatic prostate cancer. One method currently under investigation is the use of monoclonal antibodies (mAbs) as targeting agents for delivery of cytotoxic radionuclides to prostate cancer cells in the body [1]. Because mAbs do not have properties that are suited for use with some short half-lived therapeutic radionuclides of interest, e.g. bismuth-213 ($t_{1/2}$ = 46 min) or astatine-211 ($t_{1/2}$ = 7.2h) [2], we are investigating a mAb-based tumor targeting method that has been termed "pretargeting". This term is derived from the fact that a mAb conjugate is used to target the tumor prior to introduction of a radiolabeled molecule, which is capable of binding to the mAb conjugate. This two-step pretargeting method allows the use of a variety of molecules as carriers of the radionuclide. The important factors are that in vivo properties of the carrier molecule be matched with the radionuclide used, and that the carrier molecule binds with a high affinity to the mAb conjugate.

A number of research groups are investigating pretargeting using antibody conjugates of the very high affinity binding pair, biotin and streptavidin (SAv) [3]. In the present studies, we are investigating the use of succinylated SAv as a carrier of therapeutic radionuclides that can bind with biotinylated mAb (107-1A4, PSMA specific) [4] pretargeted to prostate cancer xenografts (LNCaP) in athymic mice. We are interested in succinylated SAv as the carrier, as previous studies have shown that chemical modification of recombinant SAv (pI = 7) with succinic anhydride to decrease the pI (to < 4) diminished concentration of this 52 kDa protein to its only tissue of localization, the kidney [5]. Preliminary results of our study are described herein.

2. Materials and Methods

Succinylation of recombinant SAv was conducted as previously described [5]. 107-1A4 was obtained as previously described [4] and biotinylated with Biotin-LC-NHS (Pierce) using the standard methods. Mice used in these experiments were placed on a biotin free diet for 7 days prior to injection. LNCaP tumor xenografts were implanted in athymic mice and allowed to grow for 6 weeks (range 23-534 mg) prior to the experiment. Radioiodine labeling was accomplished by the chloramine-T method [6].

3. Results

The initial step in the investigation was to biotinylate 107-1A4 (2 biotin/mAb) and evaluate the effect of biotinylation on tumor localization in LNCaP xenografts. Dual isotope labeling was performed to give biotinylated [^{125}I]107-1A4 and native [^{131}I]107-1A4. Three quantities (10, 25 and 50 μg each) were co-injected into mice bearing LNCaP prostate carcinoma xenografts. Little difference in the tissue distribution was noted at 24, 48 and 72 h post injection (pi). At 24h pi, blood concentration ranged from 14 - 17% ID/g and tumor was 15 - 22% ID/g. The blood decreased less than 30% over 72h, with tumor decreasing very little over that time.

A second experiment evaluated blood clearance of (50 μg) biotinylated 107-1A4 at 1, 2 and 4 h after avidin administration (25, 50 and 100 μg). A dramatic decrease was noted in the blood concentration (15% ID/g to 5% ID/g) within 1 h of avidin administration, with all quantities of avidin providing the same result. A third experiment evaluated tumor co-localization of succinylated [^{131}I]SAv with biotinylated [^{125}I]107-1A4. Varying quantities (1, 5 and 25 μg) of [^{131}I]SAv were administered 49 h after 50 μg of [^{125}I]107-1A4 and 1 h after 25 μg of avidin were administered. Similar tissue distributions (%ID/g) were obtained for all quantities of succinylated SAv, with increasing tumor concentrations over time. At 24h pi, the 25 μg SAv administration provided an equal quantity (pmol/g) in tumor as that of the localized mAb, 107-1A4.

In a fourth experiment, tumor co-localization of 20 μg succinylated [^{131}I]SAv with 50 μg biotinylated [^{125}I]107-1A4 was evaluated at 24, 48 and 72 h pi. Avidin clearance (25 μg) was employed 48 h pi 107-1A4 and 1 h prior to SAv injection. Good tumor localization of succinylated SAv was obtained at 24 h pi (12 - 15

%ID/g) and this was retained through 72 h. The blood concentration of SAv decreased (8 to 2 %ID/g). Unfortunately, appreciable concentrations of radioactivity were also noted in lung, kidney, spleen and liver. It is unclear what caused the localization of SAv in these tissues as succinylated SAv does not localize in them.

4. Conclusions

Radiolabeled succinylated SAv has been shown to specifically localize in prostate carcinoma xenografts when a biotinylated prostate specific monoclonal antibody (107-1A4) is prelocalized and its excess is cleared from the blood with the use of avidin. The results are encouraging for the use of this reagent as a carrier of radionuclides, but the pharmacokinetics observed indicate that a longer half-lived radionuclide (i.e. $t_{1/2}$ > 24h) should be used. The application is, however, complicated by localization in RES tissues. Future studies will be conducted to determine the cause of this secondary tissue localization.

5. Acknowledgement

We are grateful for the financial support of the Department of Defense (DAMD17-98-1-8500) and Department of Energy (DE-FG03-98ER62572) in these studies.

6. Literature Cited

(1) R.L. Vessella, in Immunotherapy of Renal Cell Carcinoma, F.M.J. Debruyne et al., Eds., Springer-Verlag, New York, pp. 38-46 (1991).
(2) D.S. Wilbur, *Antibody, Immunoconjugates, Radiopharm.* **4**, 85-97 (1991).
(3) G. Paganelli, M. Malcovati and F. Fazio, *Nucl. Med. Commun.* **12**, 211-234 (1991).
(4) L.G. Brown et al., *Prostate Cancer and Prostatic Diseases* **1**, 208-215 (1998).
(5) D.S. Wilbur et al., *Bioconjugate Chem.* **9**, 322-30 (1998).
(6) D.S. Wilbur et al., *Bioconjugate Chem.* **9**, 100-7 (1998).

PERMANENT MOULD CASTING TECHNIQUE FOR THE FABRICATION OF CESIUM-137 BRACHYTHERAPY SOURCES

K. L. N. RAO, R. N. VARMA, S. K. SAXENA, AND C. MATHEW

Radiopharmaceuticals Division, Bhabha Atomic Research Centre, Mumbai-400085, India
E-mail: isd@magnum.barc.ernet.in

Brachytherapy has been successfully employed for the intracavitary applications in the control and management of carcinoma of cervix. Reasonably long half-life of 30 a, emission of 660 KeV monoenergetic gamma radiation and availability as a byproduct from spent nuclear reactor fuels make ^{137}Cs an ideal substitute for ^{226}Ra. A simple manual after loading uterine applicator using ^{137}Cs was designed at the Bhabha Atomic Research Centre. A technology was developed by the authors for the fabrication of ^{137}Cs sources suitable for brachytherapy applicators. This consists of the following steps: (1) Immobilization of ^{137}Cs in a suitable glass matrix by melting glass-forming ingredients in a platinum-clad inconel crucible at 1050° C. (2) Conversion of the resulting ^{137}Cs glass having a specific activity of 91 GBq/g into glass spheres of ≈ 225 mg by passing through a platinum-clad inconel funnel. (3) Melting the glass spheres and loading the ^{137}Cs sources into platinum tubes of i.d. 1.3 mm and (4) Cutting the filled platinum tubes into 5 mm long sources. All the above operations can be carried out in sealed production plants provided with 100 mm thick lead shielding using remote handling tongs.

1 Introduction

The highest incidence of cancer among the Indian women is carcinoma of the cervix and accounts for 75% of all the cases of cancer among them [1]. Brachytherapy has been successfully employed for the intracavitary applications in the control and management of carcinoma of cervix. Reasonably long half life of 30 years, emission of 660 KeV monoenergetic gamma radiation and availability as a byproduct from spent nuclear reactor fuels make ^{137}Cs an ideal substitute for ^{226}Ra. A simple manual after loading uterine applicator for ^{137}Cs sources was designed at the Bhabha Atomic Research Centre (BARC). A technology was developed by the authors for the fabrication of ^{137}Cs sources suitable for these applicators and is described in this paper.

2 Materials and methods

Cesium-137 as chloride, procured from the Board of Radiation and Isotope Technology (BRIT) was converted into the nitrate form before use. All the glass-forming ingredients such as silica, boric acid, sodium nitrate, vermiculite, etc. used were of A.R. or equivalent grade. The crucible, the funnel, the tubes, the furnace, the cutting machine and the other gadgets were designed at the Isotope Division, BARC and were fabricated at the Isotope Division and Hindustan Platinum Ltd., Mumbai. Measurements of radioactivity were carried out using a well type argon-filled ion chamber supplied by the Electronics Division, BARC coupled to a Keithley 614 Electrometer after calibration. The preparation of the sources was carried out in sealed aluminium plants with 100 mm thick lead shielding.

The facile of the method is as follows.

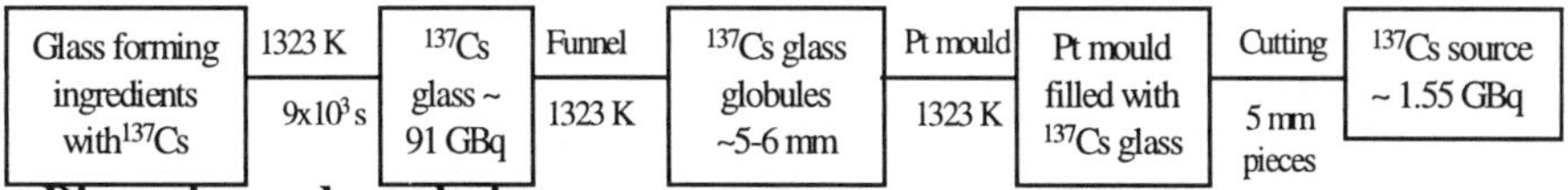

3 Discussion and conclusion

Several glass compositions from low melting to high melting with different properties were tested and ultimately a composition, containing 52% vermiculite, 13% B_2O_3 and oxides of alkali metals, along with ^{137}Cs having a glass formation temperature of about 1223 K was optimized.

The above mixture was taken in a platinum crucible with inconel casing and subjected to 1323 K for 9 X 10^3 s in the furnace. The resultant homogenous glass had a density of about 2.7 X 10^{-3} Kgm^{-3} and a specific activity of about 91 $GBqg^{-1}$. The glass was molten and passed through a platinum-cladded inconel funnel at 1323 K. The glass drops emerging out of the funnel were cooled by ambient air and formed globules weighing about 225 mg each.

To meet the requirements of the manual after loading uterine applicator designed at BARC, cesium-137 sources of activity 1.55 GBq and dimensions 1.5 mm (ϕ) X 5 mm (l) are required. Various methods like rod drawing and extrusion pulling were tried to prepare glass rods of these dimensions, but uniformity and reproducibility could not be achieved. Hence, the permanent mould casting technique was developed. It was attempted to use moulds made of different materials for shaping the molten glass into sources of required size followed by cooling to room temperature. The important criteria set for the success of the method are detailed in Table-1. Nickel, copper and stainless steel get corroded under these conditions. Graphite and inconel are more resistant to corrosion while platinum is the most resistant material. The mechanical strength of graphite is the least while that of inconel and stainless steel is the highest. In addition, the combustibility of graphite is a major disadvantage since the source has to be welded in an S.S. capsule. Considering all these parameters, platinum was selected for use as the mould.

Table-1. Materials used for mould casting of ^{137}Cs glass

Sl. No.	Material	Mechanical stability of the mould at pour point	Filling characteristics at pour point	Resistance to reaction with molten glass
1	Graphite	-	+	++
2	Nickel	++	+	+
3	Copper	++	+	+
4	S.S.	+++	+	+
5	Inconel	+++	+++	+++
6	Platinum	++	++++	++++

-: Very poor; +: Poor; ++: Moderate; +++: Good; ++++: Excellent.

The mould has a tube of i.d. 1.3 mm, o.d. 1.5 mm and height 80 mm provided with a cone of height 15 mm and i.d. 9.3 mm at the top. In the mould casting technique, the ^{137}Cs globule is placed in a mould. Four such moulds are loaded on an inconel gadget. The gadget is kept in a furnace heated to 1323 K and kept for 2.7×10^3 s. The glass melts, fills the mould and is prevented from flowing down the tip of the mould by the formation of solid glass at the tip of the mould kept outside the hot zone. The furnace is switched off and the ^{137}Cs glass in the moulds is annealed overnight. The moulds are taken out of the furnace, their conical parts are severed and the cylindrical parts are cut into 5 mm long pieces using the cutting machine. These sources after encapsulation, ultrasonic cleaning and assay are loaded into the kit for brachytherapy applications.

Permanent mould casting technique requires judicious selection of materials to meet the specifications during glass formation and casting. We have optimized a method for the regular production and supply of ^{137}Cs brachytherapy sources. The equipment, the flow sheet for the process and the design of the plant for operation have been developed indigenously.

Reference

1. Bhaskar B. K., Balakrishnan I. S., Ramakumar S., Jayaraman R. and Lakshmanan A. V., Development of indigenous afterloading units, In *Medical Physics for Human Health Care*, ed. by Bhatnagar P. K., Pradhan A. S. and Reddy A. R, (Scientific Publishers, Jodhpur, 1997) pp. 113-118.

Session F2

Production and Applications of Isotope Tracers in Industry — I

METHOD TO DETERMINE THE ACTIVITY OF PURE β^- EMITTERS IN CLOSED WASTE PACKAGE USING γ-SPECTROMETRY

S. TAKÁCS

Institute of Nuclear Research, 18/c Bem tér, Debrecen H-4026, Hungary
E-mail: takacs-s@atomki.hu

P. COVENS, G.X. EGGERMONT AND A. HERMANNE

Vrije Universiteit Brussel, 103 Laarbeeklaan 10, 1090 Brussels, Belgium,

Hospitals and research laboratories often use different radioactive materials and produce large amounts of low level radioactive waste. A method was developed to determine the activity levels of pure β^- emitters with low Z number in closed waste packages using γ-spectrometry based on a selective collection of radioactive waste. While the emitted beta particles slow down in the waste matrix, Bremsstrahlung (BS) radiation is produced. This radiation offers good possibility for quantitative determination of pure β^- emitting isotopes such as ^{32}P and ^{35}S in sealed waste packages. The fraction of the beta particle energy transformed into gamma radiation depends on the initial energy of the emitted particles and on the composition of the matrix. The shape of the resulting BS spectrum depends on the beta-decaying isotope and is influenced by the composition, density and volume of the waste matrix. An optimised region of interest (ROI) relevant to the radiation induced by ^{32}P and ^{35}S isotopes was determined in the BS spectrum. Minimal detectable activity (MDA) values were deduced for the above two isotopes in waste drums of 33 litres. Two gamma-spectrometry systems were studied based on a high purity germanium detector and a NaI(Tl) scintillation detector. Comparison was made to a 4π plastic scintillation detector system.

1 Introduction

Biomedical radioactive waste essentially consists of low-energy γ-emitting, positron emitting and pure β^-–emitting radioactive nuclides. The nuclides in the first two groups can be identified and quantitatively determined by using high-resolution gamma-spectrometry, while it is difficult to measure nuclides in the group of pure β^-–emitters.

Energetic beta particles are involved in different types of interactions with matter in which they loose their energy. The main process is the ionisation energy loss, which results in low-energy X-rays. This radiation is out of the range of the detectors used and is absorbed by the matrix. A fraction of the β^-–particle energy is transformed into electromagnetic radiation called Bremsstrahlung (BS) radiation. The BS radiation has a continuous energy spectrum with the highest photon energy close to the maximal energy of the beta particles. These photons can penetrate through the waste matrix and the wall of the waste container with limited attenuation and can be easily detected by γ-detectors. The cross section of BS radiation process for electrons is proportional to the atom concentration and to the square of the Z number of the stopping material and depends on the energy of the slowing down β^-–particle:

$$\sigma \approx nZ^2 E(\beta)$$

The probability of BS process is much lower than that of the ionisation process for beta particles but still high enough to produce measurable amounts of γ-photons. The aim of this study was to investigate in general the possibility of quantitative determination of ^{35}S and ^{32}P pure β^-–emitting nuclides in closed waste packages through BS radiation using γ-detectors.

2 Methods

The basic model makes the following assumptions:

1. Selective waste collection is required. Only one type of β^-–emitting isotope is allowed in one waste package.
2. Uniform activity distribution is assumed inside all waste drums.
3. Each component of the matrix material is evenly distributed over the whole volume of the waste drum and constant matrix composition is supposed.

The activity of a point source measured in a ROI is given as follows.

$$A = \frac{\lambda T_\gamma}{\left(1 - e^{-\lambda t_m}\right)\varepsilon(E)\beta(E)} \qquad \text{where:}$$

A is the activity of the source in Bq, λ is the decay constant, $T\gamma$ is the net counts in the ROI, t_m is the acquiring time, $\varepsilon(E)$ is the detector efficiency including geometry, scattering and absorption factors, $\beta(E)$ is the BS coefficient and gives the yield of gamma photons with energy E per decay. The $(1 - e^{-\lambda t_m})$ is a correction factor for decay during measurement. After averaging the $\varepsilon(E)\beta(E)$ quantity over the volume of the waste drum, the equation approximates the total β^- activity of the investigated waste package.

3 Results and discussion

Experiments were performed with point sources of ^{35}S and ^{32}P and two types of waste drums (small and high) filled with different types of matrices with known density. The average $\varepsilon\beta_{av}$ was measured for optimised and fixed ROI using HPGe and NaI(Tl) detectors with the Low Level Waste Assay system at VUB [1]. Then the $\varepsilon\beta_{av}$ parameter was fitted by $ae^{b\rho}$, where a and b are fitting constants and ρ is the mean density of the matrix. When the basic assumptions are fulfilled the Low Level

Waste Assay system at VUB achieves mass specific MDA values shown in figures 1 and 2. Figures show that small drums (33 l) and germanium detector (ReGe) give a better detection limit. Any deviation in the activity distribution from the ideal uniform one introduces an error in the determined activity that can be as large as 100% when the total activity is concentrated in one point near the surface of the drum. Deviation in matrix composition from the one used to calibrate the system also generates errors. Different types of materials result in different BS yield,

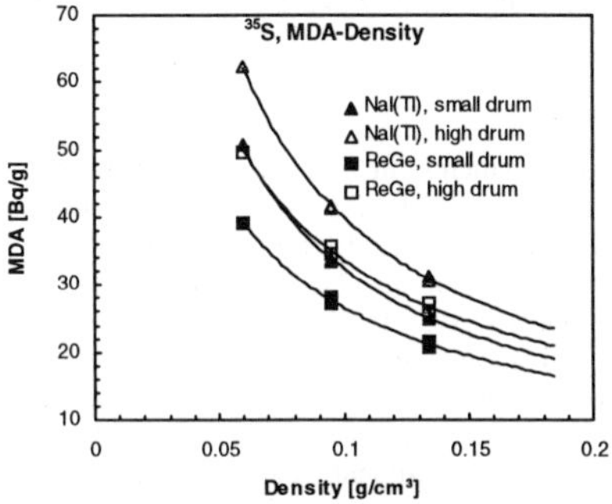

Figure 1. MDA dependence on the mean matrix density determined for ^{35}S source using paper matrix and condition of 2σ and 1000 sec acquisition time

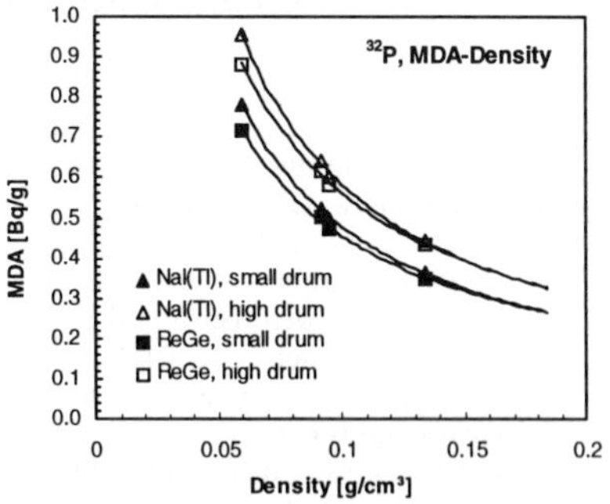

Figure 2. MDA dependence on the mean matrix density determined for ^{32}P source using paper matrix and condition of 2σ and 1000 sec acquisition time

different absorption and scattering of the photons and the averaging yields different values for $\varepsilon\beta_{av}$. The difference is about a factor of two between a paper matrix (density of 0.17 g/cm^3) and a PVC matrix. MDA values of 4 Bq/g and 5 Bq/g were reported for a system using large surface plastic scintillation detectors in nearly 4π measuring geometry for ^{32}P and ^{35}S isotopes respectively [2]. The corresponding values for our set up are 0.86 Bq/g and 53 Bq/g for a ReGe detector (1 σ above background and 30 sec acquisition time) with all the advantages of a high resolution gamma spectrometry system.

References

1 G.X. Eggermont, P. Covens, A Hermanne and D. Meert; "Decay Management of Nuclear Waste in University and Hospitals"; *Proc. of DOE Low-Level Radioactive Waste Management Conference*, Salt Lake City, USA, May (1997).

2 "Afsterfbaar radioactief afval aan de K. U. Leuven," *Report LM/hc/94.187rp*, Katholieke Universiteit Leuven, p 63, (1994).

RADIOACTIVE CORROSION MEASUREMENTS ON CLADDING AND CONTAINER MATERIALS FOR SPENT FUEL ELEMENTS IN BRINES OF PRACTICAL IMPORTANCE

J. ENGELHARDT*, A. KUPFER**, C. NEHM**, G. MARX*

**Institut für Umwelttechnologien (IUT), Rudower Chaussee 5, 12489 Berlin*

*** Freie Universität Berlin, Fachbereich Biologie, Chemie, Pharmazie, Institut für Chemie / FG Radiochemie, Fabeckstraße 34-36, 14195 Berlin*

With respect to the direct disposal of spent fuel elements in salt diapirs corrosion measurements were carried out on the cladding material of LWR and MTR fuel elements Zircaloy-4 and Al and on Ti99.8Pd, the option for a shielding material for Pollux containers, the relevant electrolyte systems being Q-brine, solution 3 and saturated NaCl solution.
The experiments were performed at various temperatures (25°C, 55°C, 80°C) applying the Radioisotope Method RIM. RIM combines neutron activation analysis (NAA) with classical electrochemical procedures as potentiodynamic and potentiostatic measurements and also impedance measurements, to be carried out simultaneously. Applying RIM a variety of important corrosion data can be obtained from the specific experiments, carried out in situ. ^{95}Zr was used as isotopic tracer for Zircaoly-4, ^{46}Sc as nonisotopic tracer for Titanium.
From the results obtained it can be clearly seen, that in brines at negative potentials the resistance of Zircaloy-4 towards corrosion is slightly better than that of Ti99.8Pd but in the range of positive potentials the corrosion resistance of Zircaloy-4 collapses. The resistance of Ti99.8Pd on the other hand still remains. Moreover the conductivity of the oxide layer of Ti99.8Pd correlates with its corrosion resistance. The kinetics of corrosion were also investigated.
In contrary to the electrochemical behavior of Zircaloy-4 and Ti99.8Pd in brines Al has only a small passive range, due to its amphoteric behavior. Detailed investigations were also performed into contact potentials of the materials aforementioned with Fe and UO_2.

1 Introduction and Experimental Details

Valve metals and their alloys are widely used in various fields of nuclear technology as for instance Ti, for Pollux containers for direct disposal of spent fuel elements in salt diapirs, moreover Zircaloy-4, the cladding material of fuel rods, and also Al, the cladding of Material Test Reactors (MTR) fuel elements. Because of their practical importance it is an urgent need to investigate the corrosion behavior of these valve metals in relevant brines.
Zircaloy-4 and Ti can be easily investigated by using RIM, ^{95}Zr and ^{46}Sc being the specific radiotracers. But there are no suitable radioisotopes available for labeling Al. So Total Reflection X-ray Fluorescence Analysis (TXRF) is used instead. This analytical method only registers the x-ray fluorescence of the material under investigation, preventing the main beam from hitting the detector. Not only the Al concentrations but also the relevant Fe concentration for investigating contact

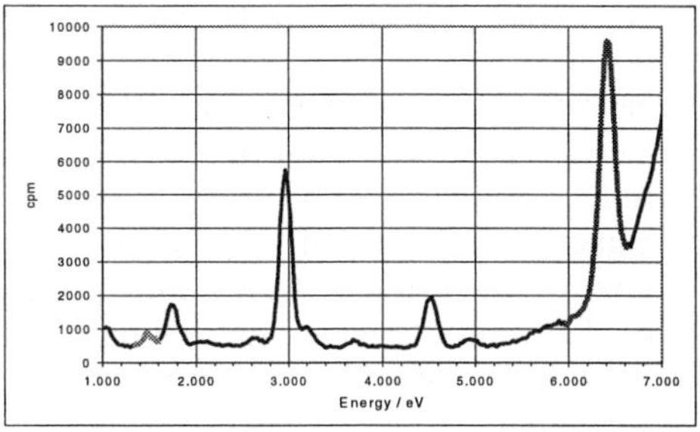

Fig.1:TXRF spectrum of 2 ppm Fe (red) and 25 ppm Al (green) in sat. NaCl solution diluted 1:20.

corrosion were determined by TXRF. For the light element Al the fluorescence efficiency is definitely decreased with respect to Auger electrons, which causes a difference of the signal height of TXRF spectra (Figure 1). The detection limit of Al therefore is only approximately 100 ppm, that of Fe on the other hand 500 ppb.

2 Results and Discussion

2.1 Titanium, TiGrade7 and TiCode12

At first our interest was focused on the investigation into the corrosion of Ti, TiGrade7 and TiCode12 in brines. The results obtained demonstrate that at the potentials applied ($-500mV \leq \Delta U \leq +500mV$) the corrosion rates of Ti, TiGrade7 and TiCode12 are smaller than 10 $\mu m \cdot a^{-1}$ under the relevant conditions. At rest potentials the corrosion rates of all three materials are smaller than 1,6 $\mu m \cdot a^{-1}$, those of the relevant welds being 4 $\mu m \cdot a^{-1}$. The corrosion rates correlate with the conductivity of the TiO_2 layer.
There is no influence of H_2O_2 on Ti corrosion at H_2O_2 concentrations smaller than 10^{-4} M, this concentration being even higher than that found under practical conditions. Hypochlorite has no influence on corrosion at all. So it can be emphatically stated that Ti and its alloys are very suitable materials for protecting plated Pollux caskets towards corrosion in brines.

2.2 Zircaloy-4

Due to the high price of Ti it was proposed to use carbon steel and stainless steel alone as materials for Pollux containers. However, in addition one might also rely on the resistance of the Zircaloy-4 cladding towards corrosion in brines.
Therefore Zircaloy-4 had to be investigated in brines and also in Bentonite Porewater within the potential range from –1500 mV to +1000 mV by applying RIM again. Figure 2 demonstrates that Zircaloy-4 is superior to TiGrade7 with respect to its resistance towards corrosion on condition that the potential applied does not reach the positive range. At higher potentials the corrosion rate of Zircaloy-4 increases abruptly. H_2O_2 does not influence Zircaloy-4 corrosion at the temperatures aforementioned on condition that the H_2O_2 concentrations do not

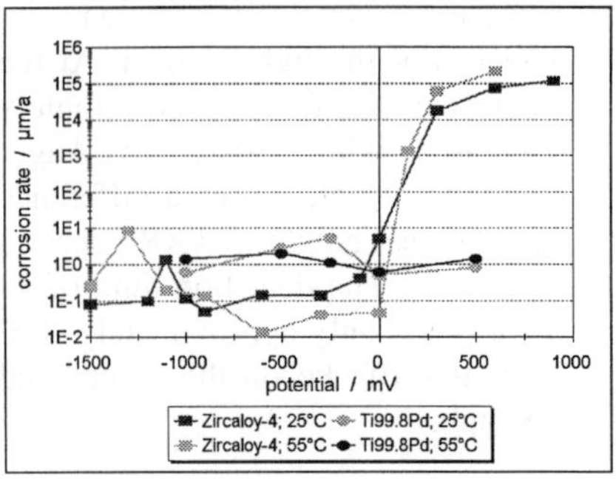

Fig.2: Corrosion on Ti99.8Pd and Zircaloy-4 at applied potentials in sat. NaCl solution

drastically exceed those existing in practice (~ 10^{-6} M). Impedance measurements demonstrate that the increase of the thickness of the ZrO_2 layer is in accordance with theory, showing a higher resistance of the ZrO_2 layer towards corrosion at 80°C, which might be due to the faster crystallization of the ZrO_2 layer, which is amorphous in the beginning.

Measurements on contact potentials show that in sat. NaCl solutions the mixed potential is dominated by the Fe (E_c = [-392 ± 25] mV). The rest potential of Fe is E_R = (-440 ± 25) mV. The corrosion rate of Zircaloy-4 is smaller than 3 $\mu m \cdot a^{-1}$ in this case. The corrosion rate of Fe, on the other hand, is 155 $\mu m \cdot a^{-1}$ under practical conditions. So the contact potential, formed by Zircaloy-4 with Fe does not increase the corrosion rate of the cladding material.

Fe^{3+} in the solution might shift the potential into the positive range enhancing Zircaloy-4 corrosion.

So Zircaloy-4 cannot be relied on as a material resisting corrosion due to its instability within the positive potential range.

2.3 *Aluminum*

The results for Al obtained from cyclovoltammetry and potentiostatic measurements and from measurements at rest potentials demonstrate that at negative potentials there is a passive range in brines (Figure 3). Under open circuit conditions the relevant rest potential of Al is just at the edge of the active and passive range.

For direct waste disposal of MTR elements therefore the resistance towards corrosion is not increased by the cladding.

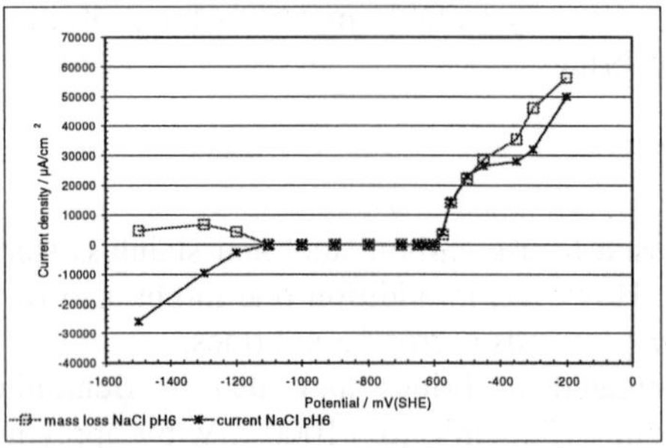

Fig.3: Current density vs. Potential for aluminum in sat. NaCl solution at pH 6.

INVESTIGATION OF FLOW PATTERN IN AN OXIDATION REACTOR IN A PTA PLANT USING RADIOTRACERS

V.N. YELGAONKAR , H.J. PANT, A.S. PENDHARKAR, AND S.V. NAVADA

Isotope Applications Division, Bhabha Atomic Research Centre, Mumbai 400 085, India
E-mail: yelgaon@magnum.barc.ernet.in

For the production of pure terphthalic acid (PTA), paraxylene is oxidised using air in an oxidation reactor. Oxidation reactor is a centrally stirred vessel with two sets of impellers in two halves. Liquid phase feed and air is introduced at the bottom and product slurry comes out through the outlet pipe from the top portion. Poor performance of the reactor necessitated a study of the residence time distribution of material in the reactor. Radiotracer (^{24}Na in the form of aqueous solution of sodium carbonate) was injected as a sharp pulse in the feed. Passage of the pulse was monitored at the inlet and outlet of the reactor using collimated scintillation detectors. From the knowledge of the theoretical and experimental mean residence times, dead volume of the reactor was computed, which varied between 20 - 33%. The tracer output function was modeled mathematically using 'tanks-in-series with recirculation' model. This revealed that the reactor behaved like a well mixed two compartments system with recirculation between the compartments. The recirculation ratio dropped during prolonged use of the reactor (it varied between 40 - 200) and increased sharply when the reactor was re-commissioned after washing during shutdown.

1 Introduction

Radiotracers are the best tools for process parameter estimation, characterisation of flow in continuously operating industrial systems and for trouble shooting [1].

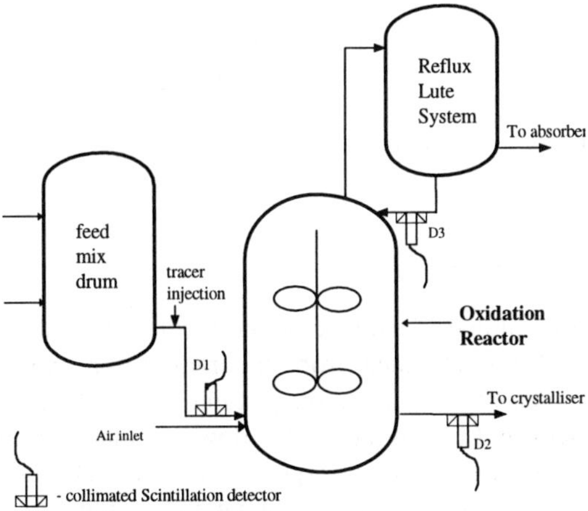

Figure 1. Block diagram of the process

In oxidation reactor investigated, for the production of pure terphthalic acid (PTA), paraxylene is oxidised by air in the reactor (a centrally stirred vessel) in presence of acetic acid and catalyst. Mixture of organic phase (paraxylene) and aqueous phase (acetic acid+water) is introduced from the feed mix drum into the reactor (Figure 1). Air is introduced into the reactor at 180 - 200^0 C. Oxidation results in the formation of PTA crystals. After removal of the crystals from the process, liquid is recycled to the feed drum. Most of the material going in to the vapour phase is condensed in the reflux lute system and fed back to the reactor.

A series of radiotracer experiments were carried out in an oxidation reactor to determine the residence time distribution of the material in the reactor.

2 Experiment

About 260MBq of ^{24}Na as aqueous solution of sodium carbonate was used as radiotracer. Radiotracer was injected at the suction end of the feed pump as a sharp pulse. It was monitored on the inlet and outlet pipes using 1" diameter scintillation detector placed in 2" thick lead collimator on the positions shown in Figure 1. Count-rates obtained for a predetermined time interval were accumulated by the data acquisition system and later transferred to a PC for analysis.

3 Analysis

Mean residence time (MRT, $\bar{t}$) of a response obtained at each detector position is calculated from the first moment of the curve[2].

$$\bar{t} = \int C(t).tdt \Big/ \int c(t)dt$$

Difference of MRTs at the outlet and inlet detector positions gives MRT of the oxidation reactor.

3.1 Computation of dead volume

From the geometric volume of the reactor and volume flow rate, theoretical MRT is calculated. If the MRT obtained from the radiotracer experiment is less than theoretical MRT, then the volume corresponding to the difference in MRTs is the volume of the reactor not taking part in the flow, popularly known as dead volume. Dead volume obtained with different experiments before cleaning the reactor was of the order of 33% and after shutdown, cleaning and re-commissioning, it was of the order of 20%.

4 Modeling

A mathematical model 'Ideal mixers in series with recirculation' was used to simulate the obtained experimental responses[3]. This model assumes cascade of N number of ideally mixed vessels with a certain recirculation ratio f. Standard deviation between experimental response and a response computed by the model for a particular set of parameters $\bar{t}$, N and f is computed. Parameters for which standard deviation is minimum are representatives of the behaviour of the system. A typical model fitting is shown in Figure 2. Value of f obtained during the different experiments before shutdown was of the order of 40 and after the shutdown it was more than 200.

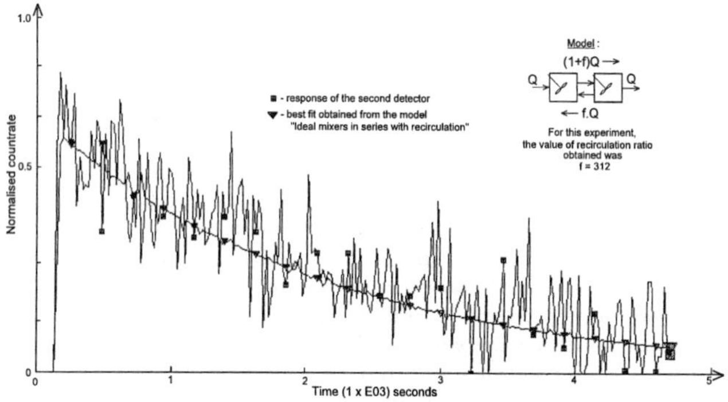

Figure 2. Best fit for a typical experimental curve

5 Conclusion

The reactor investigated behaves as a well mixed system. Quantity of dead volume and recirculation ratio was varying with the prolonged use of the reactor and it changed sharply after cleaning the reactor during shutdown and re-commissioning.

References

1. Charlton J.S., Radioisotope Techniques for Problem solving in Industrial Process plants, Leonard Hill (1986).
2. Levenspiel O., Chemical Reaction Engineering, John Wiley & Sons (1972) pp. 253-314
3. Zitny R. and Thyn J., Programs for Residence Time Distribution Analysis, in *RTD software manuals for data analysis of radiotracer experiments* (1996) pp. 104-157.

NUCLEAR DATA REQUIREMENTS FOR THE PRODUCTION OF MEDICAL ISOTOPES IN FISSION REACTORS AND PARTICLE ACCELERATORS

MARC A. GARLAND, ROBERT E. SCHENTER, AND ROBERT J. TALBERT

Pacific Northwest National Laboratory, Richland, Washington 99352, USA
E-mail: Marc.Garland@pnl.gov, re_schenter@pnl.gov, bob.talbert@pnl.gov

STEPAN G. MASHNIK AND WILLIAM B. WILSON

Los Alamos National Laboratory, Los Alamos, New Mexico 87545, USA
E-mail: mashnik@t2y.lanl.gov, wbw@lanl.gov

Simulations of isotope-production irradiations of targets in the radiation environments of fission reactors or particle accelerators require the evaluation of proton and/or neutron particle fluxes within the target and the integration of these with appropriate cross-section data. In fission reactors, proton fluxes are typically considered negligible and the approach reduces to neutron reactions at energies below the upper extents of the fission spectrum at which neutrons are born — 10 to 15 MeV. In particle accelerator target simulations, the secondary neutron flux can provide considerable production and/or collateral target activation. Where target and product nuclides are separated in Z and A, the relative contributions of neutron and proton reactions may not be apparent. Through decades of effort in nuclear data development and simulations of reactor neutronics and accelerator transmutation, a collection of reaction data is continuing to evolve with the potential of direct applications to the production of medical isotopes. At Los Alamos the CINDER'90 code and library have been developed for nuclide inventory calculations using neutron-reaction (En$\leq$20 MeV) and/or decay data for 3400 nuclides; coupled with the LAHET Code System (LCS), irradiations in neutron and proton environments below a few GeV are tractable; additional work with the European Activation File, the HMS-ALICE code and the reaction models of MCNPX (CEM95, BERTINI, or ISABEL with or without preequilibrium, evaporation and fission) have been used to produce evaluated reaction data for neutrons and protons to 1.7 GeV. At the Pacific Northwest National Laboratory, efforts have focused on production of medical isotopes and the identification of available neutron reaction data from results of integral measurements.

1 Introduction

Medical radioisotope production is receiving increased attention due to the many advances in nuclear medicine. In addition to further development in diagnostic nuclear medicine, pioneering work is being done in therapeutic applications of radioisotopes. For example, radiolabeled monoclonal antibodies are being used to treat leukemia and lymphoma, brachytherapy is being used to treat prostate cancer, radioactive stents are being used to prevent restenosis (reclogging of arteries) following angioplasty treatment of coronary heart disease, and radioisotopes are being used to palliate the excruciating bone pain associated with metastatic cancer.

Continued success in developing cures for cancer and ultimately in treating a large number of cancer patients is adversely impacted by this lack of knowledge of certain neutron capture cross sections for medically important radioisotopes. Without this data, medical radioisotope production cannot be optimized. Optimization is not only critical for economic reasons, but also for applications requiring the production of very high specific activity radioisotopes. In many cases, trial and mixed success is "de rigueur" for producing certain radioisotopes of medical significance.

2 Reactor-Spectrum Data Needs

As an example, the thermal and resonance integral cross sections are known for ^{186}W and ^{187}W , but not for ^{188}W. Thus, optimal production of the medically important radioisotope ^{188}W may not be realized since calculations related to the design of ^{186}W targets, their placements in a reactor and irradiation times cannot accurately be performed.

The design for a reactor focused on isotope production needs to consider the neutron cross sections of the medical radioisotopes to be produced so that proper neutronic conditions can be achieved for optimal radioisotope production. Neutron cross section information is needed to design targets (and hopefully a reactor) for the optimal production of medical radioisotopes.

Production of medical radioisotopes in fission reactor systems must be optimized with respect to several different parameters: position, target composition, density, configuration, etc. Research is required to determine the needed cross sections. Knowledge of these cross sections will benefit several practical applications and will also provide important modern data information for many isotopes previously unavailable.

The main objective of an initiative to address the cross section deficiencies is to access the cross sections that are of the greatest projected need. Table 1 identifies several medical radioisotopes that harbor deficiencies in cross section knowledge required for efficient, high specific activity production. In order to demonstrate how the lack of knowledge of the cross section impacts production results, calculations were made for six important medical isotope products. Table 2 shows these results comparing values with known and unity cross sections.

Table 1. Medical Isotopes with Unknown Cross-Section Data

Isotope	$T_{1/2}$ (days)	σ_{th} (barns)	R I (barns)	Medical Application
^{188}W	69.4	unknown	unknown	cancer and rheumatoid arthritis therapy, radiolabeled antibodies for cancer therapy
^{186}Re	3.72	unknown	unknown	prostate cancer and rheumatoid arthritis therapy, radiolabeled antibodies for cancer therapy, bone pain palliation
^{188}Re	0.71	unknown	unknown	medullary thyroid carcinoma therapy, bone pain palliation, radiolabeled antibodies for cancer therapy
^{194}Os	6.0y	unknown	unknown	radiolabeled antibodies for cancer therapy
^{193}Os	1.27	40	unknown	cancer therapy
^{198}Au	2.70	26E3	unknown	ovarian, prostate, brain cancer therapy
^{166}Ho	1.12	unknown	unknown	cancer and rheumatoid arthritis therapy
^{177}Lu	6.71	1000	unknown	radiolabeled antibodies for cancer therapy, heart disease therapy
^{153}Sm	1.93	400	unknown	radiolabeled antibodies for cancer therapy, bone pain palliation, treatment of leukemia
^{153}Gd	241.6	2E4	unknown	osteoporosis detection, SPECT imaging
^{127}Xe	36.4	unknown	unknown	neuroimaging for brain disorders, neuropsychiatric disorder research, SPECT imaging, lung imaging
^{125}Xe	0.71	unknown	unknown	cancer therapy
^{126}I	13.0	unknown	unknown	cancer therapy
^{131}Cs	9.69	unknown	unknown	intracavity implants for cancer therapy

Table 2. Production Results with Known and Unity Cross Section Values

Product Isotope	Target Isotope	Product 1-Group Cross Section (barns)	Target 1-Group Cross Section (barns)	Production (Ci/g-tgt) Using Known Cross Sections	Production (Ci/g-tgt) Using Unity Cross Sections	Ratio of Unity to Known Cross Section Production
^{198}Au	^{197}Au	1.62E+3	1.40E+1	688	67.4	0.10
^{60}Co	^{59}Co	3.24E-1	5.42E+0	68.1	12.9	0.19
^{125}I	^{124}Xe	1.83E+2	4.61E+1	1190	151	0.13
^{192}Ir	^{191}Ir	2.89E+1	1.53E+1	545	47.3	0.087
^{145}Sm	^{144}Sm	2.06E+1	1.47E-1	3.91	30.1	7.7
^{153}Sm	^{152}Sm	7.66E+1	3.83E+1	3420	91.5	0.027

3 Medium-Energy Data Evaluations

In the radiation environment of a proton accelerator target, neutron and proton reactions may significantly contribute to the production of the desired radionuclide. Medium-energy protons may each produce a few tens of neutrons in a high-Z target, each having a significant range and contribution to particle flux. The complexities resulting from the myriad of possible reaction paths, along with spatially varying flux magnitudes and spectra, require the evaluation of pertinent cross sections and fluxes. These are evaluated in sequential calculations with the LAHET Code System LCS[1] — the combination of LAHET and MCNP[2], or their subsequent combination in MCNPX[3] — with the CINDER'90 nuclide inventory code;[4,5] in this sequence, cross sections for reactions of protons and medium-energy neutrons are calculated with on-line nuclear models and evaluated lower-energy neutron reaction cross section are contained in the CINDER'90 library. This state-of-the-art sequence is used effectively in the analysis of medium-energy designs but requires a significant investment of CPU time.

Nuclear models have also been utilized with limited available measured cross section data to form evaluations for a growing number of target nuclides. Neutron and proton cross sections from threshold to 1.7 GeV have been evaluated for the stable isotopes of O, F, Ne, Na, Mg, Al, S, Cl, Ar, K, Zn, Ga, Ge, As, Zr, Nb, Mo, Xe, Cs, Ba, La, and Hg — or about 30% of the naturally-occuring stable nuclides. These evaluations have used available measured data from the LANL T-2 compilation,[6] the evaluations of the EAF97 library[7] for neutrons below 20 MeV, and calculations with HMS-ALICE,[8] CEM95,[9] and the BERTINI and ISABEL models of LAHET. Samples of the data and evaluations for two of nearly 700 reactions evaluated to date are shown in Figure 1. Complete results are shown in Ref. 10.

4 Conclusions

The status of simulation methods and data available for the description of isotope production is fair and improving, but many additional cross section measurements and evaluations are needed. Consequently, further research in obtaining better cross section information will have positive benefits in the field of medical science.

References

1. R. E. Prael and H. Lichtenstein, "*User Guide To LCS: The LAHET Code System,*" Los Alamos National Laboratory informal report LA–UR–89–3014 (Sept. 1989).
2. J. F. Briesmeister, Editor, "*MCNP — A General Monte Carlo N-Particle Transport Code,*" Los Alamos National Laboratory Manual LA-12625-M (November 1993).
3. H. G. Hughes et al., "*MCNPX For Neutron-Proton Transport,*" Los Alamos National Laboratory Report LA-UR-99-1935 (1999), submitted to **M&C'99**, September 27-30, 1999, Madrid, Spain.
4. W. B. Wilson et al., "*Accelerator Transmutation Studies at Los Alamos with LAHET, MCNP, and CINDER'90,*" Proceedings of the Workshop on Simulation of Accelerator Radiation Environments, January 11–15, 1993, Santa Fe, New Mexico; Los Alamos National Laboratory conference proceedings LA-12835-C (October 1994) pp.115–133; also available as Los Alamos National Laboratory informal document LA-UR-93-3080 (January 11, 1993).
5. W. B. Wilson, T. R. England and K. A. Van Riper, "*Status of CINDER'90 Codes and Data,*" Proc. Fourth Workshop on Simulating Accelerator Radiation Environments, Sept. 14–16, 1998, Knoxville, Tenn, pp. 69–79.
6. S. G. Mashnik, A. J. Sierk, K. A. Van Riper and W. B. Wilson, "*Production and Validation of Isotope Production Cross Section Libraries for Neutron and Protons to 1.7 GeV,*" Proc. Fourth Workshop on Simulating Accelerator Radiation Environments, Sept. 14–16, 1998, Knoxville, Tenn, pp. 151–162; Eprint **nucl-th/9812071** on the LANL xxx.lanl.gov server.
7. J.-Ch. Sublet, J. Kopecky, R. A. Forrest and D. Nierop, "*The European Activation File: EAF-97 REPORT File, Rev. 1,*" UKAEA Fusion report.
8. M. Blann and M. B. Chadwick, "*New Precompound Decay Model: Angular Distributions,*" Phys Rev. C57, 233 (1998).
9. S. G. Mashnik, "*User Manual For The Code CEM95,*" Joint Institute for Nuclear Research, Dubna (1995), see the Web page http://www.nea.fr/abs/html/iaea1247.html.
10. K. A. Van Riper, S. G. Mashnik and W. B. Wilson, "*Study of Isotope Production in High Power Accelerators: Detailed Report,*" Los Alamos National Laboratory informal report LA–UR–98–5379 (1998) (a 684 page report with 37 tables and 264 color figures available on the Web at: **http://t2.lanl.gov/publications/publications.html**).

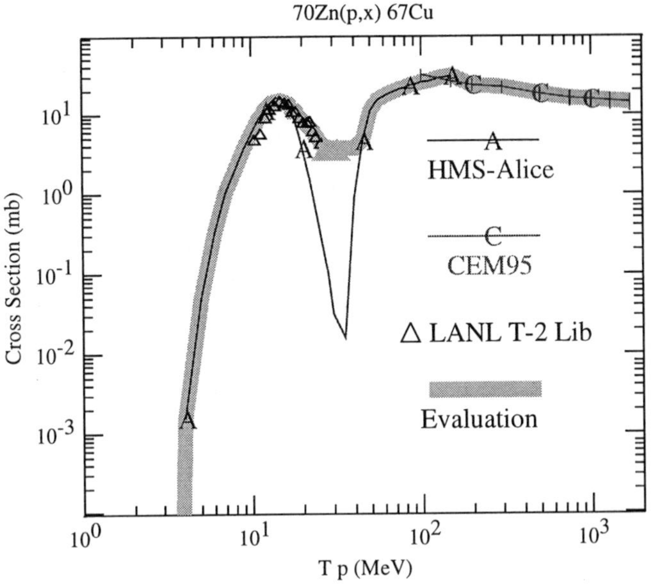

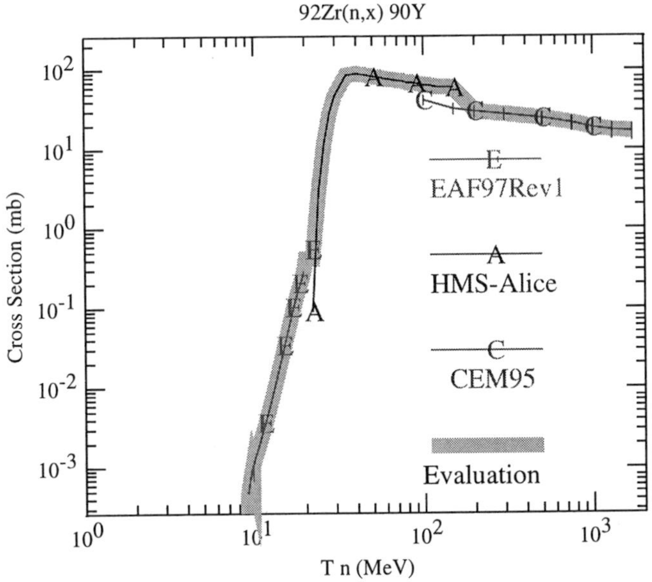

Figure 1. Samples of Data and Evaluations for (p,x) and (n,x) Reactions.

Session F3

Use of Isotopes in Environmental Studies — I

SILICON-32: DIATOMS, THE SILICON CYCLE, AND THE CLIMATE

DENNIS R. PHILLIPS, RAPHAEL M. KUDELA, VIRGINIA T. HAMILTON
Los Alamos National Laboratory, Radioisotope Research Group, MS J514, Los Alamos, NM 87545, USA

MARK A. BRZEZINSKI
Department of Ecology, University of California, Santa Barbara, CA 93106, USA

Silicon-29 and Si-30, both stable isotopes, have been used as tracers in mass spectrometric studies of the role of diatoms in the silicon cycle and their impact on the global climate as photosynthesizing phytoplankton. The methods used for the studies require substantial chemical manipulation in sample preparation, and significant skill to produce high quality data. Therefore the amount of data generated is relatively limited. We have produced high specific activity radioactive Si-32 and developed straight-forward radioanalytical methods for the study of biogeochemical cycles. We will discuss the relative merits of the mass spectrometric methods and the radioanalytical techniques, and we will summarize results from diatom studies done with Si-32.

1 Introduction

As the dominant phytoplankton species in the euphotic zone of the oceans, diatoms are the main primary producers in most coastal marine systems. It is estimated that they contribute to more than 40% of the total marine primary productivity. These photosynthesizing organisms require carbon in the form of bicarbonate and therefore represent a major sink for atmospheric carbon dioxide. The magnitude and dynamics of this sink are important to climate models. Besides the usual primary nutrients (carbon, nitrogen, phosphorus) required by autotrophic phytoplankton, diatoms also have an obligate requirement for silicon during critical parts of their cell cycle. Based upon existing data, diatoms produce 200 – 400 Tmol of biogenic silica each year.[1] Several models reveal that silicon could be the limiting nutrient for diatom proliferation in certain marine environments thereby significantly influencing the level of primary productivity in those systems. In some instances, when silicon becomes the limiting nutrient in an algal bloom, other much less benign species, especially toxin producing dinoflagellates, can become dominant. A quantitative understanding of the physiology of silicon use by diatoms and the role of silicon nutrient limitation on marine algal blooms is thus of practical environmental importance.

Laboratory and field experiments require a silicon tracer and analytical methods to study silicon uptake by diatoms. Earlier work used ^{29}Si and ^{30}Si and mass spectrometric techniques that required a significant amount of chemical pretreatment of the samples.[2] This made acquisition of data relatively time consuming and limited both the temporal and spatial resolution of oceanographic field measurements. Recently, with the availability of ^{32}Si, new radioanalytical techniques have been developed that significantly reduce the amount of analytical time required per sample and will allow acquisition of greater amounts of data.[3]

2 Production of ^{32}Si

Silicon-32 decays to ^{32}P by emission of a beta particle with a peak energy of 221 keV with a half-life of about 170 years.[4] The daughter ^{32}P decays to stable ^{32}S by emission of a second beta particle with a peak energy of 1709 keV. The half-life of ^{32}P is 14.3 days. The only effective route to production of ^{32}Si with reasonably high specific activity is by proton spallation. Polak, et al., produced a small amount of very low specific activity ^{32}Si by irradiation of vanadium metal targets.[5] In the early 1990s we determined that ^{32}Si was being made in high purity KCl targets being irradiated for spallation production of ^{26}Al, and we developed procedures to recover both of these important nuclides.[6] Our basic procedure is summarized in the following paragraph.

With 750,000 μA-hrs integrated exposure using 800 MeV protons, spallation reactions in a 100 g KCl target produce at end of irradiation about 1 μCi of ^{26}Al, 150 μCi of ^{32}Si, 500 mCi of ^{22}Na, 700 mCi ^{7}Be, 10 Ci each of ^{32}P and ^{33}P as well as traces of other nuclides. We generally allow 6 to 18 months of elapsed time prior to processing the target to allow decay of the shorter lived nuclides. We dissolve the target in water, and pass the resulting solution through a 0.45μ cellulose nitrate filter. The insoluble residues trapped by the filter contain ^{26}Al, the bulk of ^{7}Be and a fraction of the ^{32}Si. This filter is processed to recover the ^{26}Al. We process the filtrate to recover the Si-32 as follows. Using HCl we adjust the solution pH to between 1 and 2, then add a solution of ammonium paramolybdate with pH of 1.75 We then allow sufficient reaction time to form silicopolymolybdate anion $(Si^{IV}Mo_{12}O_{40})^{4-}$. The solution is transferred to a plastic column containing Sephadex (polydextran). The Sephadex quantitatively sorbs the silicopolymolybdate and passes the ^{22}Na. Silicon-32 and its ^{32}P daughter are stripped from the column using a dilute NaOH solution that decomposes the silicopolymolybdate releasing it from the Sephadex. We convert the strip solution to a solution of greater than 4.5 M HCl and pass this through an anion exchange column that retains the molybdate. We evaporate the column eluate containing the ^{32}Si to soft dryness, then reconstitute in dilute NaOH yielding a product solution containing soluble silicate having a specific activity of between 20,000 and 50,000 Bequerels ^{32}Si per microgram of silicon.

3 Comparison of Mass Spectrometric and Radioanalytical Methods

Before we comparatively discuss the mass spectrometric and radioanalytical techniques, it is useful to describe a typical oceanographic experiment. From the deck of a research vessel the investigator collects a water sample from the desired location and depth. The nutrient concentration is determined by measurement or by addition of a known amount of nutrient. Then a known amount of identifiable nutrient tracer is added to a fixed volume of the water sample. The sample is incubated for the requisite time in proper light conditions, then the sample is filtered to collect the phytoplankton. The amount of tracer taken up by the phytoplankton is measured and the data used to calculate the uptake rate.

The details of isotope dilution mass spectrometric measurement technique using stable ^{29}Si and ^{30}Si are provided in the paper by Nelson and Goering (reference 2). The radioanalytical methods using radioactive $^{32}Si/^{32}P$ are given in the paper by Brzezinski and Phillips (reference 3). The table below summarizes a comparison of the methods.

	^{29}Si, ^{30}Si	**$^{32}Si/^{32}P$**
Isotope Availability	Good	Limited
Dual Tracer	No	Yes
Isotope Cost	Low	High
Sample Processing Cost	Very High	Low
Equipment	Mass spec. and full chemistry laboratory	Gas Flow Proportional or Scintillation Counter
Analysis Time	Weeks to Months	Hours to Days
Sensitivity	Moderate	Excellent
Mass of silicon added with tracer	$\leq$ 15% of ambient	Trace (0.05 μCi adds nanomolar amount)
Shipboard Capability/Real Time Analysis	No	Yes

Table 1. Mass Spectrometry vs. Radioanalytical Method for Oceanographic Research

Though the initial isotope cost is higher for the radioanalytical methods, the amount of data that can be acquired per cruise is significantly higher and the cost of obtaining it is lower. Real-time data can be obtained with the radioanalytical method, whereas the mass spectrometric data are collected well after the cruise is completed. Since phosphorus is also an essential nutrient, if the ^{32}Si tracer is in secular equilibrium with its ^{32}P daughter at the time of addition, it is possible to gain data on both phosphorus and silicon uptake in a single incubation. The addition of ^{32}Si does not perturb the experiment in cases where ambient silica concentration is low since truly trace quantities of stable silicon are added to the incubation bottle. Probably the most serious difficulty with the ^{32}Si tracer methodology is the limited supply of the material in high specific activity. If methods can be developed to increase the supply, then both the cost and availability problems would be reduced.

4 Summary of Results from COOP 95.

In May of 1995 we field-tested our radioanalytical methods during the Coastal Ocean Processes (COOP 95) research cruise in Monterey Bay (reference 3).[7,8] During this cruise we produced the first shipboard Michaelis-Menten kinetic curves for silicon uptake using a gas flow proportional counter. After the cruise over 100 samples were analyzed by scintillation counting within three weeks of the end of the cruise. The same number of samples would have required months using the mass spectrometric methods. We completed the first systematic analysis of silicon nutrient limitation within an upwelling zone. The data from this cruise increased the measured average world-wide rate of biogenic silica production in coastal marine upwelling systems by 43% demonstrating the need for methods that will allow more rapid collection of data. In comparing the silicon data with carbon and nitrogen uptake data measured at two stations during the same cruise we observed that there may be at least a secondary control of silicon uptake by irradiance; this is the first such observation. We are currently developing a radioanalytical method that will permit the simultaneous measurement of carbon (^{14}C) and silicon (^{32}Si) uptake rates during a single incubation to further evaluate this observation in the laboratory.

---

[1] R. C. Dugdale and F. P. Wilkerson, Nutrient limitation of new production in the sea, in Primary productivity and biogeochemical cycles in the sea, ed. P. G. Falkowski and A. D. Woodhead (Plenum, 1992).

[2] D. M. Nelson and J. J. Goering, A stable isotope tracer method to measure silicic acid uptake by marine phytoplankton. *Anal. Biochem.* **78**, 139-147 (1977).

[3] M. A. Brzezinski and D. R. Phillips, Evaluation of ^{32}Si as a tracer for measuring silica production rates in marine waters. *Limnol. Oceanogr.* **42**(5), 856-865 (1997).

[4] D.E. Alburger, G. Harbottle, and E. F. Norton, Half-life of ^{32}Si. *Earth and Planetary Science Letters* **78**, 168-176 (1986).

[5] P. Polak, S. R. Garcia, W. A. Taylor, J. W. Barnes, L. Linder, H. A. O'Brien, Jr., and K. E. Thomas, A ^{32}Si Source from proton spallation of vanadium. *Radiochimica Acta* **38**, 73-77(1985).

[6] D. R. Phillips, V. T. Hamilton, D. J. Jamriska, and M. A. Brzezinski, Application of Sephadex to radiochemical separations. J. Radioanaly. Nucl. Chem., Articles, **195**(2), 251-261 (1995).

[7] M.A. Brzezinski, D. R. Phillips, F. P. Chavez, G. E. Friederich and R. C. Dugdale, Silica production in the Monterey, California, upwelling system. *Limnol. Oceanogr.* **42**(8), 1694-1705 (1997).

[8] R. M. Kudela, M. A. Brzezinski, D. R. Phillips, and F. P. Chavez, Uptake versus irradiance parameters for simultaneous measurements of C, N and Si during an upwelling event. Manuscript in preparation for submission to Limnology and Oceanography.

PERCUTANEOUS ABSORPTION OF DEUTERIUM LABELLED XENOESTROGENS BISPHENOL A AND NONYLPHENOL

MANFRED WAHREN

Institute of Organic Chemistry, University of Leipzig, Permoserstr. 15, D-04303 Leipzig, Germany E-mail: wahren@rz.uni-leipzig.de

MANFRED KIETZMANN

Institute of Pharmacology, Toxicology and Pharmacy, School of Veterinary Medicine, Bünteweg 17, D-30559 Hannover, Germany E-mail: mkietz@pharma.tiho-hannover.de

MONIKA MÖDER

Department of Analytical Chemistry, UFZ-Centre for Environmental Research Leipzig-Halle, Permoserstr.15, D-04318 Leipzig, Germany E-mail: moeder@ana.ufz.de

2,2-Bis-(4-hydroxyphenyl)-propane **1** (Bisphenol A) and a technical mixture of isomeric nonylphenols **2** are found in the environment, e.g. in surface waters. **1** is a main component of polycarbonates and epoxy resins and ethoxylated **2** was widely used in technical nonionic surfactants. **1** and **2** formed by degradation are rather persistent and a special hazard is suspected due to their estrogenic properties. We investigated the kinetics of the percutaneous absorption of these compounds. Isolated perfused udders from slaughtered cows were used as in-vitro models for human skin. As the reliability of trace analyses of possibly ubiquitous substances can be greatly improved by isotope labelling, the deuterated compounds were employed. **1-d_{14}** labelled at all carbon-hydrogen bonds was prepared by synthesis from phenol-d_6 with subsequent exchange of the phenolic OD-groups into OH. **2-d_2** labelled in the o-positions to the phenolic OH-group was prepared by an exchange procedure with the mixture **2**, which did not change its composition to a measureable extent. Udder perfusate and milk equivalent samples were taken immediately before administration of 500 mg **1-d_{14}** or **2-d_2** in ethanolic solution on an udder skin area of 200 cm^2 and then for 6 h. They were analyzed for the deuterated compounds and possible metabolites containing deuterium by GC-MS. Trace amounts of the unlabelled compounds **1** and **2** were present in all samples. A significant enrichment of the deuterated compounds in the milk equivalent produced during the experiments compared with the perfusate was noted. The results confirm the penetration of **1** and **2** into the capillary system of living mammals after skin contact.

1 Introduction

Isolated perfused bovine udders from slaughtered cows have been successfully tested as models for human skin in percutaneous penetration studies of pharmaceuticals [1]. The aim of this work was to obtain data on the significance of a possible incorporation of environmental estrogens [2] after skin contact, information which is needed for risk assessment. The selected compounds

Bisphenol A **1** and isomeric p-nonylphenols **2** are persistent degradation products of resins (**1**) or surfactants (**2**) with large scale technical use worldwide and have now become nearly ubiquitous e.g. in surface waters of industrial countries. Both **1** and **2** have at least weak estrogenic properties. Possible hazards for humans were suspected when **1** was detected in tinned foods (ref. [2], p.330) and **2** in fish, shrimp and mussels [3].

2 Experimental

1-d$_{14}$: HO-C_6D_4-$C(CD_3)_2$-C_6D_4OH was synthetized from phenol-d_6 and acetone-d_6 with 30% DCl in D_2O as catalyst. It was recrystallized 2x from toluene and then from water. It contained 93 ± 2 atom-% D at the carbon atoms only (NMR), m.p. 156^{0}C. **2-d$_2$**: p-HO-($C_6D_2H_2$)-isoC_9H_{19} was made from commercial **2** (SIGMA-ALDRICH) by 2 exchanges with 4 g of 20% DCl in D_2O (99.5 atom-% D) per g of **2** for 100 h at 105^0 C, re-exchange of the OD-group with methanol and final distillation at 155-165^{0}C / 3 x 10^2 Pa. It contained 95 ± 1 atom-% D in the positions 2 and 6 of the phenyl group.

The bovine udders were taken immediately after the slaughter of healthy cows and perfused with 80 ± 20 mL/min of gassed (95% O_2, 5% CO_2) and heparinized tyrode solution at 38.5^{0}C. More details are described in [1]. Three independent experiments, each with administration of 500 mg **1-d$_{14}$** or **2-d$_2$** dissolved in 10 ml of 70% ethanol on an udder skin area of 200 cm^2, were performed. Samples of perfusate, milk equivalent produced by the udders, and in the experiments with **2-d$_2$** also udder skin tissue from 0.02 -0.2 mm depth were taken just before (0 min) and then regularly until 360 min after application of **1-d$_{14}$** or **2-d$_2$**. They were kept below -15^{0}C.

Before any workup, known amounts of internal standards were added (**1-d$_{14}$** or **2-d$_2$** to the 0 min samples, **1** or **2** to all others). All analyses were done by GC-MS at 70 eV (**1**: HEWLETT-PACKARD 5890/II ; **2**: HP 6890 Plus /MSD 5973; GC-capillary HP-5, 30 m, ID 0.25 mm, 0.25 μm film thickness). From the experiments with **1-d$_{14}$** the perfusates were acidified with HCl, centrifuged, adjusted to pH 3, saturated with NaCl and extracted by Solid Phase Micro Extraction (SPME) with a polyacrylate coated fibre (85 μm film thickness, SUPELCO) for 60 min with stirring. The milk samples were freed from casein and fat and extracted with diethylether. The SPME-fibres were desorbed in the injection port of the GC-MSD at 280^{0}C for 2 min, the concentrated ether extracts injected in splitless mode and all measured with the GC temperature program 70^{0}C-7K/min-280^{0}C-20 min by SIM (selected ion monitoring) of m/z 240 and 222 for **1-d$_{14}$** and 228 and 213 for **1** (M^+ and M^+ -CD_3 or -CH_3). All samples from the experiments with **2-d$_2$**, which can be volatilized by steam, were purified and concentrated by steam distillation. From 70 mL sample volume with pH > 2 were distilled off 20 mL. 1 g K_2CO_3 was added to this first distillate and from it 5 mL distilled for SPME sampling. The GC temperature program was 50^{0}C-2min-8K/min-280^{0}C-20min. The ions with m/z 137 and 109 for determination of **2-d$_2$** and m/z 135 and 107 for **2** (fragment ions) were recorded in the SIM mode. In the samples taken at 0 min the following mean concentrations of environmental estrogens were found: 1.5 ± 0.5 ng **1** and 0.5 ± 0.2 ng **2** per g perfusate. All other results after due consideration of blind values, blanks and corrections for the deuterium content of **1-d$_{14}$** and **2-d$_2$** are shown in the following Table 1.

Table 1. 1-d$_{14}$ and **2-d$_2$** in udder perfusate, milk equivalent and udder skin samples (ng/g)

Sampling time (min)	**1-d$_{14}$** in perfusate	**1-d$_{14}$** in milk	**2-d$_2$** in perfusate	**2-d$_2$** in milk	**2-d$_2$** in skin
15	2.0		1.4		
30			1.7		
60	1.8	60	1.8		3.8×10^3
90			2.4		
105	6				
120			2.5	$0{,}19 \times 10^3$	
135	11				
150			2.2		
165	11				
180		3.6×10^3			5.6×10^3
195	5				
210			2.0	$0{,}22 \times 10^3$	
240	1.2		2.2		
300	1.5	1.9×10^3	2.4	$0{,}14 \times 10^3$	4.9×10^3
360	1.0				

3 Discussion

The method of reverse isotope dilution analysis used in this work (the labelled compounds are used in the experiments and the unlabelled compounds are added to the samples as internal standards) has the same advantages as the well known normal procedure with labelled compounds used as internal standards. Losses during sample processing are compensated and mass spectra are calibrated. However for analytes which are or might be present in the study objects already before the experiment, lower detection limits and enhanced reliability of results can be obtained. So the rather fast uptake of **1** and **2** from the skin surface into the skin tissue and then into the perfusate and the milk equivalent, which can be derived from the results presented in Table 1, would remain unambigous, even if the background of **1** and **2** from environmental sources had been larger.

In the perfusate as model for the blood circulation the maximum estrogen concentrations are found about 2 h after topical application. While the less lipophilic **1** seems to be washed out or metabolized soon, the concentration of **2** is not significantly changed as it is accumulated in the udder skin tissue. Both **1** and **2** are concentrated in the milk equivalent by a factor of $>10^2$, similar to the bioconcentration factor of 320 from water to mussels (ref. [3], p.3266). We are aware of the fact, that our experiments with high concentrations of **1** and **2** on a rather small area of skin with the aim of measuring dermal uptake of these substances are not simply transferable to real situations. If larger areas of skin are contaminated with much lower concentrations of **1** or **2** over longer periods of time, e.g. from dust or surface waters, the relative degree of resorption could be higher.

4 Conclusions

The data presented here are an unambiguous proof of dermal absorption of the xenoestrogens **1** and **2** with penetration into the capillary system of living mammals after skin contact. Their significant enrichment in the milk samples justifies preventive measurements of **1** and **2** in milk or milk products for children.

5 Acknowledgements

The authors thank Dr. Bettina Blume, Dr. Petra Kränke and Dipl.-Ing. (FH) Steffi Schrader for their cooperation. Financial support for M. Wahren from the Sächsisches Staatsministerium für Wissenschaft und Kunst (project 7533-70-0361/745) is gratefully acknowledged.

6 References

1. Kietzmann, M., Löscher, W., Arens, D., Maaß, P. and Lubach, D., The isolated perfused bovine udder as an in vitro model of percutaneous drug absorption. Skin viability and percutaneous absorption of dexamethasone, benzoyl peroxide, and etofenamate. *Journal of Pharmacological and Toxicological Methods* **30** (1993) pp. 75-84.
2. Soto, A. M. and Sonnenschein, C., Environmental sex hormone mimics and antagonists. *Comments Toxicology* **5** (1996) pp. 329-346.
3. Thiele, B., Günther, K. and Schwuger, M. J., Alkylphenol ethoxylates: Trace analysis and environmental behavior. *Chem. Rev.* **97** (1997) pp. 3247-3272.

Session G

Applications of Isotopes in Medical Imaging & Therapy

MEDICAL IMAGING APPLICATIONS IN PRE-CLINICAL AND CLINICAL MEDICINE

LEONARD I. WIEBE

Noujaim Institute for Pharmaceutical Oncology Research, Faculty of Pharmacy and Pharmaceutical Sciences, University of Alberta, Edmonton, Canada T6G 2N8

e-mail:leonard.wiebe@ualberta.ca

Radiopharmaceuticals are used as either imaging or therapeutic agents in clinical medicine. Radiopharmaceuticals for imaging serve not only to diagnose disease, but also to develop therapeutic drugs and drug delivery systems. Early radiopharmaceutical design was based primarily on physiology, and on the pharmacological properties of therapeutic agents. Unfortunately, chemotherapeutic drugs are usually used at high doses, whereas radiopharmaceuticals only infrequently display selective, interpretable biodistributions at these doses. Radiopharmaceuticals designed on biochemical rationale, on the other hand, give rise to superior biodistribution (selective localization in target tissue) at tracer doses of the radioactive compound. The selective, intracellular, enzymatic conversion of highly diffusible substrates into highly reactive or non-diffusible products is among the most efficacious means of achieving high relative accumulations of both diagnostic and therapeutic radiopharmaceuticals in target tissues. The molecular biology-biotechnology revolution has played an important role the development of enzyme-oriented radiopharmaceuticals, through the reduced toxicity of 'biologicals', because of high molecular (target) specificity and due to the ability deliver the drug selectively to the target site.

1 Introduction

Radiopharmaceuticals for *in vivo* scintigraphic imaging were initially used to obtain anatomical images based on physiological function. Early radiopharmaceuticals produced images attributable to vascular trapping (*e.g.* micro- and macro-aggregates) and reticuloendothelial uptake (*e.g.* colloids), while others reflected physiological functions such as renal clearance (*e.g.* iodo-aromatic acids and hydrophilic ^{99m}Tc-chelates) and hepatic clearance (*e.g.* lipophilic ^{99m}Tc-chelates). There was also early interest in using radiotracers to image metabolic processes, such as the use of 5-iododeoxyuridine (IUdR), a thymidine analogue, to delineate cell proliferation [1]. Modern imaging technologies, including x-ray CT and magnetic resonance imaging (MRI), have superior spatial resolution compared to radioscintigraphy (planar and tomographic imaging) using single photon and positron emitting radionuclides.

Approaches to imaging molecular targets in membranes, and in extracellular, cytoplasmic and nucleoplasmic compartments, continue to exploit the sophisticated techniques developed in the basic sciences of molecular biology and genetic engineering.
A number of fundamental barriers challenge the intracellular delivery of targeted, macro-molecular radiopharmaceuticals. Innovative *in vivo* delivery of 'biological' radiopharmaceuticals, whether they are monoclonal antibodies (MABs), MAB fragments, peptides or oligonucleotides remains a major challenge for scientists in this field. The *in vivo* instability of these substances, their exaggerated clearance kinetics and their difficulty in permeating biological membranes are among their drawbacks. Although today's macromolecular radiopharmaceuticals are unlikely to disappear, there is a revival of small-molecule radiopharmaceuticals for diagnostic imaging.

2 Small-molecule radiopharmaceuticals to target enzymes expressed by transferred genes (gene therapy)

Gene therapy represents the ultimate approach to the intracellular delivery of macromolecules, bypassing the normal barriers by genetically coding for the intracellular production of the desired substance within the target cell. Although this does not solve the problems of delivering a macromolecular radiopharmaceutical, it does pave the way for creation of unique molecular targets (*e.g.* enzymes) for radiopharmaceuticals.
Gene transfer therapy, directed *in situ* at the root-cause of disease within the host genome, has already been applied to a number of pathologies, with the first clinical trial in 1990. The Recombinant DNA Advisory Committee of the US FDA has subsequently approved more than twenty gene therapy protocols that support upwards of 200 clinical trials world-wide. The current consensus is that the process needs substantial improvement, that there are risks, and that there have not been any unequivocal cures [2].
International efforts to map the human genome have greatly contributed to the identification of the genetic basis of many diseases. None-the-less, there are major technical impediments to successful implementation of gene transfer, challenges similar to those inherent in the delivery of classical intracellularly-targeted drugs. Site-specific delivery, and control of gene catabolism, gene expression and gene deletion are major challenges that remain to be resolved.
Replacement gene therapy is particularly attractive for treating congenital disorders arising from absence or deficiency in expression of a gene, whereas introduction of new genes is a common objective in gene therapy of cancer. Gene transfer can be effected *in*

vitro or *in vivo*. For *in vitro* transfer, the therapeutic gene is inserted into target cells (e.g. cord blood stem cells for a blood cell disease) isolated from the patient; after *in vitro* replication these homologous transformed or transfected cells are injected back into the donor patient at the appropriate tissue site. For *in vivo* delivery, the gene-carrying vector is injected (plasmid or engineered virus) or implanted (producer cell) into the target tissue *in situ*.

2.1 Imaging in gene therapy

Imaging is the only contemporary clinically acceptable method to establish *site-specific delivery* and *site-specific expression* of a therapeutic gene *in vivo*. These criteria are important, for if the transferred gene is not expressed, there can be no therapeutic benefit, and expression at the wrong (non-target) site could introduce severe complications for the patient. Whole-body imaging ensures that all sites are accessed simultaneously, subject only to signal intensity. The major role of imaging is to monitor the gene therapy protocol, to alert the therapist of impending risk, to predict therapeutic efficacy and to indicate when no further improvement is to be expected.

Two imaging modalities are currently of interest for gene therapy imaging. Beta-galactosidase has been investigated [5] as an enzyme target marker for Magnetic Resonance Imaging (MRI), and other MRI target enzymes such as tyrosinase are certain to emerge [6]. Nuclear medicine approaches, using planar, SPET and PET imaging, are currently the most clinically feasible technologies. Molecular targets for gene therapy imaging include therapeutic enzymes, receptors and nucleic acids encoded by the gene.

The virus encoded HSV-1 tk^+ enzyme was first selected as an enzyme target for imaging herpesvirus infections, using radiolabelled nucleoside analogues [5-7]. Virus-encoded selective phosphorylation of the tracer nucleosides in HSV-1-infected tissue has been adapted as the underlying principle for HSV-1 tk^+ gene therapy imaging. The radiopharmaceuticals (enzyme substrates) used to detect the gene product (enzyme) are small, chemically-defined molecules that are highly diffusible and readily prepared, making them ideal for practical application in the clinic. 1-(2-Fluoro-2-deoxyribofuranosyl)-5-[^{123}I]iodouracil (FIRU; Fig. 1) is one of the radiopharmaceuticals currently being developed for use in a clinical glioma tk^+ gene therapy study [8]. A second gene therapy enzyme target proposed was fungal cytosine deaminase, using ^{18}F-5-FCT as the reporting radiopharmaceutical [9]. This system is compromised by substrate catabolism and is therefore not expected to be viable with currently available radiopharmaceutical substrates. A bacterial reductase enzyme [10] for use in combination with the hypoxia-imaging agent ^{123}I-IAZA has been recently proposed [11].

3 Small-molecule radiopharmaceuticals to target enzymes expressed under specific pathologies (hypoxia)

Pathological disease usually predicates specific biochemical perturbation. Pathologies that are characterized by local hypoxia are a classical example. Many drugs are reductively bioactivated under hypoxic conditions, and consequently undergo binding that results in their hypoxia-selective accumulation [12]. This 'adduct' formation by nitroimidazole radiopharmaceuticals, for example, is the basis for radiotracer accumulation in target tissues; selectivity of accumulation is attributable to oxygen-sensitive, reversible reduction in viable but oxygen deficient cells. Enzymes that effect reduction (reductases), including flavin-dependent xanthine and aldehyde oxidases, and quinone oxidase and cytochrome P-450 reductase, are active in hypoxic tissue (e.g. rapidly proliferating tumours, cerebral ischaemia, myocardial ischaemia).

The radioiodinated 2-nitroimidazoles (azomycin derivatives) comprise the main body of literature that deals with agents for scintigraphic detection of tissue hypoxia. Their electron reduction potentials lie within the range for O_2-reversible reductions to occur, so the main challenge in molecular design is to adjust metabolic (non-reductive) and pharmacokinetic properties. Misonidazole remains the reference compound against which the sensitizing properties of radiosensitizers are compared. The radiohalogenated azomycin nucleosides were selected for on the basis of their relatively low lipophilicity, with the benefit of several side-chain hydroxyl groups to counteract the lipophilic effects of iodine. Of these, [^{123}I]IAZA (Fig. 1) is among the most widely evaluated radiopharmaceuticals for delineation of tissue hypoxia [13-19].

[^{125}I]IAZA undergoes hypoxia-dependent adduct formation, with minimal deiodination and metabolic degradation *in vivo* in the EMT-6 murine tumour model [20]. [^{123}I]IAZA has been used to image a variety of experimental and clinical pathologies. In photodynamic therapy (PDT) of a rat Dunning prostate tumour model, [^{123}I]IAZA localized with inverse correlation to ^{99m}Tc-HMPAO uptake, in regions of PDT-induced hypoxia [21]. A dual radionuclide autoradiographic and MRI study of cerebral occlusion in rats depicted uptake of [^{125}I]IAZA in ischaemic areas and mutually exclusive uptake of ^{99m}Tc-HMPAO in well-perfused regions [22].High diffusibility into poorly-vascularized (ischemic) tissues, high reductive binding rate, moderately rapid clearance from blood, rapid total body clearance and minimal loss of radiolabel have been used to rationalize the utilization of [^{123}I]IAZA for clinical hypoxia imaging. ^{123}IAZA-based investigations of regional hypoxia in cancer [13-15], diabetes [16], arthritis [17], brain trauma [18] and exercise-stress [19] pathologies have been reported.

The initial application of hypoxia imaging agents was to predict radioresistance in (hypoxic) tumours. There is current interest in applying this diagnostic procedure not only to detect hypoxic tissues for radiotherapy planning, but to differentiate salvageable hypoxic tissue from dead tissue in myocardial ischaemia and stroke. This is important for prognosis and for monitoring therapeutic intervention. Demonstrations of the widespread influence of hypoxia on metabolic processes such as drug resistance continue to create interest in hypoxia imaging to address basic scientific questions.

4 Small-molecule radiopharmaceuticals to target tissue-specific enzymes

Malignant neural crest tumors and malignant melanomas are aggressive once they develop metastases, and current therapies have low probabilities of success. Radioiodinated MIBG has been found to be selectively stored in neuro-secretory granules, and has been used in both imaging and radiotherapeutic management of pheochromocytoma and neuroblastoma [23,24]. Tumors of neural crest origin and malignant melanomas share at least two common features. Biochemically they express a metabolic pathway that leads from the oxidation of tyrosine to dopa to either catecholamine or melanin, and biologically they produce secretory granules, catecholamine vesicles and melanosomes. At high concentrations of tyrosine, cell pigmentation is enhanced through these processes, and at sufficiently high concentrations, tyrosine exerts selective toxicity towards melanin producing cells. Since tyrosine is a natural substrate, however, it is virtually impossible to attain these toxic levels *in vivo* [25].
A series of cysteaminyl phenols (Fig. 1) have been developed to act as potential antimelanoma agents *via* interaction with tyrosinase, the branching enzyme that directs the oxidation of dopa to melanin precursors. These lipophilic sulphur homologues of tyrosine [26,27] are competitive substrates of tyrosine for tyrosinase. They undergo selective accumulation *in vitro* in melanoma cells as compared to non-pigmented cell lines, and they cause depigmentation in new growing hair follicles in C57BL black mice *in vivo*, without apparent acute toxicity. The development of tyrosinase-specific radiopharmaceuticals that act as either substrates or as strong/irreversible binders of tyrosinase is an important objective in the development of antimelanoma diagnostics and therapeutic drugs.

5 Summary

Contemporary nuclear medicine focuses on molecular targets, using both biological and chemical radiopharmaceuticals to obtain functional images. Gene therapy is a new treatment modality that can initiate or enhance the expression of unique molecular targets. Imaging these gene therapy targets will provide essential prognostic information critical to the assessment of both efficacy and untoward effects. At the same time, there are tissue-specific and pathology-specific molecular targets, particularly enzymes, which have the potential to serve as markers of disease that can be detected and exploited by nuclear medicine imaging.

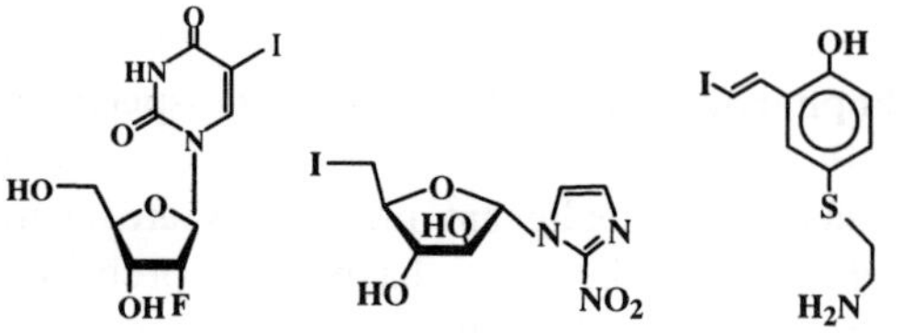

Figure 1. Enzyme targeted radiopharmaceuticals for enzyme-based imaging in gene therapy (FIRU), hypoxia (IAZA) and neural crest tumors (IV-SCAP).

6 References

1. Silvester D. J. and White N. D. Nature 200 (1963) pp. 65-66.
2. Friedman T. Nature Medicine 2 (1996) pp. 144-147.
3. Moats R. A., Fraser S. E. and Meade T. J. Ange-wandte Chemie Int. Ed. 36 (1997) pp. 726-731.
4. Jimbow K. personal communication.
5. Saito Y., Price R. W., Rottenberg D. A. et al. Science 217 (1982) pp. 1151-1153.
6. Saito Y., Rubenstein R., Price R. W. et al. Ann. Neurol. 15 (1984) pp. 548-559.
7. Tovell D. R., Samuel J., Mercer J. R. et al. Drug Des. Del. 3 (1988) pp. 213-221.
8. Haberkorn U., Oberdorfer F., Gerbert J., et al. J Nucl Med. 37 (1996) pp. 87-94.
9. Cox P. H. Personal communication.
10. Bridgewater J. A., Springer C. J., Knox R. J. et al. Eur. J. Cancer 31A (1996) pp. 2362-2370.
11. Wiebe L. I., Knaus E. E., Morin K. W. and McEwan A. J. U.S. Provisional Patent (1997).
12. Chapman J. D., Franko A.J. and Sharplin J. Brit. J. Cancer 43 (1981) pp. 546-550.
13. Urtasun R. C., Parliament M. B., McEwan A. J. et al. Br. J. Cancer 74 (1996) pp. S209-S212.
14. Groshar D., McEwan A. J. B., Parliament M. B. et al. J. Nucl. Med. 34 (1993) pp. 885-888.

15. Parliament M. B., Chapman J. D., Urtasun, et al. Br. J. Cancer 74 (1992) pp. 90-95.
16. Al-Arafaj A., Ryan E. A., Hutchinson K., et al. Eur. J. Nucl. Med. 21 (1994) pp. 1338-1342.
17. Skeith K. J., McEwan A. J. B., Schmidt R. and Wiebe L. I. unpublished.
18. Vinjamuri S., O'Driscol K., Maltby P., et al. Clin. Nucl. Med. (1999) in press.
19. Cwik V. A., McEwan A. J. B. and Brooke, MH. Neurology 45 (Suppl 4) (1995) pp. A446.
20. Mannan R. H., Somayaji V. V., Lee J. et al. J. Nucl. Med. 32 (1991) pp. 1764-1770.
21. Moore R. B., Chapman J. D., Mercer J. R. et al. J. Nucl. Med. 34 (1993) pp. 405-413.
22. Lythgoe M. F., Williams S. R., Busza A. L. et al. Mag. Res. Med. 41 (1999) pp. 706-714.
23. Wieland D. M., Swanson D. P. and Brown L. E. J. Nucl. Med. 20 (1979) pp. 155-158.
24. Mozely P. D., Kim C. K., Mohsin J. et al. J. Nucl. Med. 35 (1994) pp. 1138-1144.
25. Wick M. M. and FitzGerald G. B. Pigment Cell 10 (1990) pp. 214-227.
26. Miura S., Ueda T., Jimbow K. et al. Arch. Dermatol. Res. 279 (1987) pp. 219-225.
27. Iwashina T., Jimbow K. and Wiebe L. I. Appl. Radiat. Isot. 45 1994) pp. 703-705.
28. Singh S., Jimbow K., Kumar P. et al. J. Label. Comp. Radiopharmaceuticals XLI (1998) pp. 355-361.

Session H1

Radiopharmaceutical Applications and Medical Imaging — II

COMPARISON OF VARIOUS RHENIUM-188 LABELED DIPHOSPHONATES FOR THE PALLIATION OF METASTATIC BONE PAIN

SHYH-JEN WANG AND WAN-YU LIN

Department of Nuclear Medicine, Taichung Veterans General Hospital, Taichung 40705, Taiwan
E-mail: sjwang@vghtc.vghtc.gov.tw

BOR-TSUNG HSIEH AND GANN TING

Institute of Nuclear Energy Research, Lung-Tan, Taiwan.

F.F (RUSS) KNAPP JR

Oak Ridge National Laboratory (ORNL), Oak Ridge, U.S.A.

Many diphosphonates has been used as bone scan agents. In addition, diphosphonates have been labeled with beta-emitted isotopes and developed into useful therapeutic drugs for palliation of metastatic bone pain. However, it is not clear which diphosphonate is the best choice when labeling with Re-188. In this study, we labeled methylene diphosphonate (MDP), hydroxyethylidene diphosphonate (HEDP), and hydroxymethane diphosphonate (HDP) with Re-188 and analyzed the bone uptakes after I.V. injection in rabbits to assess their potential for clinical applicability.

After I.V. injection 37 MBq Re-188 labeled diphosphonate preparations, whole body scans were performed using a large-field gamma camera equipped with a high resolution collimator. Bone to soft tissue ratios (B/S ratio) were calculated using a computer program. Our data showed that Re-188 HEDP with carrier (10^{-4} M carrier) could accumulate in the skeletal system while very little absorption by bone was observed in the rabbits which were injected with carrier free Re-188 HEDP. In addition, no significant bone uptake was demonstrated for Re-188 MDP or Re-188 HDP, with or without carrier. The B/S ratio was 25.06 in the Re-188 HEDP with carrier group but less than 3 in the other groups.

In conclusion, our preliminary results indicate that HEDP is the best choice among these three bone seeking agents when labeled with Re-188. But, it is necessary to add carrier when preparing Re-188 HEDP for the palliation of metastatic bone pain.

1 Introduction

Systemic administration of radiation therapy with radiopharmaceuticals plays an important role in pain control in patients with malignant bone metastases. In the past 40 years, many bone-seeking radiopharmaceuticals, including P-32, Sr-89, Sm-153 EDTMP, and Re-186 HEDP, have been used in attempts to treat patients with widespread skeletal metastases. Recently, Re-188 was introduced as a very good potential candidate for radiotherapy because it is available from an in-house generator system similar to the current Tc-99m generator, which makes it very convenient for clinical use. In addition, it emits both beta and gamma rays and has a short physical half-life of 16.9 hours. The short physical half-life reduces the problems of radioactive waste handling and storage .

Since 1975, methylene diphosphonate (MDP) has become the most common agent for imaging the skeleton. MDP cold kit is cheaper than HDP cold kit. Following an experimental comparison of bone agents in 1976, several new diphosphonates were introduced . Re-188 HEDP has been reported to be useful in the palliative treatment of bone metastases . However, it is of interest to compare diphosphonates to determine which is the best choice for labeling with Re-188. In this study we labeled MDP, HEDP, and hydroxymethane diphosphonate (HDP) with Re-188 and analyzed the biodistributions and bone uptakes following intravenous injection in rabbits to find the best diphosphonate for labeling with Re-188.

2 Methods

2.1 Preparation of Re-188 perrhenate solution

Re-188 was obtained from an alumina-based W-188/Re-188 generator .

2.2 Preparation of Re-188 HEDP

A kit formulation was designed to contain exactly 15 mg of Na_2HEDP, 4.5 mg of $SnCl_2.2H_2O$ and 4.0 mg of gentisic acid in a 15ml vial. After being dissolved in an appropriate amount of distilled water, the vials were subject to complete freeze drying. To make Re-188 HEDP, 1.0 ml of NH_4ReO_4 solution (which contained 0.01-0.1 mg of NH_4ReO_4) was added to the kit, followed by 1.0 ml of 1.5 - 2.5 mCi Re-188ReO4-. The whole mixture was heated for 15 min in boiling water, and allowed to cool by running water to room temperature. Another 1 ml of 0.3 M sodium acetate trihydrate solution was added to adjust the pH to a range of 5-6 then filtrated with 0.22u millipore.

Preparation of Re-188 MDP

A lyophilized formulation was designed to contain exactly 10 mg of MDP, 3 mg of $SnCl_2 \bullet 2H_2O$ and 4 mg of gentistic acid with freeze drying as HEDP kit preparation. To make Re-188 MDP, 0.1 mg of potassium perrhenate which dissolved in 0.1 mL of distilled water and 1.0~2.0 mL of Re-188 perrhenate was added to the vial. The reaction mixture was heated for 30 min in boiling water, then allowed to cool to room temperature and stood at least 30 min before using the labeled product. For pH adjustment to neutral, 1 mL of 0.3 M sodiumacetate trihydrate was added and then filtrated with 0.22u millipore.

2.3 Preparation of Re-188 HDP

Lyophilized HDP kit was purchased from Mallinckrodt Medical, Inc., and additional 4 mg of $SnCl_2 \bullet 2H_2O$ was obtained from Sigma, minimum purity of 95%. The procedure of preparation of Re-188 HDP was similar to the preparation of Re-188 MDP.

2.4 Scintigraphy

Rabbits were used to evaluate the clinical potential of three Re-188 radiopharmaceuticals. Each radiopharmaceutical was further evaluated in two conditions, with and without carrier (carrier concentration = 10^{-4} M). Each rabbit was injected with 1 mCi Re-188 diphosphonate compounds via tail vein. Four hours after the injection, the animals were imaged using gamma camera. The images were used for the determination of bone to soft tissue uptake ratio (B/S ratio).

2.5 Bone to soft tissue uptake ratio (B/S ratio)

For determination of bone to soft tissue ration, regions of interest (ROI) were drawn over the L-spine and adjacent soft tissue. The ratio of the mean counts in the ROIs was taken as bone to soft tissue uptake ratio (B/S ratio).

3 Results

The B/S ratios are shown in Table 1. The B/S ratio was 25.06 in the Re-188 HEDP with carrier group but was less than 3 in the other groups. Our study revealed that HEDP is better than MDP and HDP as a bone seeking drug when labeled with Re-188. However, it is necessary to add carrier when preparing Re-188 HEDP for the treatment of bone metastases.

Table 1. Bone to soft tissue ratios in rabbits at 4 hours after infection of various Re-188 diphosphonate preparations

Re-188 diphosphonate preparations	Bone/Soft tissue	
	With Carrier	Without Carrier
Re-188 MDP	1.47	1.89
Re-188 HEDP	0.89	25.06
Re-188 HDP	2.42	1.15

ISOTOPE TRACER TECHNIQUES IN CELL CULTURES FOR THE STUDY OF PLANT DRUGS IN RELATION TO RECEPTORS AND POST-RECEPTOR SIGNAL TRANSDUCTION *

HU YAER, FAN GUOHUANG, HE KAI,FANG CAILONG, XIA ZONG

Isotope Laboratory, Shanghai Sec Med Univ, Shanghai 200025, China

Primary cultures of neonatal rat neuronal cells and cardiac cells have been used for the study of actions of plant drug components on certain cell regulation mechanisms. It was found that during the aging of cultured cells the M receptor density, the inhibitory effect of carbachol on cAMP, and the coupling of G protein to M receptors were all decreased. Among the sapogenins studied, one was found to have the effect of reversing the above defects. The M receptor density was increased by about 20% and the drop of cAMP after carbachol stimulation was also increased by about 20% in drug treatment than control. The coupling of G protein to M receptors was also improved by the drug as shown by the increased carbachol induced incorporation of ^{35}S-GTPγS into G protein.

1 Introduction

Cultured cell lines or primary culture of cells are commonly used for screening of agonists and antagonists of receptors [1]. However, the cultured cells have important uses in other studies of receptors in addition to screening of agonists and antagonists. In previous works, we found the decreased M-receptor density in aged animals were elevated by certain compounds from plant drugs, but these compounds were not agonist and antagonist[2]. Since it was somewhat difficult to distinguish direct or indirect actions of these compounds in vivo experiments, culture cells were used for the mechanistic studies of the plant drug components. Here is a brief report of these studies.

2 Materials and methods

Cell Culture: The cells from neonatal rat brain or heart were cultured in MEM medium[3,4]. The morphological changes were observed with phase-contrast microscope. The M receptor densities of brain or cardiac cells was measured with ^{3}H-QNB by the method of Yamamura[5] with slight modifications and was expressed as fmol/mg protein, the protein was measured by Lowry's method. The coupling of G-protein to M receptor was estimated by incorporation of ^{35}S-GTPγS [6]. The level of cyclic AMP was measured by RIA.

---

* The authors gratefully acknowledges the support of K.C. Wong Education Foundation, Hong Kong

Drugs: The compound was added to the cultured medium after several days' incubation and samples were prepared and measured after 48 hours' action of the compound. The compound used was a sapogenin from a plant drug commonly used in traditional Chinese medicine. DMSO was used as the vehicle of this fat-soluble sapogenin and was at a final concentration of 0.5% in all groups.

3 Results

(1)Time course of changes in morphology and M-receptor density: In the early stage of primary culture of rat brain, the cell bodies enlarged and the neurites extended. By the end of the second week, the cell bodies began to shrink, and some processes began to rupture. After then these changes became more and more marked and by the end of about 4 weeks. The morphological change of cardiac cells was not as characteristic as neuronal culture, but the beat of the cells became weaker and cell bodies shrank by the end of about 2 weeks. The M receptor density increased gradually in the first few days, reaching a peak at about 10-12 days for both neuronal and cardiac cells, and then began to decline slowly. Therefore, the decline of M-receptor density was slightly earlier than the observable morpholo-gical changes. In order to examine whether the drug had an effect of delaying the decline of M-receptor density (i.e., an effect of increasing the density when it was decreased), addition of drugs was started on the 12 day in most experiment, except where indicated.

(2)Action of the sapogenin on M receptor densities in neuronal and cardiac cells: It was found that the M receptor density was dropped markedly in 12 days and was elevated by sapogenin. The analytical results are showed in Table 1.

Table1. Action of the sapogenin on M receptor densities in cells

	Neurons (fmol/mg prot.)	Cardiac cells (fmol/mg prot)
Cells cultured for 12 days with DMSO	217± 49 (6)	86± 23 (7)
Cells cultured for 12 days with sapogenin	315± 91 (6) *	119± 21 (7) *

Data were expressed as mean±SD (n); * $p<0.05$

(3)Effect of the sapogenin on the carbachol-stimulated binding of ^{35}S-GTPγS: The effect of sapogenin on the carbachol-stimulated binding of aged cells is shown in Table 2. The sapogenin had significant elevation effect on stimulated binding.

Table 2. Effect of the sapogenin on bindingof ^{35}S-GTPγS in neuronal cells

	Carbachol-stimulated binding (fmol/mg prot)	Carbachol-stimulated binding (% of basal binding)
Cells cultured for 12 days with DMSO	139± 38 (8)	21.6± 7.9 (8)
Cells cultured for 12 days with sapogenin	213± 59 (8) *	38.2± 14.3 (8) *

(4)Effect of the sapogenin on the muscarine induced inhibition of cyclic AMP production in cultured cardiac cells: When Forskolin was added to the cultured cells it stimulated the AMP cyclase and hence raised the cyclic AMP to certain amount. When muscarine was added simultaneously, the elevation of cyclic AMP was lowered down because stimulation of M2 receptors induced an inhibition effect on the cyclase via Gi protein. It was found that in normal cultured cardiac cells (i.e., harvested after 9 days' culture), the effect of muscarine was significant, but in aged cultured cardiac cells (i.e., harvested after 14 days' cullture), the action of muscarine was very weak. The sapogenin showed the effect of raising the responsiveness of M receptors to muscarine. The data were shown in Table 3.

Table 3. Effect of the sapogenin on the muscarine inhibition of cyclic AMP level

	cAMP (pmol/10^6cells) Forskolin	cAMP(pmol/10^6cells) Forskolin+ Muscarine
Cells harvested on 9 day	15.6 ± 2.6 (4)	9.8 ± 3.3 (4) *
Cells harvested on 14 day with DMSO	13.7 ± 4.9 (4)	11.6 ± 2.5 (4)
Cells harvested on 14 day with sapogenin in DMSO	14.9 ± 2.0 (4)	10.1 ± 1.6 (4) *

4 Discussion

The above results indicate that the cultured cells, if suitably selected, are useful tools for study of plant drugs which have effects in regulating the receptor numbers and/or the post-receptor signal transdution other than receptor agonists or antagonists. The advantage of using cultured cells is that many indirect action can be abolished and hence are useful to give information about the direct action of the drugs. The key point is to choose the suitable cells and suitable culture time. In this work, we were studying a drug which may be benefit in anti-aging, the culture time were thus selected when the cells begins to show some phenomena of aging. It should be mentioned that the morphological change of aging happened somewhat later than biochemical changes, the choose of cultured time based on the latter seemed to be more sensitive for the study of drug actions.

References

1. Pei XF, Gupa TH, Badio B et al. J. Med. Chem. 1998;41:2047-2055.
2. Yi NY, Hu YE and Xia ZQ. Pharma. and Clinics of Chinese Materia
3. Medica. Vil4, Beijing:Military Medicine Publisher. pp.180-192.
4. Dichter MA. Brain Res 1978; 149:279
5. Galper JB, et al. Proc. Natl. Acad. Sci. USA. 1978; 755:5831-9.
6. Yamamura HI and Snyder SH. Proc. Natl. Acad. Sci. USA. 1971; 1725.
7. Wang HY, et al. Neurosci. Lett. 1994; 173: 37-39.

Session H2

Production & Applications of Isotope Tracers in Industry — II

POLONIUM-210 ALPHA-PARTICLE SOURCES FOR SOLVING OF ION BEAM TASKS

A.Z. ILIASSOV
RITVERC GmbH, 28, 2-nd Murinski Ave., St-Petersburg 194021, Russia
E-mail: ritverc@atom.nw.ru

S.L. FADDEJEV
V.G. Khlopin Radium Institute, 28, 2-nd Murinski Ave., St-Petersburg 194021, Russia
E-mail: fedvv@atom.nw.ru

Ion beam technologies are widely used for production of modern microelectronics devices. For solving new arising technologic problems alpha particles accelerated up to energy of 2÷3 MeV are used. Such accelerators can also be used for various analyses performing, viz. by Rutherford backscattering, PIXE and nuclear reaction methods.
Visible disadvantages of accelerators (high cost, need in special building and in highly qualified stuff) can be overcome with special set-ups using alpha-particle sources instead of an accelerator designed in collaboration of Russia and Uzbekistan.
One of them is a plant for irradiation of semiconductor devices. In this plant FAKT-1 alpha-particle sources with ^{210}Po are used. The uniqueness of this source is in its spectroscopic features under activity of 10^9 Bk. and negligible gamma-ray background (10^{-5} gammas per decay). The temperature of irradiation may be increased up to 350^0C. The plant is fully automated. Overall dimensions of this plant are 0.9x0.9x2.0 m^3. and weight is 150 kg.
Another one is a set-up for Rutherford backscattering (RBS set-up). Our RBS set-up is equipped with FAKT-1 alpha-particle source, too. Special non-traditional geometry without any collimation of alpha-particles stream was suggested. Application of a radioactive source allows decrease the cost of the set-up sharply while all another parameters would changed till slightly. The set-up is mounted on the special frame 1m^2 in area and operated by one person.

1 Introduction

Alpha-particles are a powerful tool for modification and investigation of solids: semiconductors, metals and polymers. They are widely and effectively used in the production of integrated circuits and semiconductor diodes, triodes etc. Their application permits to shorten the number of technology operations, obtain unique properties of semiconductor devices unachievable with traditional technologic procedures and increase their firmness to radiation. Alpha-irradiation of metals increases their mechanical properties, such as fatigue, corrosion resistance etc. Ion irradiation is widely used for radiation modification of polymers, too.

From other hand high-energy alpha-particles are successfully used for analyses of composition and concentration of admixtures and determination of deposited layers thickness. These analytical methods are based on the phenomena of Rutherford backscattering (RBS), particle induced X-rays emission (PIXE) and nuclear reactions (NR) [1].
Generally ion accelerators for these purposes are used. Visible disadvantages of accelerators application are high cost, need in special building and in highly qualified stuff. As alternative decision application of alpha-particle sources can be offered. But their intensity is too low. The situation became more optimistic after a FAKT-1 alpha-particle source with ^{210}Po nuclide was designed [2].

In this paper, we describe features of this source and give several applications of such sources for irradiation and analysis.

2 FAKT-1 alpha-particle source

A FAKT-1 alpha-particle source was designed in the V.G. Khlopin Radium Institute [2]. High intensity of this source is achieved by using of ^{210}Po isotope with half-life of 138.38 days. The attractiveness of this nuclide is in its extremely low (about 10^{-5} gammas per decay) gamma-background. All these features show the Polonium-210 radionuclide is very convenient for mentioned tasks solving.

Polonium-210 was used for alpha-particle sources production with thin foils multilayer system protecting active layer [3]. But specific activity of such sources were too low and less than $5 \cdot 10^6$ Bk·cm^{-2}.

In the Radium Institute this demerit was overcame by using low absorbing alpha-particles oxide layer filled with Polonium-210.

Our FAKT-1 alpha-particle source has following specifications:

- specific activity $< 4 \cdot 10^9$ Bk·cm^{-2} ;
- the full width at the half maximum of ^{210}Po line 150 keV;
- alpha-particles energy 4.9 MeV;
- active spot diameter 7 mm;
- overall diameter 12 mm;
- height 5 mm;

Some applications of this unique source are described below.

3 The plant for irradiation of semiconductor devices with alpha-particles

A plant for alpha-particles irradiation of semiconductor devices using FAKT-1 sources instead of an accelerator was designed in collaboration of Russia and Uzbekistan.

Two plates with about 60 tightly packed sources in each are used for irradiation. To get homogeneous field of irradiation a special gear impelling these plates was designed ensuring homogeneity of irradiation within ±5%. Diameters of irradiated chips may be varied from 40 to 100 mm. Two cassettes loaded with 200 chips are mounted on the rotating carrousel. Two plates with chips extracted from cassettes are transferred with special device into irradiation zone and after irradiation and radiation control are put into another two cassettes with irradiated chips. The temperature of irradiation may be varied from 20 to 350^0 C. The plant is fully automated. Overall dimensions of this plant are 0.9x0.9x2.0 m^3 and weight is 150 kg. The plant compactness allows including it into existing technologic lines.

4 A set-up for Rutherford backscattering analyses

Rutherford backscattering method is powerful mean for investigation of thin films and deposited layers [1]. This method is based on the analysis of energy spectra of backscattered alpha-particles. The energy spectrum shape depends on the atomic weight of target atoms and their depth of bedding. From the analysis of such spectrum one can obtain information about thickness of a layer and its position. The accuracy of RBS method is about $3\cdot10^{-7}$ cm.

Of course, the activity of our FAKT-1 source is lot lower than the intensity of an accelerator's beam and therefor the traditional geometry of measurements is not acceptable. In the proposed geometry of measurements we put our detector not behind but before the investigated target [4]. This nontraditional geometry gave us opportunity to increase count-rate and curtail measuring time down to about 2 hour which comparable with measurements with using of an accelerator (~0.5 hour). Depth resolution is slightly lower, too (10^{-6} cm with FAKT-1 source against $3\cdot10^{-7}$ cm when accelerator is used). But this disadvantage is compensated with simplicity, reliability and cheapness of our set-up.

References

1. Ion beam handbook for material analysis, Ed. Mayer J.W., Rimini E.N. (Acad. Press, New York, 1977).
2. Author's certificate SU No. 1085422 A, G 21 G 4/04
3. USA Patent No. 2476644, G 21 G 4/04 (252-301.1).
4. Iliassov A.Z., Mikhailin A.V. and Sadikov R.A. Device for Analyzing of Impurity Composition and Distribution in Solids, Uzbek Patent No. 473.

APPLICATION OF ^{113M}IN RADIOTRACER TO WELL LOGGING

SUN WENHAO SONG DAOHAI HUANG YONG
Nuclear Power Institute of China

P.O.Box 291-107 Chengdu,China 610005

A new radiotracer ^{113m}In developed for well logging is presented in this paper. The radiotracers are synthesized with ^{113m}In eluted from ^{113}Sn-^{113m}In generator and various carriers, they can be used for production profiles of oil, gas production wells and injection profiles, and good results will be achieved in well logging. Moreover, the half-life of ^{113m}In is very short (99.5min) and ^{113m}In can be obtained on the well site constantly, this new method for well logging is easily and safely to be performed and won't result in poisonous and radioactive contamination to the environment

1 Introduction

Radioisotope tracing technique has been applied to production log for a long time in China [1]. It can be used for production profiles of oil and gas production wells and injection profiles. The radioisotopes used before for well logging were ^{110m}Ag, ^{51}Cr, ^{65}Zn, ^{59}Fe, ^{124}Sb, ^{131}I and ^{131}Ba, and the radioisotope tracing flowmeter were $C_6H_5{}^{131}I$, $Na^{131}I$ and $Na^{131}IO_3$ solution. Because all these radioisotopes have long half-life, so serious radioactive contamination is hardly to be canceled in short time. In addition, the density of $C_6H_5{}^{131}I$ is not suitable for well logging, and moreover, $C_6H_5{}^{131}I$, $Na^{131}I$ and $Na^{131}IO_3$ are miscible with the fluid, which can bring serious errors in well logging. So some radiotracers with short half-life, different density and solubility are need to be developed.

Radioisotopes ^{113m}In and a series of its radiotracers presented in this paper fully satisfy the above requirement.

2 Advantages of ^{113m}In raditracer

^{113m}In is very suitable for well logging, it has following advantages:

1. It has short half-life (99.5min), hardly contaminates well, well logging instruments and well site environment.
2. It can be eluted from ^{113}Sn-^{113m}In-generator on the well site constantly and easily [2] and the half-life of ^{113}Sn is long (115d) enough to be used more than one year.

3. It belongs to low poisonous group, and has no chemistry toxicity and contamination to the water source underground.

3 In -113m radioracing flowmeter

^{113m}In and a series carriers can be used as a flowmeter for measuring flow rate of the fluid in well.The radiotracer is released into the fluid and moves with the fluid synchronously.Velocity of the radiotracer measured by scanning with • -probe multiplied by the annular area of well pipe is equal to the flow rate of the fluid. It is quite evident that the accuracy of measured results is depended on the degree of synchronous moving between the radiotracer and the fluid, et.al. So the density of the radiotracer should as the same as that of the fluid, otherwise great measuring error will be produced, moreover, the accuracy is depended on the sharp line of • -scanning pattern, too. In our experience, the sharper the line is, the more accurate the measuring results are. The quality of the• -scanning pattern is depend on the intermiscibility between the radiotracer and fluid.

To solve these problems, a series of ^{113m}In radiotracers [3] with density various from 0.8g/cm^3 to 1.03 g/cm^3 have been developed by our laboratory, and have been used for well logging with good results in China.

4 Measuring injection profiles by 113mIn-Charcoal

The injection profiles in oilfields must be monitored constantly to decrease the water in well and increase the rate of extracting oil. As mentioned above, ^{131}I-charcoal and ^{131}Ba microsphere were used before, but they are not suitable because of their long half-life and great desorption from the carriers. We found that if charcoal treated by a special chemistry technology, ^{113m}In are adsorbed quickly and effectively. The procedure of well logging by ^{113m}In: 2mCi ^{113m}In solution and 80mLNaOH solution (0.1mol/L) were added into a 200mL beaker and stir, then, pour the solution in another 300mL beaker which contains 40g charcoal, finally, the mixture was transferred into a radiotracer releaser and released at 10m above injection layer. The injection profiles is arrive or not through comparing the • -scanning curve with background• -scanning curve. Thousands of wells have been logged by this procedure and satisfied results with 95% successful rate achieved.

5. Conclusion

The procedure used ^{113m}In as radiotracer for well logging, is simple, safe and effective, and deserve to be popularized and applied widely.

^{113m}In-Charcoal has various diameters, high thermal stability. But, because of its short half-life, the injection profiles should be calibrated with well logging time.

When ^{113m}In are used as a radioisotope tracing flowmeter, ^{113m}In carriers with the same density and opposite solubility as the fluid in well should be selected so as to get accurate well logging results.

Reference

1. Jiang Wenda, Application of isotope logging technique to oilfield development, Isotope,1989,2(4):245.
2. Sun Wenhao et al., Study on production technology of high activity ^{113}Sn-^{113m}In generator, Nuclear Power Engineering, 1988,9(3): 49.
3. Sun Wenhao et al., A radioisotope tracing logging method, China Patent

RESEARCH & DEVELOPMENT ON MCA-BASED COAL ASH GAUGE

WU ZHIFANG, XIANG XINCHENG, WANG LIQIANG

INET, Tsinghua University, Beijing 102201, P.R.China
E-mail: zhouly@mail.tsinghua.edu.cn

This paper describes how to use multi-channel analysis (MCA) technique to determine the ash content of coal in pipe. It analyzes the problems existing in the previous work. A new-type apparatus is designed with ^{137}Cs and ^{241}Am sources and a fast data acquisition system of one million Hz sample rate. Several high efficient methods are adopted to smooth spectrum data, to locate peak, to calculate peak area and to modify counter rate with two-level pile-up probability. Therefore, the measuring accuracy is dramatically improved and the measure period is greatly shortened. The auto-location of peak edges makes the gauge interference-resistant, the spectrum-fixed equipment and temperature-keeping equipment unnecessary. This paper also discusses the influence of several factors to the measuring system and gives out the way to overcome them. The base error of this system is better than ±0.5%, the stability is better than ±0.25%, the measuring period is about 100s(for medium-level ash content).

1 Introduction

The ash content is closely associated with the key indexes of coal, e.g. heat-producing ability. It is one of main criteria on evaluating coal quality and deciding the coal price. Fast determination of ash content of coal also plays a guiding function in process control, product quality and saving energy etc. in some industrial domain. The previous methods have some defaults such as long measuring period, long delay time, poor accuracy or stability etc.

In this paper, we adopt dual-energy •-rays transmission method and multi-channel analysis (MCA) technique to determine the ash content of coal. It overcomes the disadvantages mentioned above, and features good accuracy, high speed and strong interference-resistant ability. It can be widely applied in the industrial fields such as metallurgy, coal power-generating factory, coal mine and harbor etc.

2 Methods

This apparatus measures the ash content of coal by dual-energy•-rays transmission method [1]. The two kinds of •-rays transmit the specified coal sample. The intensity attenuation of low energy•-rays indicates the ash content of coal, while that of high energy•-rays is used to eliminate the error caused by the thickness variation of the specified coal sample. MCA technique is made full use of to measure the intensity attenuation of •-rays in this apparatus.

1. Detector and Pre-amplifier

NaI(Tl) crystal is used as•-rays detector. Fast pre-amplifying and shaping circuits are designed with a count rate of $4x10^5$/s. The high counting rate is the essential factor to realize fast determination of ash content of coal.

2. Acquisition of spectrum data

The foreground and background parallel process is adopted. The hardware circuit realizes AD conversion and data buffer in the background with a sample rate of 1.26MHz. The computer fetches the data in buffer regularly or randomly to form energy spectrum, and process it to work out ash content of coal in the foreground.

3. Process of spectrum data

After the energy spectrum of•-rays has been attained, it becomes a key problem to work out the ash content of coal.

We divide the entire spectrum into two segments to proceed Am peak and Cs peak respectively. Each peak follows four steps: 1) to use 11-points filter to smooth the spectrum data for three times, 2) to locate the full-energy peak by first differential of the spectrum, 3) to calculate the net area of peak, 4) to figure out the counting rate of the peak.

The process is auto-adaptive, having no interference from peak movement and shape distortion. The base and the contribution of Compton scatter and the derived x-rays can be well discounted. Furthermore, the spectrum data is smoothened for several times before they are used to count the peak area, so the random noise is greatly reduced.

4. Rectification of count rate

The intensity of •-rays is not exactly linear with the count rate, so the count rate must be rectified. The rectification method is designed in consideration of dead time and two-level pile-up probability etc.

5. System calibration and auto-adaptive measurement

A math model is built to calibrate the measuring system in three segments of the ash content respectively. The measuring period is auto-adaptively decided by the size of ash content of coal. Therefore, good accuracy can be got in the low, medium and high segments of ash content.

Generally speaking, the accuracy and the speed of measuring the ash content of coal are dramatically improved, and the inference-resistant ability of this apparatus is well enhanced because of auto-track of peak position.

3 Measuring Apparatus

In the coal electricity-generating factory, the ash content of coal powder need to be inspected before it enters the combustion furnace. It is necessary to guarantee the furnace to work safely and economically.

According to this situation, we design the following apparatus to measure the ash content of coal in pipe on-line. It's shown in Fig.1.

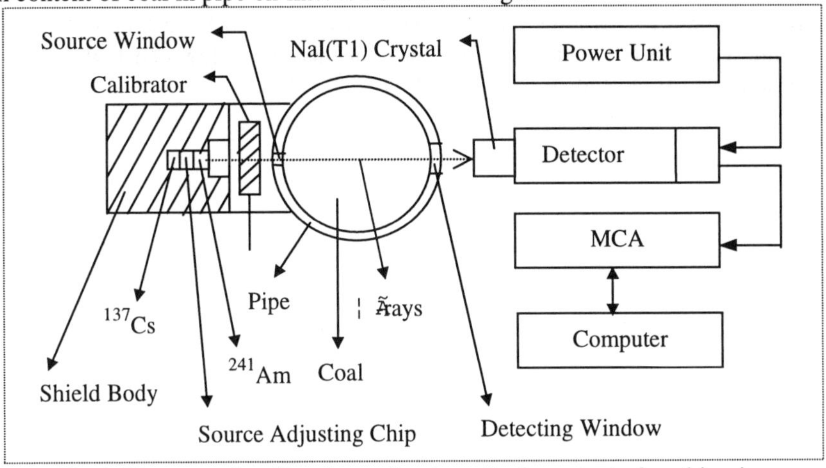

Fig.1 Diagram of on-line determination of ash content of coal in pipe

4 Basic Behavior and Advantages

This apparatus realizes on-line determination of ash content of coal in pipe. It may also inspect the given coal sample statically. The main features are as the following:

a) Basic error: better than ±0.5%.
b) Stability: better than ±0.25%.
c) Measuring period: short to 100-180s (related to the size of ash content of coal).
d) Good measuring accuracy due to MCA technique.
e) Strong interference-resistant ability. This apparatus isn't sensitive to the factors such as temperature, working voltage, detector's stability, derived x-rays, Compton scatter, count rate and base etc.
f) Simple equipment without spectrum-fixing and temperature-keeping devices. It's easy to install and to maintain.
g) Auto-adaptive timing of measurement, ensure to meet statistic requirement for coal samples of different ash content.
h) Having almost the same accuracy to coal samples of different ash content.

References

1. R.A.Fookes, V.L.Gravitis etc. On-line determination of the ash content of coal using a "Siroash" gauge based on the transmission of low and high energy•-rays, Int. J. Appl. Radiat. Isot. Vol.34, pp63-69 (1983)

Session H3

Use of Isotopes in Environmental Studies — II

THE ROLE OF RADIOTRACER TECHNIQUES IN OFF-SHORE CONTAMINANT TRANSPORT STUDIES

PETER AIREY, MYINT ZAW, JIYUAN TU, TOM KLUSS, RON SZYMCZAK AND QING HE*

Australian Nuclear Science and Technology Organisation, MENAI 2234 Australia
**State Key Laboratory for Estuarine and Coastal Research, Shanghai, China*

E-mail: pla@ansto.gov.au

Recent advances in off-shore tracer technology have enabled the implementation of increasingly sophisticated investigations. Applications are being increasingly directed towards the validation of contaminant transport codes. Research interest is focusing on the fate and behaviour of particulates, because they are little understood, they play a significant role in the transport of environmental pollutants and there are currently no effective alternatives to radiotracing. Examples of recent research into the transport of fine sewage particles from the sewage outfall at Burwood Beach, Newcastle, and into the migration of sand at depth will be presented.

1 Background

International support for projects designed to enhance an understanding of the coastal zone is a reflection of national priorities. For instance, the IAEA/UNDP Regional Cooperative Project: *Management of the Marine Coastal Environment and its Pollution* is a vehicle for building specific capabilities underpinning the application of nuclear techniques to sustainable development in the coastal zone. Again, the pollution of regional seas from land based activities is the first key issue for large-scale sustainability in the South East Asian region listed in the SARCS1999 *Integrated Study Science Plan* [1].

2 Role of radiotracer techniques in model validation:

Modelling and tracing are complementary approaches to the study of complex environmental systems. Numerical modelling uses an holistic approach, but is inevitably constrained by a range of simplifying assumptions. Tracing enables more detailed experimental studies of specific components of such systems. ANSTO and the Unisearch Water Research Laboratory (WRL) were involved for a number of years in the radiotracer evaluation of predictive models for the transport of sewage from the deep ocean outfalls off Sydney[2].

More recently, the two groups have been involved in tracer studies at the North West New Territories (NWNT) sewage outfall, Hong Kong. The work was undertaken for the Environmental Protection Department through a consultancy managed by Montgomery Watson HK and involving the survey company EGS(Asia) and the University of Hong Kong. One of the outcomes of this work was a comparison of observed effluent dilutions from tracer studies with those predicted by the near-field numerical model JETLAG [3,4] .

This study also involved a comparative evaluation of the radiotracer gold-198 and the fluorescent dye tracer rhodamine-WT. The tracers were released simultaneously by ANSTO and EGS (Asia) respectively. The correlations were excellent. Under the conditions of the experiment, the sensitivities of the two techniques were comparable.

Research is therefore focusing on the fate and behaviour of particulates, because they are little understood, they play a significant role in the transport of environmental pollutants and there are currently no cost-effective alternatives to radioisotopes for tracing. In this report ^{198}Au is used to study particulates and ^{192}Ir incorporated in sized glass is used for bedload transport. Tritium (HTO) is used as a tracer for the receiving water.

3 The transport of particulates from a sewage outfall; tracer studies

A study of the influence of salinity on the aggregation of sewage particles discharged to a coastal marine environment was undertaken at the Burwood Beach Wastewater Treatment Plant, Newcastle, NSW [5] by ANSTO in collaboration with WRL. The aim of this field oriented research was to carry out both bench scale and *in situ* studies of the aggregation of natural and sewage particulates.

Detailed studies of the transient size distributions and structures of sewage particles under simulated estuarine mixing conditions ware made in the laboratory. Complementary studies of the dilution and transport kinetics of the sewage plume were obtained using ^{198}Au and tritium tracer techniques. Two properties of the particles were measured on samples collected from within the plume: the particle size distribution and the fractal dimension. Fractal theory provides a new approach to the characterisation of many natural and engineered systems that have no definite form or regularity [6]. The most important numerical parameter in fractal theory is the fractal dimension D. The lower the value of the fractal dimension, the higher the porosity.

Major findings of this research include:

(1) In the laboratory, the rate of aggregation of sewage particles increases with increasing salinity, with equilibrium being reached within a few minutes. A particle aggregation model which accounts for salient features of the laboratory data is being developed [7].
(2) This contrasted with field observations, where equilibrium floc size was not achieved during the 5 hour study. The particle size distributions were very similar and independent of mixing time and salinity.
(3) The tracer data showed definitively that particles discharged from the Burwood Beach Treatment Plant did not settle during the 5 hour period of this study. This is based on direct observations of ^{198}Au and of the ^{198}Au/HTO ratios
(4) Sewage particle sizes increased while fractal dimensions decreased ie the porosity increased with the distance traveled by the plume from the diffuser. Hence, the more aggregated the particles, the greater the entrainment of the surrounding water within the particle structure. Since the particles are moving from lower density sewage effluent into higher density marine water, there will be a net buoyancy effect which inhibits particle settling.

In order to develop a more complete understanding of the fate and behaviour of effluents released to the sea from engineered structures, two complementary modelling approaches are being progressed. The first is a two-phase turbulent flow model which has been applied to the near field of a prototype of the Malabar outfall near Sydney [8,9].

Far field modelling at Burwood Beach is being progressed [7] using a code developed by the Virginia Institute of Marine Science [10]. As indicated above, the sewage plume rose to the surface during this study. Consequently, the tracer plume could only be modelled satisfactorily by taking into account the impact of the wind field on the current patterns in the upper layers of the water column. The authors gratefully acknowledge the contribution of the Hunter Water Corporation and the Water Research Laboratory to the tracer study.

4 The migration of larger particles: bedload transport

Although there is a significant quartz component in sewage sludge, much of the interest of sand migration relates to impact of storm events on the fate of beach sands in the short term, and on the impact of sea level rise on beach dynamics in the longer term. Precise position fixing linked to stable instrumentation has enabled bedload transport studies to be extended to greater depth with enhanced resolution. This is illustrated with a study of sand migration , 2 km from shore at the 68m contour off Bondi, NSW. Observations were made over a nine month period. Some dispersion was observed, but no systematic migration despite three significant storm events.

5 Acknowledgements

The contributions of collaborating organisations are acknowledged in the text. Special gratitude is due to Mr J Walker WRL and Mr G Spelman ANSTO.

References

1. South East Asian Regional Committee for START 1999 *Integrated Study Science Plan,* Chap. 1 "The Changing Environment of South East Asia" Section 'Sustainable Development Strategies'
2. T. R. Prichard, R. S. Lee and A. Davison, "Sydney Deepwater Outfalls: In situ Observations of Plume Characteristics", *Proc. Eleventh Conference on Coastal and Ocean Engineering*, Townsville pp 53 - 58 (1993).
3. Horton P.R and Wilson J R (1996) 'NWNT Sewage Outfall, Urmston Road, Hong Kong, Comparison of Near-Field JETLAG Modeling with Tracer Experiments' Australian Water and Coastal Studies Report 96/29.
4. Horton, P R, Airey P L and Wilson J R (1997) 'Radioactive and Dye Tracer Studies for the NWNT Sewage Outfall, Hong Kong and Comparison to Near-Field Modelling' Proc.Second Int Conf Isotopes (2ICI) Sydney 1997
5. M. Zaw, R. Szymczak and P. Airey, "Influence of Salinity on the Aggregation of Sewage Particles Discharged to a Coastal Marine Environment", *Proc. of the Australian Water and Wastewater Association, 18th Federal Convention,* Adelaide, CD ROM, April 8 pp (1999).
6. Li Da-hong, and Ganczarczyk J (1989) 'Fractal Geometry of Particle Aggregates Generated in Water and Wastewater Treatment Processes'. *Environ. Sci. Technol.*, **23**, pp. 1385-1389.
7. Peter Airey, Myint Zaw, Ron Szymczak, Thomas Kluss, Jiyuan Tu and Jerard Barry 1999, 'Application of Radiotracer Techniques to the Validation of Contaminant Transport Models at Sewage Outfalls' Proceedings of the Conference Paper 145 *Global 99* (Jackson Hole Wyoming, August 1999).
8. Tu J Y (1997) ' An Efficient Method for Modeling Two-Phase Flow' ISAC'97 High Performance Computing on Multi-Phase Flow (Ed Y Matsumoto, A Prosperetti) p105
9. Ye S Z and Tu J Y ' Numerical Modeling of Mixing Flow near an Ocean Outfall' *ibid*
10. Hamrick J M 1992 *A three-dimensional environmental fluid dynamics computer code: Theoretical and computational aspects* Special Report in Applied Marine Science and Ocean Engineering (Virginia Institute of Marine Science, Glouscester Point, VA 23062)

DYNAMIC CHANGES IN GROUNDWATER CONDITIONS CAUSED BY TUNNEL EXCAVATION AT ASPO HRL, SWEDEN- CHANGES IN ISOTOPE HYDROLOGY

Y. MAHARA, T. IGARASHI, K. MIYAKAWA, K. KIHO, Y. TANAKA AND T. HASEGAWA

Central Research Institute of Electric Power Industry,1646 Abiko, Abiko-shi, Chiba, 270-1194, JAPAN

E-mail: mahara@criepi.denken.or.jp

Groundwater environmental changes caused by tunneling have been investigated over a period of two years by measuring dissolved noble gases, stable isotopes, chloride concentration and chlorine isotopes. We can categorize groundwater into three groups based on the correlation between sampling depth and the dissolved helium content: the undisturbed zone which has not yet been affected by the intrusion of Baltic seawater, the mixing zone which has been gradually exchanged by Baltic seawater, and the disturbed zone which the primary condition of the dissolved helium has already lost. The residence time of groundwater which ranges from 900 to 9×10^5 years increased when increasing the sampling depth from - 60 m deep to - 460 m. However, very old groundwater (more than one million years) is trapped somewhere at - 350 m depth in the mixing zone. On the other hand, the residence time in the disturbed zone turned extremely into too young at around - 400 m depth.

1 Introduction

Central Research Institute of Electric Power Industry (CRIEPI) in Japan has joined the international cooperation research organized by SKB at Aspo HRL in Sweden. CRIEPI conducted groundwater surveys at Aspo HRL in 1995 and 1997, for the purpose of investigating the origin and residence-time of groundwater and estimating changes and stability in groundwater environments during tunnel excavation [1]. Groundwater was investigated in noble gas hydrology, isotope hydrology and chemical hydrology by collecting groundwater samples at 21 points in the tunnel and one deep borehole outside the tunnel. Dissolved noble gas contents and isotope ratios, stable isotopes, $^{36}Cl/Cl$ ratios, tritium concentration and chloride concentration were measured to collect palaeohydrological information.

2 Study Area

The study area is the tunnel of the Aspo HRL, constructed by SKB for R&D into the geological disposal of HLW, located about 450 km southwest of Stockholm, and facing the Baltic Sea. The HRL has a tunnel of 3600 m length and was built to - 460 m depth below the ground surface. Aspo island and the surrounding area are part of the Precambrian basement in SE Sweden. The rocks are between 1760 and 1840 million years old (dated by U-Pb method [2]) and are composed of Smaland granite and Aspo diorite, which are highly fractured and water bearing. There are various fracture zones and fracture swarms surrounding the site. Furthermore, many minor fractures are found in the rock mass between the fracture zones. Groundwater might mainly circulate through the fracture zones and gradually move through fractures in the rock mass.

3 Results and Discussion

Distributions of dissolved He content and chloride concentration were heterogeneous in the entire tunnel. The highest content of He and chloride were located surrounding KA2862, which is 380 m below the ground surface and is not the deepest location in the tunnel. There was little change in the content of He and chloride at KA2862 in the past two years. The area surrounding KA2862 is estimated to be isolated from groundwater disturbed by tunneling. This zone is defined in the undisturbed zone. On the other hand, the areas including KA3105 and KA3110 have already been lost the primary geohydrological conditions because of the intrusion of much Baltic seawater via fractures. The He content in samples collected here is 50 times less than that measured in the undisturbed zone. There were no changes in stable isotopes, chloride concentration and He content in this zone, which was defined as the disturbed zone. Dissolved He content at KR0013, KA3010 and KA3067(g) changed drastically in the past two years because of the tunneling. However, other indicators (stable isotopes and chloride concentration) remained little changed. Furthermore, the He content is mid-level compared with that of samples collected at the same depth in the undisturbed zone and the disturbed zone. This area is called the mixing zone where it has gradually been mixing with Baltic seawater intruding through micro fractures or fissures.

The changes in the stable isotopes changes for two years indicate that groundwater in the tunnel has consisted of a mixing between groundwater which has not lost the primary geohydrological conditions in the undisturbed zone and Baltic seawater intruding by tunnel excavation. The mixing line was expressed in eq.•D = 6.38•^{18}O - 9.97 having high saline water and Baltic seawater as both end-members. KA3358 moved greatly to a direction of Baltic seawater on this mixing line for two years. On the other hand, KA2862 and KA3010 also moved greatly to a direction of Laxemar Deep groundwater on a different eq. •D = 11.3•^{18}O + 56.2. This suggests that deep high-saline water at Aspo has not completely been isolated from groundwater flow but has gradually circulated in a regional groundwater flow including the Laxemar region.

^{36}Cl content and He content were measured to estimate groundwater residence time. However, since the in-situ ^{36}Cl production is not negligible, groundwater residence time can not be estimated from a decrease in ^{36}Cl atoms produced by cosmic irradiation. This result is supported by Louvat et al. [3]. The residence times were predicted from dissolved He content and He dispersion effects in stratum. The oldest groundwater with a high salinity of more than 14000 ppm of Cl^- is estimated to be more than 1.8×10^6 years old. The groundwater residence time ranges from 900 to 9×10^5 years in the gradually mixing-zone. Groundwater in the disturbed zone where a violent intrusion of Baltic seawater occurred is too young to date reliably and the primary paleohydrological characteristics have already been lost.

4 Conclusions

On Aspo island, very old groundwater of which the residence time is over one million years has partially been remained and it might have kept palaeohydrological conditions such as stagnant water. However, this is still open for discussion. On the other hand, some parts of the tunnel have been losing primary conditions of groundwater by the violent intrusion of Baltic seawater via fractures from the tunneling. The investigation for the past two years indicates that the groundwater flow is very complex in crystalline rocks and depends on the characteristics of fractures. The in-situ production rate of ^{36}Cl in the granite rocks is not negligible for that of the cosmogenic ^{36}Cl.

References

1. Mahara Y., Igarashi T., Miyakawa K., Kiho K., Tanaka Y. and Hasegawa T., Dynamic changes in groundwater conditions caused by tunnel construction at the Aspo Hard Rock Laboratory, Sweden., *SKB ICR* **98-04**, (1998).
2. Johansson A., The age and geotectonic setting of the Smaland Varmland granite proryprv belt. Geologiska Foreningens I Stockholm, Forhandlingar, (1988).
3. Louvat D., Michelot J. L. and Aranyossy J. F., Salinity origin and residence time of the Aspo groundwater system, *Proceedings of the second Aspo International Geochemistry Workshop,* ed. By Laaksoharju M. and Wallin B., *SKB ICR* **97-04**, (1997) #5.1-7.

THE IDENTIFICATION OF PLASMA CATALYTIC REACTIONS OF N_2O, NO, NO_2 BY MEANS OF LABELLED COMPOUNDS

J. W. LEONHARDT, K. P. FRANCKE, H. MIESSNER, R. RUDOLPH

I.U.T. Institut für Umwelttechnologien GmbH, Volmerstraße 9B, 12489 Berlin, Germany
E-mail: info@iut-berlin.com

Cold plasma in combination with an appropriate catalyst can reduce the NO_X emission of car engines. The knowledge of the different pathways of removal and new formation of NO_x in engine exhaust system is essentially for any further development. Isotopically labelled compounds as ^{15}NO, $N^{18}O$, $^{15}N_2$ and ^{14}C-hydrocarbons help to clarify the situation in detail. Plasma-chemical reaction channels can be identified.

Measurements have been done in a dielectric barrier discharge reactor with cylindrical geometry. Various catalysts have been placed downstream to the discharge. The plasma conversion of labelled compounds was analyzed by means of on-line FTIR spectroscopy and chemiluminescence but also by GC-MS (off-line).

Using of labelled ^{15}NO it could be shown that:

- The plasma-chemical formation of NO in oxygen containing exhaust gases is proportional to the energy density in the plasma, it is not influenced by the presence of a catalyst.
- NO_2 is synthesized by NO-oxidation.
- N_2O is synthesized plasma-chemically both from N_2+O_2 and from NO_2.
- NO_x can be reduced by an appropriate catalyst.
- The reduction of NO_x by NH_3 in plasma catalytic SCR-reactions requires the partial intermediate oxidation of NO to NO_2 and results in the formation of N_2.

Tracer experiments by means of ^{14}C-labelled hydrocarbons are in progress.
Research is sponsored by the German Federal Ministry of Education an Research.

1 Introduction

The emission of NO_x into the atmosphere is dominated by power stations and traffic. Car engines produce it in relatively large amounts. Typical concentrations of NO_x in exhausted air are in the range 150-500 mg/m³. A significant reduction of this concentration is demanded by environment authorities. The knowledge of production and reduction reactions of N_2O, NO, NO_2, HNO_x is essential for any progress. One of the technical options for car engines exhaust gas cleaning is the plasma catalytic treatment. The plasma in this context is non-thermal or „cold".

2 Methods

The plasma reactor is designed as shown in Fig. 1. It consists of a glass tube as dielectric with a central stainless steel rod with equidistant discs as high-voltage electrode and a copper grid as outer ground electrode. The space between the discs can be filled with a catalyst to be studied. The power supply delivers 5 - 20 mA and

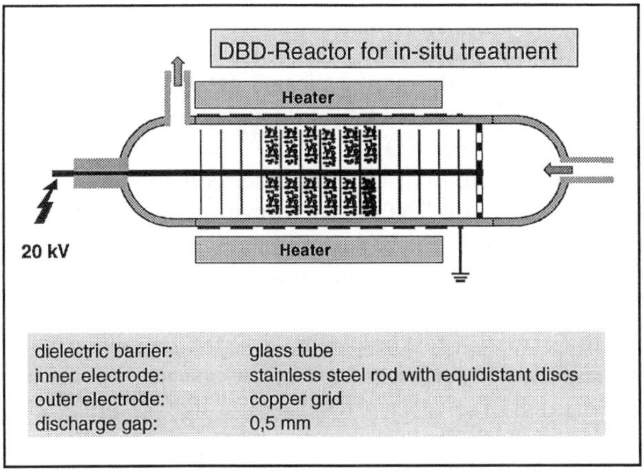

20 kV. The exhaust gas stream is simulated by means of purified nitrogen, oxygen and added NO-gas, which was available also as isotopic labelled ^{15}NO.

Figure 1. Principle scheme of a dielectric barrier discharge reactor for plasma-catalytic treatment of NO_X and VOCs

The plasma treated gas mixture passes several analytical devices: a Perkin Elmer FTIR with a 20 m gas-cell, a chemiluminescent NO_x-meter and, sometimes, an IUT-IMS and GC-PID for on-line determination of organic compounds. This setup is also suitable to check the efficiency of VOC-removal (Volatile Organic Compounds) in a plasma catalytic reactor[1-3]. In that case carbon-14 labelled compounds are used to characterize reduction reactions. Labelled chlorinated benzene (100 mCi/mMol) for example is evaporated into the gas. Isotope ratios are determined by GC-MS and injection of the collected gas samples by means of a head space. The organic content after leaving the reactor is burnt to CO_2, which is converted to $Ba^{14}CO_3$.

On-line isotope analysis of nitrogen-15 was possible by means of isotope shift in the infrared spectra of NO, N_2O and NO_2. The isotopic content of an actual gas mixture is then determined by arithmetical spectrum fitting procedures. The isotope ratio used is defined as the δ-15 value:

$$\delta = \frac{^{15}N}{^{15}N + ^{14}N}$$

The analysis of carbon-14 rations were carried out by means of liquid scintillation counting and GC-MS.

3 Results and Discussion

Experimental data of the production of NO_x in air in a cold plasma equipment was collected on-line by means of an IMS. Measuring the time profiles of NO, NO_2, HNO_x it became clear, that the oxidation of N will give NO, which is a step to get NO_2.

These results were confirmed by labelled compounds. For the synthesis of NO_2 by oxidation of NO according $NO + \frac{1}{2}O_2 \rightarrow NO_2$ it should be expected that both

NO and NO_2 have the same δ-15 values. In fact, that is what we have found over the whole range of energy applied.

The analysis of δ-15 for N_2O formed during the plasma catalytic treatment of ^{15}N labelled NO indicates the formation of N_2O by two independent pathways: directly from N_2 and O_2 and by destruction of NO_2.

Labelling the NO by ^{15}N, it is possible to discriminate the reductive removal of ^{15}NO and the oxidative new formation of ^{14}NO. The formation of ^{14}NO in the plasma is proportional to the energy density in the system and cannot be influenced by a catalyst. The reductive removal of ^{15}NO can be increased, on the other hand, by appropriate catalysts.

Using a NH_4-loaded zeolite as catalyst for the plasma-assisted removal of NO_x we were able to obtain a significant synergetic effect, which is due to the combination of the oxidative potential of an electric discharge in oxygen-containing gases and the catalytic reaction with the NH_3 in the zeolite [4-6]. The formation of $^{15}N_2$ during the plasma catalytic reaction of ^{15}NO with $^{15}NH_4$-zeolite (Fig. 2) proves a mechanism consisting of a partial intermediate oxidation of NO to NO_2 and the subsequent reduction with NH_3.

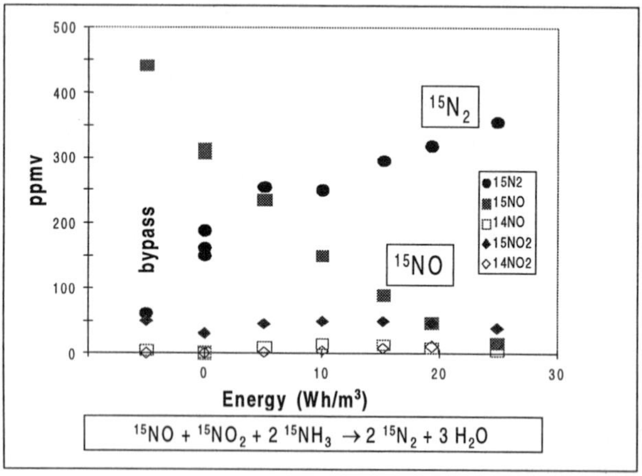

Figure 2. Formation of $^{15}N_2$ during the plasma-catalytic reaction of ^{15}NO with $^{15}NH_4$-zeolite

4 Conclusions

The chemistry of NO_x-removal in plasma catalytic processes was studied using isotopically labelled NO_x compounds as tracers. It was possible to discriminate the different pathways of NO_x-removal and -formation in dependence of the plasma energy introduced and the catalysts used. The design of plasma catalytic systems for car engines can be improved on the basis of these results.

The use of carbon-14 labelled ring systems opened up new prospects to characterise the removal of extremely low concentrations of toxic compounds like dioxins.

References

1. Francke K.-P., Rohrbeck W., Rudolph R. and Volkmann H., 1 kW Pulse Corona Reactor for VOC Destruction. In *Proceedings of the 12th Inter-national Conference on Gas Discharges & Their Applications*, vol. 1, pp. 308-311, Greifswald 1997.
2. Bensch H., Francke K.-P. and Rudolph R., Oxidation of VOCs in Air Flows by Pulse Corona Reactor and Heterogeneous Catalyzer Hybrid Systems. *The 1998 European Workshop on Water and Air Treatment Technologies: Innovative and Commercial Applications* EPFL, Lausanne, Switzerland, October 11-14, 1998.
3. Francke K.-P., Miessner H., Rudolph R. and Rutkowsky J., Removal of VOCs by plasmacatalysis. *11th Symposium on Elementary Processes and Chemical Reactions in Low Temperature Plasma*, June 22-26 1998, Liptovsky Jan, Low Tatras.
4. Miessner H., Rudolph R. and Francke K.-P., Plasmacatalytic low-temperature conversion of NO_x to N_2 by ammonium-loaded zeolites in a dielectric barrier discharge. *J. Chem. Soc., Chem. Commun.* (1998) pp. 2725-2726.
5. Miessner H., Rudolph R. and Francke K.-P., Plasmakatalytische Tieftemperatur-Umwandlung von NO_x zu N_2 durch Ammonium-Zeolithe in einer dielektrischen Entladung. *32. Jahrestreffen Deutscher Katalytiker,* 17.-19. 3. 1999, Friedrichroda
6. Francke K.-P., Miessner H. and Rudolph R., Untersuchung des plasmakatalytischen NO-Abbaus in sauerstoffhaltigen Abgasen mittels isotop-markierter Ausgangsstoffe. *9. Bundesdeutschen Fachtagung Plasma-technologie*, 3.-5. März 1999, Stuttgart

CARBON-14 ACTIVITY OF PORE WATER OBTAINED BY VACUUM-DISTILLATION AND ONE-DIMENSIONAL COMPRESSION METHODS ON YUCCA MOUNTAIN CORES

I. C. YANG

U. S. Geological Survey, MS-421, Lakewood, Colorado 80225, USA

G. R. DAVIDSON

Department of Geology and Geological Engineering, The University of Mississippi, University, Mississippi 38677, USA

K. M. SCOFIELD

Pacific Western Technologies, Lakewood, Colorado 80215, USA

The vacuum-distillation method is a valuable tool for obtaining pore-water carbon-14 data from rock cores where moisture contents are too small (less than 6 percent) to be obtained by the one-dimensional compression methods. Dissolved inorganic carbon was extracted directly from the pore water of Yucca Mountain rock cores by vacuum distillation. These data were compared to carbon-14 values from pore water extracted indirectly by one-dimensional compression. The carbon-14 activities of pore-water samples collected from the same rock core by both methods differed by 6.8 percent modern carbon. The carbon-14 activities of samples collected by the vacuum distillation and one-dimensional compression from cores separated by 1.7 to 6.3 meters ranged from 8.7 to 24.8 percent modern carbon, and the delta carbon-13 values ranged from 0.5 to 5.1 per mil (relative to Vienna Pee Dee Belemnite reference standard). The carbon-14 activities of samples collected from nearby core, separated by 8 meters, using the one-dimensional compression method alone differed by 10.7 percent modern carbon and the delta carbon-13 values differed by 2.4 per mil.

1 Introduction

The pore waters of many nonwelded tuffs from cores at Yucca Mountain, Nevada have been analyzed for ^{14}C composition of the dissolved inorganic carbon (DIC) by extraction of pore water using one-dimensional compression. However, welded tuff, such as the Topopah Spring Tuff of Tertiary age at Yucca Mountain (moisture content less than 6% by weight), yields no water even at high compression loads [6]. Therefore, no water can be obtained for ^{14}C analysis. Davidson et al.[2] developed a vacuum-distillation method for extracting DIC from the pore water of tuff samples similar to the Topopah Spring Tuff. This method is not applicable to ^{14}C dating of pore water if DIC contains carbon from marine carbonate, calcite or dolomite (dead ^{14}C) because age correction by the δ^{13}C value of pore water cannot be accurately assessed (carbon- isotope fractionation during vacuum distillation).

In order to validate the distillation method for its applicability at Yucca Mountain, comparative analyses were performed on several adjacent or nearly adjacent core samples from Yucca Mountain boreholes. Core samples were analyzed for ^{14}C activity obtained by compression-extraction and by distillation of pore water at the U.S. Geological Survey (USGS) in Denver, Colorado.

2 Methods and Analyses

Distillation of dissolved inorganic carbon was performed on preserved cores from Yucca Mountain, Nevada, in laboratories at the U.S. Geological Survey in Denver, Colorado. Core samples came from the Topopah Spring Tuff and the Prow Pass Tuff in boreholes USW NRG-7a, SD-7, SD-9, SD-12, WT-24, and from the North Ramp Exploratory Studies Facility (ESF). All cores were drilled using atmospheric air as the drilling medium. Sulfur hexafluoride was injected into the drilling air as a tracer, but was not found in significant concentrations in preserved core samples. Preservation was completed in the field at the time of drilling in 1994 by sealing the core inside a lexan liner and then again inside a heat-sealable aluminum laminate (Protec Core®)[1].

Vacuum-distillation procedures are outlined in Davidson [1] and Davidson et al. [2]. Briefly, the preserved core sample was opened and immediately transferred into an argon-filled quartz-glass (at the USGS laboratory) connected to a high-vacuum line. The cylinder was heated to 180°C and extracted vapor and gases were drawn through a series of cryogenic traps to capture both water (dry-ice/alcohol traps) and carbon dioxide (CO_2) (liquid nitrogen traps). Total distillation time ranged from 6 to 12 hours.

Calcite (no ^{14}C activity) of the Harding Pegmatite from central New Mexico was acidified in the USGS distillation system to determine the ^{14}C-background activity. The analyzed ^{14}C value was 0.4 ± 0.04 percent modern carbon (pMC), a value close to the pMC value of zero for the ^{14}C activity of calcite from the Harding pegmatite, indicating that the sample-preparation system was leakproof, with no contamination from the environment.

Pore water was extracted from the cores by using a high-pressure one-dimensional compression procedure [4, 5, 3]. A pressure of 103.4 MPa initially was applied and gradually increased to a final pressure of 827 MPa. Additional pore water was expelled by injecting dry nitrogen gas (greater than 99.99 % N_2) into the pore space to force out pore water at the end of compression. The extracted water was filtered through 0.45μm Nucleopore filters into glass bottles for ^{14}C analysis.

Pore-water DIC samples for ^{14}C and $\delta^{13}C$ analysis were sent to Beta Analytic, Inc. in Miami, Florida, for target sample preparation and to the Lawrence Livermore National Laboratory in California for analysis by Tandem Accelerator Mass Spectrometry (TAMS). Uncertainty for $\delta^{13}C$ values was ±0.2‰ (2σ) and for ^{14}C

[1]The use of trade name in the paper is for identification purposes only and does not imply endorsement by the U. S. Geological Survey.

was ±1.0 pMC (2σ).

3 Comparison of compression method versus vacuum-distillation method

The compression and vacuum-distillation methods were compared directly for only one sample from Yucca Mountain. A sample from the ESF was first compressed, then the compressed core was distilled (table 1). The ^{14}C activity of carbon extracted by the two methods differed by 6.8 pMC. The $\delta^{13}C$ value (relative to Vienna Pee Dee Belemnite reference standard) of the distilled sample was lighter by 8.9‰ relative to the compressed sample. These results are consistent with the expectation that precipitation of carbonate during distillation will result in depletion of the heavier isotopes (^{13}C and ^{14}C) in the extracted CO_2.

Core samples separated from the Calico Hills Formation in borehole WT-24 by 0.91 m (midpoint of each core) differed by 7.2 pMC (table 1). The $\delta^{13}C$ value of the distilled sample was heavier by 0.6‰ relative to the compressed sample, contrary to prediction by fractionation. The ^{14}C values were also inconsistent with the $\delta^{13}C$ values, that is, lighter $\delta^{13}C$ values should correspond with smaller ^{14}C activities. All other comparisons were made between similar parameters from distilled and compressed core samples that were collected at different depths, some as much as 6.3 m apart. First, the differences observed between the ^{14}C activity of samples obtained by distillation and compression methods from cores separated by 1.7 m to 6.3 m ranged from 8.7 to 24.8 pMC, and the $\delta^{13}C$ values ranged from 0.5 to 5.1‰. Two core intervals from SD-7, separated by 7.6 m, were compressed; ^{14}C activities between the two samples differed by 10.7 pMC and the $\delta^{13}C$ values differed by 2.4‰ (table 1). Differences between the ^{14}C activity of samples from SD-9 core obtained by distillation compared to compression was 8.7 pMC; and from SD-12, 17.9 pMC. Second, the observed ^{14}C activity of a distilled sample was not always smaller than that of samples obtained by compression. Minor fractionation is expected during distillation which would lower the activity by a few percent, but kinetic fractionation is not the only phenomena controlling the ^{14}C composition of distilled carbon. Davidson et al. (written communication) found that distillation also recovered carbon from a secondary source, such as adsorbed CO_2 on mineral phases, that was depleted in ^{14}C relative to DIC in the bulk pore water.

4 Conclusions

The vacuum-distillation method is a valuable method for obtaining pore-water carbon from rock cores where moisture contents are too low (<6%) to obtain it by more conventional methods, such as one-dimensional compression. The ^{14}C activities of two pore-water samples collected from the same rock core by both

Table 1. Comparison of carbon isotopic data obtained by pore water extraction using vacuum distillation Comparison of vacuum-distillation data between the two laboratories

Borehole names	Pore-water carbon collected by distillation-extraction method			Pore-water carbon collected by compression-extraction method			
	Core Interval (Feet)	^{14}C (pMC)	$\delta^{13}C$ (‰)	Core Interval (Feet)	^{14}C (pMC)	$\delta^{13}C$ (‰)	Geologic Unit
ESF core North Ramp	Moisture Study #13/4.3-5.1	61.8	-15.2	Moisture Study #13/4.3-5.1	68.6	-6.3	Paintbrush Tuff Nonwelded unit
WT-24	1812.5-1813.6	54.6	-16.6	1809.6-1810.5	61.8	-17.2	Calico Hills Formation
SD-7	1503.9-1504.5	48.7	-20.2	1498.2-1498.9 1524.6-1525.0	73.5 62.8	-15.1 -17.5	Calico Hills Formation
SD-9	1468.1-1468.9	94.6	-14.6	1452.6-1453.0	85.9	-14.1	Topopah Spring
SD-12	1498.6-1499.7	87.8	-18.2	1517.0-1517.9	69.9	-17.4	Calico Hills Formation

methods differed by 6.8 pMC . The ^{14}C activity of samples collected by the vacuum distillation and one-dimensional compression methods from cores separated by 1.7 to 6.3 m ranged from 8.7 to 24.8 pMC, and the $\delta^{13}C$ values ranged from 0.5 to 5.1‰. The ^{14}C activities of samples collected from nearby core, separated by 8.0 m, using the one-dimensional compression method differed by 10.7 pMC and the $\delta^{13}C$ values differed by 2.4‰.

5 Acknowledgements

This study was conducted by the U.S. Geological Survey, in cooperation with the U.S. Department of Energy, under Interagency Agreement DE-AI08-97NV12033. The ^{14}C analyses at the U. S. Geological Survey laboratory in Denver were performed by K. M. Scofield.

References

1. Davidson, G. R., 1995 Geochemical and isotopic investigation of the rate and pathway of fluid flow in partially welded fractured unsaturated tuff. Ph.D dissertation, The University of Arizona, 208 p.

2. Davidson, G. R., Hardin, E. L. and Bassett, R. L., 1995 Extraction of C-14 from pore water in unsaturated rock using vacuum distillation. *Radiocarbon* 37: 861-874.
3. Higgins, J. D., Burger, P. A. and Yang, I. C., 1997 The one-dimensional compression method for extraction of pore water from unsaturated tuff and effects on pore-water chemistry. *U.S. Geological Survey Water-Resources Investigations Report* 96-4281, 34 p.
4. Mower, T. E., Higgins, J. D. and Yang, I. C., 1991 Pore-water extraction from unsaturated tuffs using one-dimensional compression. *In* Proceedings of the 2nd International High-Level Radioactive Waste Management Conference, Las Vegas, Nev., April 28-May 3, 1991. *American Nuclear Society*, La Grange Park, Ill.: 989-1006.
5. Mower, T. E., Higgins, J. D., Yang, I. C. and Peters, C. A., 1994 Pore-water extraction from unsaturated tuff by triaxial and one-dimensional compression methods, Nevada Test Site, Nevada. *U.S. Geological Survey Water-Resources Investigations Report* 93-4144, 78 p.
6. Yang, I. C., Rattray, G. W. and Yu, P., 1996 Interpretations of chemical and isotopic data from boreholes in the unsaturated zone, Yucca Mountain, Nevada. *U.S. Geological Survey Water-Resources Investigations Report* 96-4058, 58 p.

Session I1

Radiation Safety Aspects at Isotope Facilities

RADIATION SHIELDING AND SAFETY OF HIGH ENERGY PARTICLE ACCELERATOR FACILITIES

Takashi Nakamura
Cyclotron and Radioisotope Center, Tohoku University
Aoba, Aramaki, Aoba-ku, Sendai 980-8578, Japan
E-mail: nakamura@cyric.tohoku.ac.jp

1. Introduction

High-energy proton and heavy ion accelerators are now in wide use and are planned for cancer therapy, intense spallation neutron source and radioactive beam facility. The shielding design becomes important both for the radiation protection of radiation workers and nearby inhabitants and for the reasonable estimation of facility building cost. The radiation shielding for the accelerator facility design necessitates to estimate radiation source term, bulk shield, skyshine, streaming through maze and ducts, induced radioactivities in air, water, accelerator and building materials. As the radiation source term, neutron production data from thick targets are especially important and the attenuation length of deep penetration is essential for bulk shield calculation. The skyshine, streaming and air activation can be usually estimated by using several simple analytical or semiempirical formulas. The radiation safety problem is induced/residual radioactivities in air, accelerator and building materials, underground water and soil. The former two are important for external and internal exposures of radiation workers during maintenance, and the latter two give an environmental impact to nearby inhabitants. The low activation materials for concrete and accelerator components are investigated to use.

The recent studies on these topics will be given in this review talk.

2. Source Estimation

As the radiation source of the shielding calculation, secondary neutrons produced by various projectile-target reactions are the most important because of their strong penetrability. The double differential neutron yield in energy and angle, DDY, produced from thick target is required. DDY can also be estimated by integrating the double differential neutron production cross section, DDX, from thin target, considering the projectile energy loss in thick target under the neglection of external cascade reaction. For projectiles of energy higher than 100MeV/u, several experimental results on DDY have published on neutron production from proton incidence [1-5], however for heavier ions, there have been the only one experimental data on DDY for 160 and 177.5MeV/u helium ions[6]. Recently, Heilbronn et al. published two reports on DDY for 155MeV/u He, 155MeV/u C and 272, 435MeV/u Nb ions[7,8]. Our group also performed a systematic study on DDY and published three papers on DDY for 100, 180MeV/u He, 100, 180, 400MeV/u C and 100, 180, 400MeV/u Ne ions[9-11]. We further obtained new results for 400MeV/u Ar, 400MeV/u Fe, 400MeV/u Xe and 800MeV/u Si ions.

Neutron spectra measured in the forward direction have a broad peak at high-energy end, especially a large bump at 0 degree and this peak due to a break-up process becomes more prominent for lighter target and for higher projectile energy. The peak energy of this bump is about 60 to 70% of the projectile energy per nucleon. The high-energy neutrons in the forward direction spread up to about the energy which is about the twice of the incident particle energy per nucleon. The whole neutron spectra can be well fitted to the two-term moving source model that represents the intra-nucléar cascade neutron components and the evaporation neutron components.

The total neutron yields above 5MeV were integrated over a hemisphere from 0 to 90 degree, and they are shown in Fig. 1. The total neutron yields become slightly larger with increase of the target mass, but their dependence on the target mass is very small compared with the difference of neutron numbers of the target. The difference of neutron yields between He, C and Ne ion incidences is also very small, but the yields of Ar and Fe ion incidences are larger than the former three data. These differences of neutron yields might be caused by the neutron production cross section, the thickness of the target and neutrons produced by secondary charged particles.

The total neutron yield per incident projectile of proton and helium is summarized as a function of projectile energy Ep based on experimental and calculated data [12]. The yields in Fig. 1 and Ref. 12 are roughly proportional to the square of Ep below 500MeV/u and to Ep above it. In the case of no experimental data of DDY, they are usually calculated by using various Monte Carlo codes treating hadron cascade, such as NMTC, HETC, FLUKA, LAHET, MARS and so on, but these codes can not treat the transport of heavy ions heavier than He ion.

3. Bulk shielding calculation

Most shielding design for high energy accelerator has been generally performed by using a point kernel method, that is the Moyer model, which is based on single exponential attenuation of neutron dose equivalent after penetrated through a thick shield to be in a spectral equilibrium state. In this deep penetration problem, the dose attenuation length that is ruled by high-energy neutrons above about 100MeV is essentially important in the Moyer model.

In the energy lower than GeV, the slightly modified Moyer model has been developed by several workers that considered the energy dependence of neutron attenuation length[12]. The original Moyer model can only be applied for lateral shield calculation that corresponds to 90 degree neutron transmission from the neutron producing target on the projectile beam line. For shielding calculation in other directions, the angular dependence of the attenuation length is also given in some cases for concrete shield[12]. In this context, the attenuation length is of basic importance.

Neutron shielding experiments to obtain the dose attenuation length at several accelerator facilities have ever been performed to get the attenuation length, but the reliable experimental data penetrated through a very thick shield are still very scarce and dispersed as seen in Fig. 2. Since 1992 the shielding experiments of deep penetration have been performed at an intense spallation neutron source facility, ISIS, of the Rutherford Appleton Laboratory(RAL) [13]. We measured the neutrons penetrated through an additional shielding of iron and concrete that were put upon the top of neutron source shield consisting of approximately 3m iron and 1m ordinary concrete to obtain the attenuation length of high-energy neutrons.

In this experiment, we measured the attenuation profiles of the ^{11}C production rates by the $^{12}C(n, 2n)^{11}C$ reaction with the iron and concrete thickness. The attenuation length of this reaction rates correspond to those for neutron flux above 20MeV, and are 155g/cm^2 for iron (density of 7.8g/cm^3) and 125g/cm^2 for concrete (2.33g/cm^3) , as is shown in Fig. 2.

References

[1] J.W.Wachter, W.A.Gibson, W.R.Burrus, Phys. Rev., C6, 1496 (1972)
[2] M.A.Lone et al., Nucl. Instr. and Meth., A256, 135 (1987)
[3] M.M.Meier et al., Nucl. Sci. Eng., 102, 310 (1989)
[4] M.M.Meier et al., Nucl. Sci. Eng., 104, 339 (1990)
[5] J.V.Siebers et al., Nucl. Sci. Eng., 115, 13 (1993)
[6] R.A.Cecil et al., Phys. Rev., C21, 2471 (1980)
[7] L.Heilbronn et al., Nucl. Sci. Eng., 132, 1 (1999)
[8] L.Heilbronn et al., Phys. Rev., to be published
[9] T.Kurosawa et al., Nucl. Sci. Eng., 132, 30 (1999)
[10] T.Kurosawa et al., J. Nucl. Sci. Technol., 36, 41 (1999)
[11] T.Kurosawa et al., Nucl. Instr. and Meth., A430, 400(1999)
[12] T.Nakamura, Radiation Physics and Accelerator Safety Engineering, Chijin Shokan, Tokyo (1995), in Japanese
[13] Y.Uwamino, Proc. of OECD/NEA specialists meeting on Shielding aspects of Accelerator, Target and Irradiation Facilities (SATIF-1) Arlington, Texas (1994)

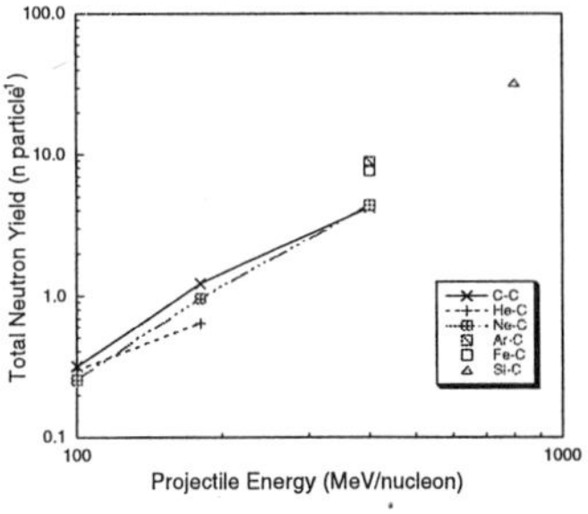

Fig. 1 Total neutron yields above 5MeV integrated for hemisphere from 0 to 90 degree

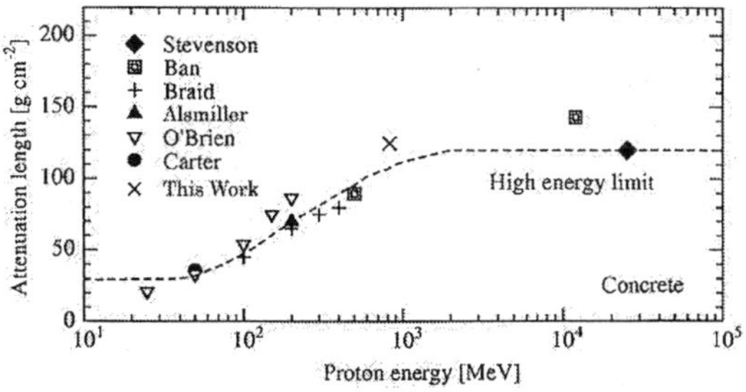

Fig. 2 Attenuation length of concrete for neutrons produced by various proton energies

RADIATION SAFETY CONSIDERATIONS FOR ISAC, THE ACCELERATED RADIOACTIVE ION BEAM FACILITY AT TRIUMF

LUTZ E. MORITZ

TRIUMF, 4004 Wesbrook Mall, Vancouver, B.C., Canada V6T 2A3
E-mail: lutz.moritz@triumf.ca

There has been a steadily growing worldwide interest in generating accelerated beams of unstable nuclei for use in a variety of applications such as nuclear physics, nuclear astrophysics, atomic and condensed matter physics, and medicine. At ISAC, an intense proton beam from the TRIUMF 500 MeV cyclotron bombards a thick target which is raised to high temperature. Radioactive atoms are in this way driven off into an ion source and the ions are extracted at a suitable potential. A high-resolution mass analyzer then separates the undifferentiated ion beam and a particular species is selected. These low-energy ions can be delivered to experiments directly or a post-accelerator can further accelerate them. The most severe radiological problems arise out of the deliberate generation of volatile activities in the target/ion source only a small fraction of which are extracted as a usable ion beam. A wide spectrum of activity must be addressed as it is desirable to use a variety of target materials up to and including Z=92. This paper describes the methods for calculating the radioactivity production, the means of limiting and containing the contamination and the equipment and procedures for handling the highly radioactive targets.

1 Introduction

ISAC (**I**sotope **S**eparator and **AC**celerator) is the new radioactive ion beam facility at TRIUMF. Radioactive ion beams are produced by bombarding a thick target with 500 MeV protons. The target is heated to high temperature and the atoms liberated in this way are transported by diffusion or effusion to an ion source contiguous with the target. The beam of mostly radioactive ions generated in the target/ion source is separated by a magnetic spectrometer and a selected species of ions is transported to experiments or to be further accelerated to ~ 1.5 MeV/u by an accelerator complex consisting of an RFQ and a drift tube linac. A wide variety of radioactive ions may be produced through judicious choice of the target material. There is some incentive to use high mass targets such as ThC or UC_2 to serve as general purpose generators of a wide spectrum of ions.

The goal is to eventually bombard targets with up to 100 μA of approximately 500 MeV protons, i.e. with a beam power of 50 kW. For a typical target this would lead to a saturated source strength of approximately 500 TBq and a residual γ-radiation field of up to 10 Gy h^{-1} at 1 m. It is clear that such targets must be handled remotely .

An associated radiological issue is the possibility of high levels of contamination that may be generated by operating such targets. The greatest concern is for those high-Z targets that can produce α-emitting radioisotopes.

2 Calculation of target radioactivity inventories

In order to make estimates of the radiation fields from the irradiated targets and to evaluate the possible consequence of releases of radioactive volatiles, the radioactive inventories have been calculated for typical targets using the Silberberg and Tsao semi-empirical cross section formulae [1,2] and the Monte Carlo code FLUKA [3] as well as the LAHET code system (LCS) [4]. For critical radioisotopes such as the radio-iodines, experimental data were used wherever possible. Fig. 1 shows a comparison between the Silberberg and Tsao esti-

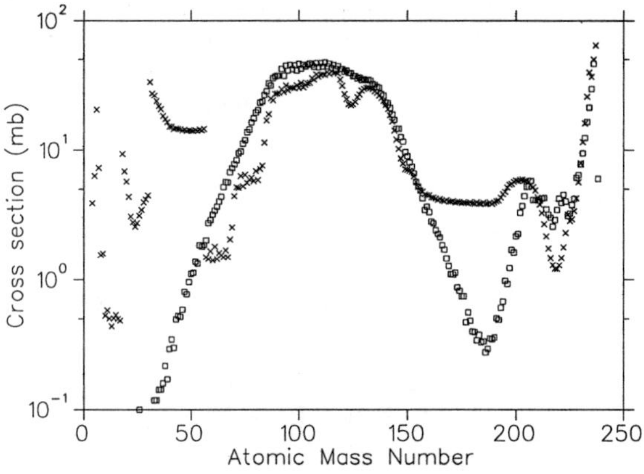

Figure 1. Radioisotope production cross sections summed over Z as a function of atomic mass number A for 500 MeV protons bombarding a U target. Open squares are calculated using LCS, crosses using the semi-empirical formulae of Silberberg and Tsao.

mates and calculations using LCS. The agreement is poor in some regions of radiological significance.

3 Handling of Targets

Because the γ-radiation fields from the targets are expected to be as high as 10 Gy h^{-1} at 1 m, ISAC has been designed with the capability to perform target servicing remotely. The target stations, two hot cells, and a service area are located inside the Target Maintenance Hall which is shielded on all sides by 1.2 m of concrete. This hall is kept at a negative pressure with respect to the outside world to minimize the spread of loose radioactive material. The targets are transported from the irradiation position to the service hot cell using a remotely operated crane (Fig. 2).

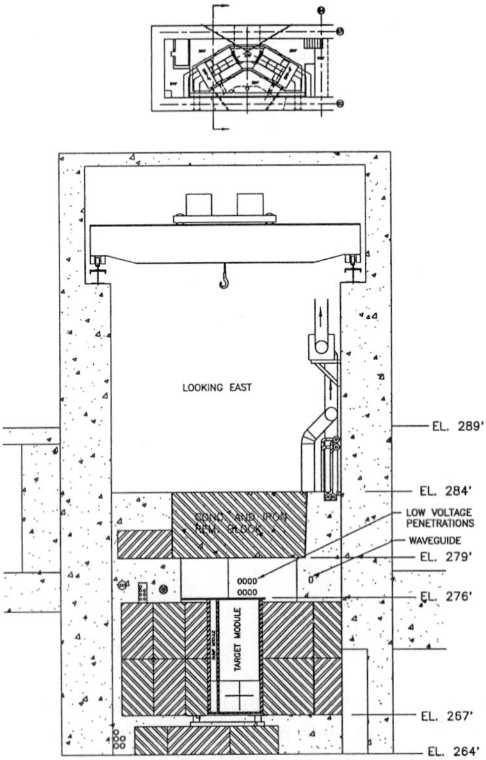

Figure 2. Cross section through the heavily shielded ISAC Target Maintenance Hall showing the remotely driven crane used to transport radioactive components of the target/ion source.

4 Loose Radioactivity Containment

The requirement of the experiments to efficiently extract the greatest possible quantity of radioactivity from the targets is at odds with the requirement to minimize the spread of radioactive material. Typically targets consist of an 'oven' made of a refractory material such as tantalum, filled with the target material which may be heated to temperatures up to 3000° C during bombardment. Optimizing such a target/ion source to yield the maximum ion beam intensity for a particular elemental species will also optimize it for chemically and physically similar species. The most severe problems will be encountered for targets of high atomic number such as Th and U because they are capable of producing α-emitting radioactivity.

The volatile species generated present the more severe problem. For high-mass targets all noble gas, iodine and radon isotopes may be produced. Of these the radio-iodines and radon progeny are the most radiologically significant. The radio-toxicity of alpha-emitting radioactivities is generally two orders of magnitude greater than that of the radioactivities usually encountered at accelerators. The solution at ISAC has been to capture and hold all the gases exhausted from the vacuum system during operation. The targets and front-end electrodes are housed in "containment boxes" with water cooled aluminum windows for the proton beam. The penetrations for the ion beam can be closed by shutters during transport of the target and front-end modules.

5 Internal Exposure of Workers

The threshold for the need to monitor the internal exposure of workers is usually defined by the national regulatory authorities in terms of the Annual Limit on Intake (ALI). Useful guidance is also provided by the ICRP Publication 54 [5]. In Canada the Atomic Energy Control Board gives regulatory guidance [6] for bioassay monitoring according to the scheme illustrated in Table 1. One of the most critical α-emitting radioactivities that can be produced by the bombardment of U by 500 MeV protons, both directly and via the decay of several Rn precursors, is ^{210}Po. At ISAC the equilibrium inventory of ^{210}Po in the facility is expected to be $\sim$ 35 GBq corresponding to $400,000\times$ALI [7] for this radioisotope alone. The need to set up a bioassay program for those workers involved in servicing the targets therefore seems unavoidable.

It will be necessary to be able to detect several 10's of Bq of ^{210}Po in a day's urine [8]. This is achievable but monitoring of a large number of people

Table 1. Requirements for Bioassay (adapted from *General Bioassay Guide*).

Confinement	Gases, volatile liquids and powders	Non-volatile liquids and solids
None	> 5 ALI	$> 5,000$ ALI
Fumehood	> 10 ALI	$> 10,000$ ALI
Glovebox	> 500 ALI	$> 500,000$ ALI
Pseudo-sealed	> 50 ALI	$> 10,000$ ALI

will require significant investment in time and resources.

6 Conclusion

Commissioning of ISAC is now proceeding. There have been two brief production runs at a proton beam intensity of 1 μA using a CaO target and a one month-long run at a proton beam intensity of 10 μA on a Nb target. The latter established ISAC as the most intense radioactive beam facility in the world. So far there have been no unexpected radiological problems although much work remains to prepare the facility for routine operation at full intensity.

References

1. R. Silberberg and C.H. Tsao, *Astroph. J. Supp.* Series No. 220(I), 25, 315 (1973).
2. R. Silberberg and C.H. Tsao, *Astroph. J. Supp.* Series No. 220(I), 25, 335 (1973).
3. A. Fassò et al., *FLUKA98 Users Manual*, (1998).
4. R.E. Prael and H. Liechtenstein, LA-UR-89-3014, Los Alamos National Laboratory, (1989).
5. ICRP Publication 54, *Annals of the ICRP* Vol. 19, No. 1-3, (1988).
6. Atomic Energy Control Board of Canada, *General Bioassay Guide*, Report of the AECB Working Group on Internal Dosimetry, Draft, (July 1997).
7. ICRP Publication 61, *Annals of the ICRP* Vol. 21, No. 4, (1988).
8. J.R. Johnson, Bioassay Monitoring Requirements for the TRIUMF ISAC Facility, *TRIUMF Internal Report*, (1998).

CONTINUOUS MONITORING OF POSSIBLE FUEL FAILURE OF A POOL-TYPE REACTOR

KEN-ICHI SASAKI AND CHIKAYOSHI YAMADA

College of Science, Rikkyo University, 3 Nishi-Ikebukuro, Toshima-ku, Tokyo 171-8501, JAPAN
E-mail: ksasaki@mail.com

SHU A. HAYASHI AND TATSUO MATSUURA

Institute for Atomic Energy, Rikkyo University, 2-5-1 Nagasaka, Yokosuka 240-0101, JAPAN

A simple and sensitive continuous monitoring system for the possible fuel failure of a pool-type nuclear reactor was successfully developed. Fuel failure detection (FFD) can be made by the detection of fission products (FP's) leaking from fuel element containers. In our system, fission rare gases (^{88}Kr, ^{89}Kr, and ^{138}Xe) in the cooling water are collected with bubbled nitrogen gas, and then they are continuously led into a vertical empty column, where they decay to yield daughter nuclides (^{88}Rb, ^{89}Rb, and ^{138}Cs) which are extracted to aqueous phase. The radioactivity of daughter nuclides in the aqueous phase are continuously measured using a flow-through Ge-detector with excellent signal-to-noise ratios, because the collected fission rare gases are completely isolated from non-volatile background radioactive nuclides in the reactor coolant. The newly developed *continuous* FFD system responded with so high sensitivity that an anomalous (twice) change in the concentration of FP rare gases in the coolant would be detected within about 30 minutes using the short-lived ^{89}Kr- ^{89}Rb pair.

1 Introduction

A more sensitive and simpler continuous monitoring system for possible fuel failure in a nuclear reactor must be established for safe reactor operation and quick response for an accident, although some techniques have been developed [1, 2, 3].

Nuclear reactor coolant contains fission originating rare gases, ^{89}Kr (3.15 min) and ^{138}Xe (14.08 min), which are easily collected by purging with bubbled nitrogen gas [3] and decay to yield ^{89}Rb (15.15 min) and ^{138}Cs (33.41 min), respectively. ^{88}Kr (2.48 hr), whose daughter is ^{88}Rb (17.78 min), is collected at the same time. The collected gases are completely isolated from non-volatile radioactive nuclides in the coolant, so that they make FFD easy with excellent signal-to-noise ratios.

In a previous work [4], we proposed a simple, economical, and fully-automated continuous monitoring system, which consists basically of a column-like large chamber, where the collected rare gases decay and their daughters are extracted into a washing water flowing counter-currently to the gases. All these processes proceed continuously, and the effluent solution is led to a radiation detector for monitoring. The present paper reports the results obtained using the proposed column systems.

2 Layout and Operation of Monitoring Systems

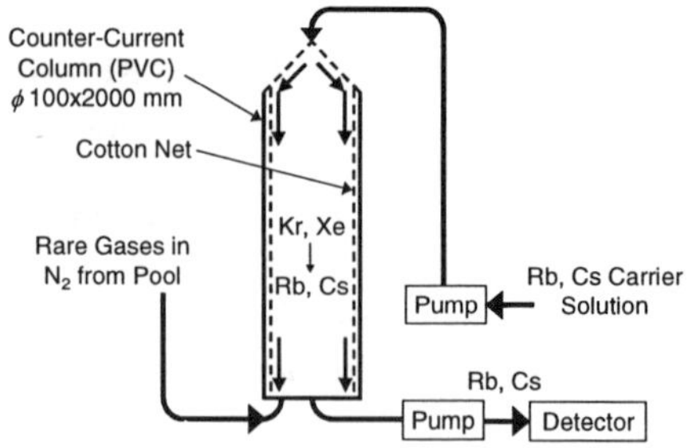

Fig. 1 Main part of FFD system consists of counter current column for the decay of rare gases and for the extraction of daughter nuclides into water.

Figure 1 shows the design of the major part of the most successful FFD system, we tried, using a counter current column without packing materials except cotton net covering inner wall for wetting. A nitrogen gas bubbler and a gas collector, a large funnel, in the pool (both not illustrated) collect rare gases with bubbled nitrogen, and supply them to the column continuously, where the gases fed into from the bottom, flow upward, and decay to produce daughters, ^{88}Rb, ^{89}Rb and ^{138}Cs. On the other hand, a washing solution containing rubidium and cesium chlorides as carriers is fed into the column from the top, and flows downward along the hydrophilic cotton net counter-currently to the gas to extract and wash the daughters out of the column. ^{41}Ar coming with Kr and Xe is removed from the effluent using a small nitrogen bubbling unit, and then the effluent is led into a flow through cell, which is made of long polyvinyl chloride tubing (20 m in length and 5 mm in inner diameter; 340 mL) in the shape of coil wound around a Ge-detector.

3 Results and Discussion

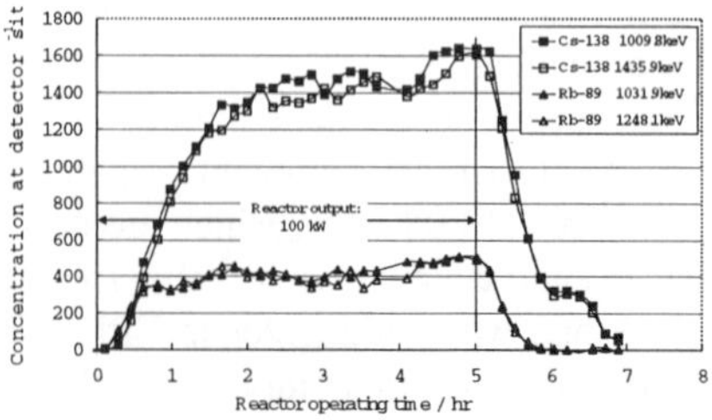

Fig. 2 Typical response curve obtained using **counter column**

In Fig. 2, the concentrations of the rare gas daughters are plotted against reactor operating time. The corresponding counting rates are normally a few hundred counts per minute, which are the background radioactivity level in the TRIGA Mark II reactor, Rikkyo University.

Although the empty column allows the convective flow of gases inside, it showed the best sensitivity among several "decay-extraction chambers" tried, such as packed column [3] and a rotary drum.

According to Fig. 2, the radioactivity of shorter-lived ^{89}Rb rises and saturates more rapidly than ^{138}Cs. The activity of ^{88}Rb (not shown) does not saturate during the reactor operation and is not helpful for FFD.

There are many parameters affecting response curves such as Fig. 2. Major parameters are a sampling position in the pool, the flow rates of nitrogen gas and of the washing carrier solution, a column volume, and the volume of connecting units.

In the present work, we varied the sampling position and the flow rate of nitrogen gas, and kept all the other parameters constant.

Figure 3 shows the effects of the flow rate of nitrogen and of a sampling zone upon the counting rate of ^{89}Rb. Similar results were obtained for ^{138}Cs. In any zone, the counting rates showed maxima at 2-3 L min^{-1}. The higher flow rate reduces the more traveling (decay) time between the pool and the column, but the volume of their connecting unit is only 0.3 L, so that the marked increase in counting rate is not caused by the traveling time reduction but by the increased rare gas concentration in the sampling zone owing to the accelerated circulation of water with bubbling. The decrement of the counting rates in the higher flow rate is caused by the reduction of the residence time of rare gasses in the column for their decay.

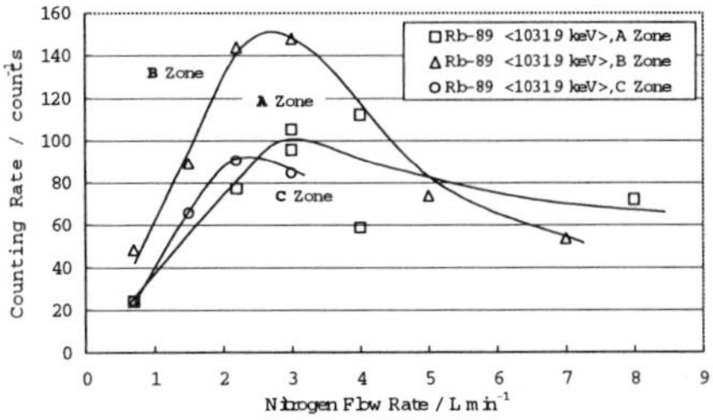

Fig. 3 Effect of sampling position and nitrogen gas flow rate for sampling.

As is seen in Fig. 3, the position of gas sampling affected the observed counting rate critically. Three zones locating horizontally in the cylindrical pool are selected as sampling positions; A-zone, a central part, the upper position of fuel elements; B-zone, the midway of the center and the pool wall; C-zone, the closest position to the wall. Interesting to say, the maximum counting rate was not observed in A-zone but in B-zone. These results show that the distribution of fission originating rare gases is complicated and critically influenced by various parameters such as natural convective flow, circulating flow controlled by the coolant maintenance system, and the additional circulation by bubbling for sampling.

References

1. Ando, T., Kaneko, and Y., Fujisawa, *M., FAPIG*, No. 94 (1980) pp. 21.
2. Dodd, B., and Johnson, A. G., Searching for a possible fuel element leak, In *Tenth Biennial U. S. TRIGA Users& Conference, College Station, Texas, April 6-9, 1986,* TOC-18 (1986) pp. 3.5-3.10.
3. Hayashi, S. A., Matsuura, T., Tomura, K., and Harasawa, S., Effective extraction of fission-originating rare gas nuclides in a cooling water of a research reactor and its use as a sensitive monitor of fuel failure, In *Proc. SERINA (Symp. on Env. Rad. Nucl. Impact in Asia), Taipei,* (1996) pp 177-185.
4. Sasaki K., Hayashi S. A. and Matsuura T., Proposal of Novel Method of Continuous Monitoring of Possible Fuel Failure of a Pool-Type Reactor, In *Conference Handbook, The Second International conference on Isotopes*, Paper 125 (1997) pp. 53.

Session I2

Applications in Agriculture and Nutrition

VEGETATION CHANGES DURING THE QUATERNARY INFERRED FROM CARBON ISOTOPES ANALYSES ON CHARCOAL AND SOIL ORGANIC MATTER, STATE OF SÃO PAULO, SOUTHEAST BRAZIL

L.C.R. PESSENDA[1]

1 ^{14}C Laboratory, Centre for Nuclear Energy in Agriculture, University of São Paulo, Piracicaba, SP, Brazil
E-mail: pessenda@cena.usp.br

S.E.M. GOUVEIA[1], R. BOULET[2], R. ARAVENA[3] AND R. SCHEEL-YBERT[4]

2 IRD, France, Geosciences Institute, University of São Paulo, São Paulo, SP

3 Department of Earth Sciences, University of Waterloo, Waterloo, Ontario, Canada

4 Laboratoire de Paléoenvironnements, Anthracologie et Action de l'Homme, Institut de Botanique, Université de Montpellier II, Montpellier, France

This paper presents an attempt to reconstruct vegetation changes and inferred climatic conditions in the Jaguariúna and Botucatu regions (state of São Paulo, southeast of Brazil) during the late Pleistocene and Holocene, based on δ^{13}C values of soil organic matter (SOM), ^{14}C dating and analysis of charcoal present in the soil profiles. Sampling sites were located under natural forest vegetation ("cerradão"), along the slopes of small hills. Charcoal was found predominantly between 50 and 150 cm deep, indicating a period of greater frequency of fires between 3000 and 6000 years BP. This suggest the existence of drier conditions during this period. The presence of charcoal at depth on the tops of the hills sites suggests sedimentation could not be the major process responsible for burying the charcoal within the soil. Biological activity (termites, ants, earthworms) played an important role in this process. For the Jaguariúna site, enriched δ^{13}C values of SOM were observed from the late Pleistocene until mid Holocene, indicating the presence of grasses and probably a drier climate when compared with present-day conditions. For the Botucatu site, the δ^{13}C profile of SOM suggests the predominance of forest vegetation during the entire Holocene. Structural analyses of buried charcoal confirmed the presence of savanna species in Jaguariuna supporting the interpretation of the δ^{13}C data.

1 Introduction

Many studies have demonstrated that past climatic variations caused alterations in the ecosystems. These variations can remain historically registered in several matrix, such as in the sediments deposited in lacustrine environment [1] and in the soil organic matter [2, 3, 4]. The study of charcoal fragments found in these materials also supplies information about palaeoclimatic changes [5, 6].

Charcoal datings present great reliability concerning the establishment of chronology on paleoclimate studies [9]. Its presence can only be explained by the occurrence of paleofires during drier episodes and/or the human disturbance [3, 5, 6].

In this paper we report $\delta^{13}C$ data of the soil organic matter (SOM), microscopic analysis and ^{14}C dating of charcoal from four soil profiles located under natural vegetation (cerradão), along the tops and slope of small hills. The aim of the project was to study probable vegetation and climate changes in the Botucatu and Jaguariúna regions (state of São Paulo, southeast of Brazil), during Pleistocene/Holocene.

2 Study sites

The study sites, Botucatu and Jaguariúna, are located in the São Paulo State. The soils of Botucatu are developed on basalt, with interference of sandstone, and the soils of Jaguariúna on diabase. The two soil profiles collected at Botucatu were obtained in the top of two neighboring hills and the soil is Oxisol. The soil samples at Jaguariúna were collected on a slope (Figure 1). One profile is located near of the top, immediately downstream of the emergent rock and the soil is an Oxisol. The other is 75 m downstream of the first profile and it corresponds to an Acrisol. In both sites the vegetation is a subtropical semideciduous forest (cerradão).

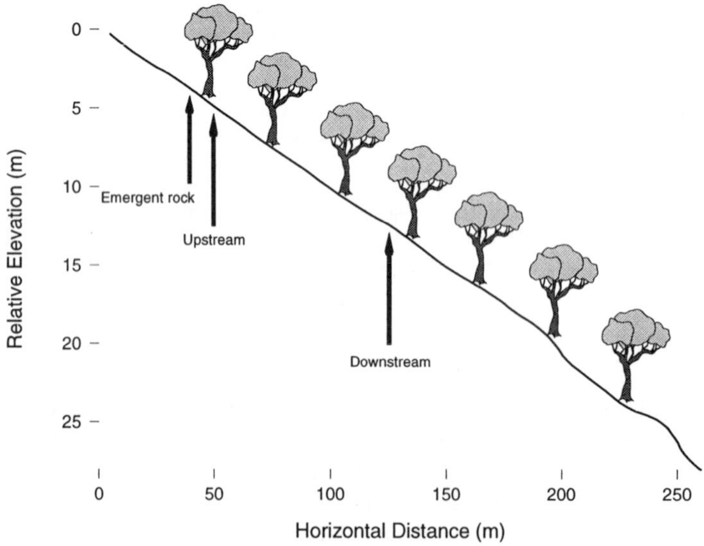

Figure 1 - Location of excavations **at** Jaguariúna.

3 Material and Methods

3.1 SOM

Soil samples were collected from excavations of 250 cm depth. Sampling involved the collection of up to 10 kg of material at 10 cm intervals.

The $\delta^{13}C$ analysis and total organic carbon on soil samples were carried out at the Environmental Isotope Laboratory, University of Waterloo (Canada). $^{13}C/^{12}C$ data are expressed in δ (‰) units relative to the PDB standard and organic carbon in percentage.

3.2 Charcoal

Charcoal were collected during the sieving of soil samples. The ^{14}C analyses were carried out at the Radiocarbon Laboratory (Brazil) using the benzene method and liquid scintillation counting [8] and, for small samples, at the Isotrace Laboratory (Canada), employing the AMS technique. Radiocarbon ages are reported as years BP. Charcoal fragments were analyzed at the Laboratoire de Paléoenvironnements, Anthracologie et Action de l'Homme (France).

4 Results

The Botucatu soils are clayey or clay-sandy. The total carbon concentration varied from 2.8% to 0.5% in both locations [9].

The highest amount of charcoal was observed between 50 and 150 cm soil depth, related to the period 4000-5600 years BP, probably corresponding to a period of frequent fires and drier conditions [9].

The $\delta^{13}C$ values of SOM (Figure 2) on the Botucatu I range from -26.0‰ to -24.7‰ and are characteristics of C_3 plants. The data of Botucatu II from -26.1‰ to -22,2‰, indicated a probable influence of C_4 vegetation, but with predominance of C_3 type. These data suggest that a significant (drastic) climate and vegetation change didn't occur in the Botucatu area during the studied period. The more depleted $\delta^{13}C$ values (up to 2‰) and the higher amount of charcoal observed in the interval 50-160 cm in Botucatu I suggests a higher density of trees in the past.

The charcoal analyses indicated a important contribution of semideciduous forest taxa (*Sebastiania* sp, cf. *Acacia,* cf. *Chlorophora*) and wooded savanna taxa (*Peltogyne* sp, *Qualea* sp, *Vochysia* sp, *Gramineae*, *Palmae*), supporting the isotopic data.

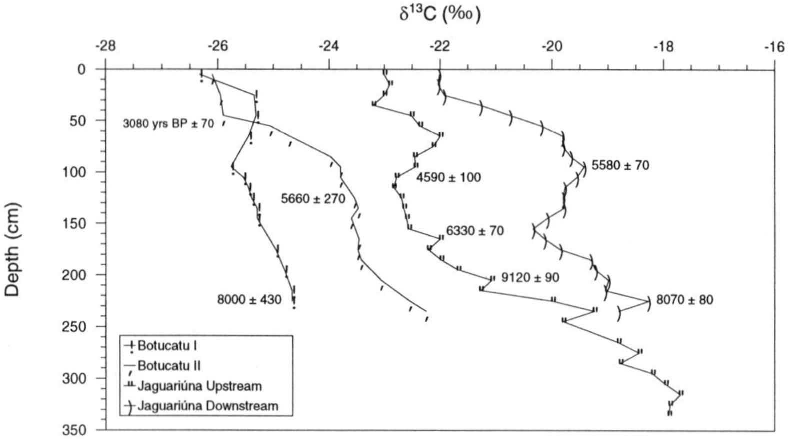

Figure 2 - $\delta^{13}C$ values of soil organic matter of Botucatu and Jaguariúna areas.

At the Jaguariúna site, the soils are clayey and the carbon content decreased from the surface to deeper layers (2.1% to 0.4%). The period of higher frequency of charcoal (60-220 cm) and probable paleofires correspond to the 4,000 and 9,000 years BP age range [9].

The $\delta^{13}C$ values of SOM are more enriched than the Botucatu soils (Figure 2) indicated a mixture of C_3 and C_4 vegetation. A trend toward more enriched $\delta^{13}C$ values is observed below 4,000 year at both profiles suggesting a major influence of C_4 plants, typical of savanna vegetation. Assuming that the charcoal ages increase regularly with depth, the predominance of C_4 plants (-17.7‰) at 240 cm indicates a drier climate in the late Pleistocene.

The charcoal analyses indicated a important contribution of forest taxa (*Patagonula americana, Centrolobium aff. Tomentosum)* in the upper part of the profile (younger than 4000 years BP). In the lower part (from 4000 to 9000 years BP) it was detected the presence of savanna taxa (*Acosmium* sp, *Bowdichia* sp, *Plathymenia reticulata, Qualea* sp, *Gramineae).* These analyses support the interpretation of the isotope data.

5 Conclusions

The soil carbon isotopes data and the ^{14}C dating and charcoal analyses, indicated the predominance of C_3 vegetation during the whole Holocene in the Botucatu area, suggesting no significant climate change occurred in this region.

In contrast, the data collected at the Jaguariúna sites indicated the predominance of C_4 grasses during the late Plestocene to the mid Holocene, suggesting a drier climate. From mid Holocene to the present, the data suggest the presence of more C_3 plants, indicating a change to more humid conditions.

6 Acknowledgements

We gratefully acknowledge financial support from São Paulo Foundation for Research (FAPESP) grants no.95/5047-5, 96/12777-2 and PRONEX/FINEP no. 41.96.0938.00. We also thank M.V.L. Cruz for sample preparation and ^{14}C analyses.

7 References

1. Ledru, M.-P. 1993. Late Quaternary environmental and climatic changes in central Brazil. *Quaternary Research,* 39:90-98.
2. Desjardins, T.; Carneiro Filho, A.; Mariotti, A.; Chauvel, A.; Girardin, C. 1996. Changes of the forest-savanna boundary in Brazilian Amazonia during the Holocene as revealed by soil organic carbon isotope ratios. *Oecologia,* 108:749-756.
3. Pessenda, L.C.R., Aravena, R., Melfi, A.J. and Boulet, R., The use of carbon isotopes (^{13}C, ^{14}C) in soil to evaluate vegetation changes during the Holocene in central Brazil, *Radiocarbon* **38** (1996) pp.191-201.
4. Pessenda, L.C.R., Gomes, B.M., Aravena, R., Ribeiro, A.S., Boulet, R. and Gouveia, S.E.M., The carbon isotope record in soils along a forest-cerrado ecosystem transect: implications for vegetation changes in the Rondonia state, southwestern Brazilian Amazon region, *The Holocene* **8** (1998) pp. 631-635.
5. Soubiès, F. 1980. Existence d'une phase sèche en Amazonie brésilienne datée par la présence de charbons dans les sols (6000-3000 ans BP). *Cahiers ORSTOM. Géologie,* 11(1):133-148.
6. Scheel, R., Vernet, J.-L., Wengler, L., Fournier, M. 1995. Carvões do solo em São Pedro, Estado de São Paulo, Brasil: Datação, notas sobre o paleoambiente no Quaternário recente, condições de depósito e origem do fogo e proposta de estudos antracológicos. In: Congresso da ABEQUA, 5, Niterói, Anais, p.169-175.
7. European Science Foundation, 1985. Handbooks for archaeologists. 1.ed. Strasbourg; ESF, 65p. (Radiocarbon Dating, 3).
8. Pessenda, L.C.R. and Camargo, P.B., Datação radiocarbônica de amostras de interesse arqueológico e geológico por espectrometria de cintilação líquida de baixa radiação de fundo, *Química Nova* **14** (1991) pp.98-103.
9. Gouveia, S.E.M.; Pessenda, L.C.R.; Boulet, R.; Aravena, R.; Scheel-Ybert, R. 1999. Isótopos do carbono dos carvões e da matéria orgânica do solo em estudos de mudança de vegetação e clima no Quaternário e da taxa de

formação de solos do estado de São Paulo. *Anais da Academia Brasileira de Ciências* (aceito para publicação).

NITROGEN FIXATION IN CHICKPEA AND GROUNDNUT GENOTYPES AS INFLUENCED BY BRADYRHIZOBIAL INOCULANT USING ^{15}N TECHNIQUE

M.A. SATTAR AND A.K. PODDER
Bangladesh Institute of Nuclear Agriculture, Mymensingh-2200, Bangladesh

E-mail: bina@bdmail.net

Field experiments were conducted to determine the extent of N uptake derived from soil (Ndfs), fertilizer (Ndff) and atmosphere (Ndfa) in several genotypes of chickpea and groundnut using ^{15}N isotope technique as influenced by bradyrhizobial inoculation. Variations were observed among the genotypes as regard to N uptake from above sources. The Ndfs by chickpea ranged from 15-26 kg/ha under uninoculated condition and 16. 9-32.9 kg/ha under inoculated condition. The ranges of Ndfa was 34.6-48.5 kg/ha in uninoculated condition and 79. 8-87.3 kg/ha with inoculation. Of the four genotypes, G-97 recorded highest amount of Ndfa under inoculated condition and Ndfs under both inoculated and uninoculated condition while Hyprosola recorded higher Ndfa under uninoculated condition. In case of groundnut, the ranges for Ndfa were 57-9-115.0 kg/ha and 111-192 kg/ha under uninoculated and inoculated condition, respectively. The ranges for Ndfs were 41-100 kg/ha in uninoculated condition and 53-143 kg/ha with inoculation. The genotype 43-17 recorded highest amount of Ndfa at both the conditions. However, the Ndfs in plant was highest by genotype 62-30 in uninoculated condition and by 86-54 in inoculated condition. The study also indicated that the yield and yield contributing characters were positively correlated with the Ndfa but not with the Ndff and the Ndfs.

1 Introduction

Legumes have the capacity to derive a considerable proportion of their N requirement from the atmosphere through symbiosis with rhizobia [4]. Rhizobia, however, differ in their abilities to survive and to nodulate legumes under adverse soil conditions and on different varieties [1]. Differences in nitrogen fixation potential of the cultivars of the same legume species occur [2].

Chickpea (*Cicer arietinum* L.) and groundnut (*Arachis hypogaea* L.) are the two important legumes widely grown in Bangladesh, but little is known as to the amount of N fixed by these legumes in this part of the region. Therefore, studies were conducted to ascertain the proportion and the amount of N fixed by several promising genotypes of chickpea and groundnut with and without inoculation.

2 Materials and Methods

Two field experiments with chickpea and groundnut were conducted during November 1993 to June 1994 at the experimental field of Bangladesh Institute of Nuclear Agriculture (BINA), Mymensingh, Bangladesh. The soil was a sundy loam

(Old Brahmaputra Floodplain/Haplaquepts, Sonatala series) with C 0.67%, total N 0.05%, P 10.5 ppm, pH 6.8 (Soil : water). Soil was fertilized with a basal dose of triple super phosphate (@ 25 kg P ha^{-1}) and muriate of potash (@ 33 kg K ha^{-1}) at the time of land preparation one day before sowing.

Four genotypes of chickpea(Hyprosola, G-97, G-299 and commercial variety Nabin) and six genotypes of groundnut(43-17, 62-30, 86-54, 64-82, 16-90 and commercial variety Dacca-1) were used in the experiments. The genotypes were selected on the basis of extensive field trials conducted for several years. Seeds of chickpea and groundnut were inoculated with peat based mixed culture bradyrhizobial inoculant, BINA-CP-2 and BINA-GN-2, respectively. Sugarcane molasses (40 gm kg^{-1} seed) was used as an adhesive [5].

A solution of ammonium sulphate containing 5% ^{15}N atom excess (a.e.) was uniformly sprayed one week after planting @ 20 kg N ha^{-1} to the legume plots, while the reference crops (wheat for chickpea and maize for groundnut) received a 100 kg N ha^{-1} containing 1% ^{15}N a.e. $^{15}N/^{14}N$ ratio analysis of samples was performed at Seibersdorf Laboratory of FAO/IAEA, Vienna, Austria using a mass spectrometer. The "A value" approach of Fried and Broeshart and the equations used in Hardarson et. al. [3] were followed to calculate the amount and proportion of N fixed.

The design of the each experiment was a split plot with varieties in the main plot. Isotopic plots of 1m x 3m were established to the yield plot of 3m x 4m. The treatments were replicated three times. Analysis of variance was performed on the data and means were classified according to Duncan's Multiple Range Test ($P < 0.05$) within each parameter.

3 Results

Significant effects of chickpea genotypes, rhizobial inoculant and their interaction were detected for almost all parameters excepting varietal effect on nitrogen derived from air (Ndfa) and interaction effect on total N were non-significant. The proportion and total amount of N obtained from atmosphere was always higher in inoculated treatments (Table 1). The ranges of Ndfa was 34.6 - 48.5 kg/ha in uninoculated condition and 79.8-87.3 kg/ha with inoculation. The nitrogen derived from soil (Ndfs) ranged from 15 - 26 kg/ha under uninoculated condition and 16.9 - 32.9 kg/ha under inoculated condition. Of the four genotypes G - 97 recorded highest amount of Ndfa under inoculated condition and Ndfs under both inoculated and uninoculated condition while Hyprosola recorded higher Ndfa under uninoculated condition.

The data for groundnut as regard to total N, nitrogen derived from air, soil and fertilizer showed that these characteristics were significantly influenced independently by the *Bradyrhizobium* inoculant and the genotypes as no genotype x inoculant interaction was observed. The amount of total N and the proportion of atmospheric N fixed were also different for different genotypes (Table 2). The ranges for Ndfa were 57.9 - 115.0 kg/ha and 111 - 192 kg/ha under uninoculated and inoculated condition respectively. The ranges for Ndfs were 41 - 100 kg/ha in uninoculated condition and 53 - 143 kg/ha with inoculation. The genotype 43-17 recorded highest amount of Ndfa at both the conditions. The Ndfs was, however, highest by genotype 62-30 in uninoculated condition and by 86-54 in inoculated condition.

Correlation studies showed that Ndfa was highly correlated with nodule number (r = 0.680/0.707) and mass (r = 0.882/0.598) and dry matter (r = 0.642/0.554) at mid-flowing stage and seed yield (r = 0.852/0.578) at harvest (data not presented) in both crops (chickpea/groundnut) studied. Significant negative correlation between Ndfa and % Ndfs were also recorded.

Table 1 : Total N, nitrogen derived from fertilizer (Ndff), soil (Ndfs) and fixation (Ndfa) in kg ha^{-1} (or percent) of chickpea genotypes as influenced by inoculation.

Treatments					
Genotype	**Inoculant**	**Total N**	**Ndfs**	**Ndff**	**Ndfa**
G-97	Uninoc.	74.9c[a]	26.1b (34p)	4.34b (5.9p)	44.3bc (59s)
	Inoc.	126a	32.9a (26qr)	5.55a (4.4qr)	87.3a (70qr)
G-299	Uninoc.	69.2cd	24.3bc (39p)	4.27b (6.7p)	34.6c (54s)
	Inoc.	109b	23.8bc (23r)	4.25b (3.9rs)	79.8a (73q)
Nabin	Uninoc.	58.2d	16.9d (29q)	2.93c (5.0q)	38.4bc (66r)
	Inoc.	102b	18.3cd (18st)	3.17c (3.1st)	81.0a (79p)
Hyprosola	Uninoc.	68.2cd	15.0d (23rs)	2.51c (3.8rs)	48.5b (74q)
	Inoc.	105b	17.3d (16t)	2.99c (2.8t)	84.4a (81p)
	CV (%)	10	15	13	9

[a]Values in the same column followed by the same letter are not significantly different ($P < 0.05$)

4 Discussion

Total N and the proportions of N from all the three sources viz. air, soil and fertilizer were found strongly influenced by the inoculant in both legumes studied. The non-significant genotype x inoculant interaction in groundnut is probably due to narrow genetic base of the genotypes investigated. Some of the genotypes tested in this programme were better than the commercial variety. Genotypes that use less soil N and fix more atmospheric N may be utilized in the crop rotation programme to reduce further degradation of the soil.

Table 2 : Total N, nitrogen derived from fertilizer (Ndff), soil (Ndfs) and fixation (Ndfa) in kg ha^{-1} (or percent) of groundnut genotypes as influenced by inoculation.

Treatments					
Genotype	Inoculant	Total N	Ndfs	Ndff	Ndfa
43-17	Uninoc.	170def[a]	53.2cde (31st)	1.95b-e (1.1qr)	115cd (68pq)
	Inoc.	281a	87.4b (30st)	2.18bcd (0.8s)	192a (69pq)
62-30	Uninoc.	197cde	100b (51pq)	2.47b (1.3pq)	93.8de (48st)
	Inoc.	276a	90.2b (33st)	2.22bc (0.8rs)	183a (66pq)
86-54	Uninoc.	153ef	93.3b (61p)	2.29bc (1.5p)	57.9e (38t)
	Inoc.	257ab	143ab (56p)	3.52a (1.4pq)	111cd (43t)
64-82	Uninoc.	140f	48.3de (34st)	1.18de (0.8rs)	90.4de (65pq)
	Inoc.	261ab	79.3bc (30st)	1.92b-e (0.7s)	180ab (70pq)
16-90	Uninoc.	159ef	72.0bcd (45qr)	1.77b-e (1.1qr)	85.3de (54rs)
	Inoc.	236abc	89.3b (37rs)	2.17bcd (0.9rs)	144bc (62qr)
Dacca-1	Uninoc.	125f	41.0e (33st)	1.00e (0.8rs)	82.8de (66pq)
	Inoc.	216bcd	53.1cde (25t)	1.31(0.6s)	162ab (75p)
CV	(%)	13	21	28	17

[a]Values in the same column followed by the same letter are not significantly different ($P < 0.05$)

5 Acknowledgements

We thank Dr. S.K.A. Danso of IAEA for his kind help in ^{15}N analysis and Ms. Jana Thomson of TRIUMF for restructuring the manuscript.

References

1. Bromfield E.S.P. and Ayanaba A., The efficacy of soybean inoculation on acid soil in Tropical Africa. *Plant and Soil* 54 (1980) pp. 95-106.
2. Duc. G., Mariotti A. and Amargar N., Measurements of genetic variability for symbiotic nitrogen fixation in field grown fababean (*Vicia faba* L.) using a low level ^{15}N-tracer technique. *Plant and Soil* 106 (1988) pp. 269-276.
3. Hardarson, G., et al., Measurements of nitrogen fixation in fababean at different N fertilizer rates using the ^{15}N dilution and 'A-value' methods. *Plant and Soil* 131 (1991) pp. 161-168.
4. Senaratne R., et al. Effect of combined nitrogen on nitrogen fixation of soybean (*Glycine mox* L Merill.) as affected by cultivar and rhizobial strain. *Plant and Soil* 103 (1987) pp. 45-50.
5. Vincent J.M., A Manual for the Practical Study of the Root Nodule Bacteria. 1BP Handbook No. 15. Blackwell Scientific Publications Ltd. Oxford (1970) pp. 168.

ELEMENTAL COMPOSITION OF COMMERCIAL BRAZILIAN COFFEE USING NEUTRON ACTIVATION ANALYSIS

F.S. TAGLIAFERRO, E.A.N. FERNANDES AND E. J. FRANÇA

Centro de Energia Nuclear na Agricultura, Universidade de São Paulo
P.O. Box 96, 13400-970, Piracicaba SP, Brazil
E-mail: fstaglia@carpa.ciagri.usp.br

Some Brazilian coffee brands were studied using instrumental neutron activation analysis (INAA). As roasted ground coffees are normally blended, in order to compare mixed with pure coffee, commercial brands were selected from the local market, five whole beans and seven roasted grounds. Additional information on the differences between conventional and organic coffee is supplied. After irradiation, the induced radioactivity was detected in germanium hyperpure detector and the elemental concentration was analyzed by the k_0 method. ANOVA was used to verify the hypothesis of equality of elemental composition among the different brands and kinds of coffees for all determined elements.

1 Introduction

Currently, one of the greatest challenges is the production of food, not only in sufficient amounts, but also with adequate nutritional quality, assuring nutritional balanced and a contaminant free food supply. In this context, interest in studies related with toxic and essential elements have increased in last years, stimulating use of refined techniques for quantification. The evaluation of coffee quality in relation to elemental composition has great and special interest, particularly because of the importance of coffee in nutrition and international trade [1].

Brazil is the world largest coffee producer with 1,69 x 10^9 t in 1998, which corresponds to approximately 26% of the world production. Internal consumption has been increased from 384 x 10^6 kg in 1985 to 750 x 10^6 kg in 1998. Despite this fact, there are few studies on the quality of Brazilian commercial coffees related to elemental composition. Nowadays there are about 1800 roasting enterprises in the country, which are responsible for about 2500 coffee brands. Thus, to evaluate the elemental composition in a comprehensive way, some brands of coffee were selected based on participation and tradition in the market and analyzed by INAA. This study should be seen as an initial investigation of Brazilian coffees, and should not be extrapolated as a national mean estimate.

2 Experimental

Coffee samples were obtained from local market, selecting 5 whole beans and 7 roasted grounds. Among roasted grounds, one was made with organic coffee. After particle size reduction in alumina grinder, samples were homogenized and 3 replicates of approximately 350 mg placed into special polyethylene capsules. Certified reference materials SRM-1515 (NIST), RM-8433 (NIST) and V-10 (IAEA) were also analyzed. Irradiation was carried out in the research reactor IEA-R1m, of Instituto de Pesquisas Energéticas e Nucleares (IPEN), with a neutron flux of about 10^{13} n cm^{-2} s^{-1} during 8 hours. Induced radioactivity was measured with a hyperpure germanium detector (45% relative efficiency) after 4, 8, 17 and 40 days. Quantification was performed on basis of the k_0 method.

3 Results and discussion

Elemental concentration in food is a consequence of many factors, such as environmental conditions, production and processing methods, pesticides and fertilizers exposure, besides soil composition of plantation area. The mean concentrations and coefficients of variation of three replicates for Br, Ca, Co, Cs, Fe, K, Na, Rb, Sc and Zn are shown in the Table 1. The highest levels of Br were observed for the *Pelé* brand (2.58 µg g^{-1}), about 3.5 times higher than the levels encountered in the brands *Morro Grande* (0.670 µg g^{-1}) and *Do Ponto* (0.711 µg g^{-1}), both whole beans. The highest concentrations found by Orli• *et al.* [2] and Krivan *et al.* [1], both studying green coffees from different countries, were 1.2 µg g^{-1} and 1.7 µg g^{-1}, respectively. Among roasted ground coffee brands, the lowest concentration was observed for *Ituano* (0.825 µg g^{-1}), which uses an organically produced coffee.

No significant differences were observed for Ca, with mean concentrations between 1142 and 1408 µg g^{-1}, which agree with those found by Orli• *et al.* [2] and Martín *et al.* [3], but were greater than those published by Krivan *et al.* [1]. In an analogous way, the levels of K (20416 to 24300 µg g^{-1}) and Zn (5.67 to 8.76 µg g^{-1}) present small variations probably due to the fact that these elements are essential in plant mineral nutrition. Cobalt concentrations were between 0.221 and 0.327 µg g^{-1}, except for the organic coffee, where the concentrations were 3 to 4 times lower with a mean of 0.086 µg g^{-1}. All the Cs concentrations were bellow 0.2 µg g^{-1}, with the extremes of 0.0785 and 0.185 µg g^{-1} for the brands *Kühl* and *Do Ponto*, respectively. The values agree with those of Krivan *et al.* [1], who encountered values between 0.015 and 0.190 µg g^{-1}.

The largest variations were observed for Fe (42.1 to 1602 µg g^{-1}) and Sc (0.00313 to 0.270 µg g^{-1}). It is interesting to note the existence of a high correlation ($r=0.99$) between the concentrations of Fe and Sc. This fact may be occurred due to the presence of soil in the coffee, because the largest levels of Fe and Sc found by Krivan *et al.* [1] did not pass 48.0 and 0.00192 µg g^{-1}, respectively, without any suggestion of a correlation. Such correlation is commonly found in soils, and Sc and Fe have been used as tracers of the presence of soil in sugar cane consignment [4].

Lower concentrations of Na were found in whole beans when compared to roasted ground, except for the *Ituano* brand, where the incorporation of the element during processing was suggested.

Table 1. Mean elemental concentrations (μg g^{-1}) of three replicates of the selected Brazilian coffee brands. Coefficients of variation are present between parentheses.

	Brand	Br	Ca	Co	Cs	Fe	K	Na	Rb	Sc	Zn
WHOLE BEANS	Do Ponto	0.711^{i} *(6)*	1178^{a} *(19)*	0.327^{a} *(7)*	0.0850^{ef} *(6)*	50.2^{b} *(7)*	21483^{bc} *(6)*	6.78^{cd} *(15)*	45.9^{de} *(6)*	0.00313^{d} *(14)*	7.34^{abc} *(10)*
	Morro Grande	0.670^{i} *(3)*	1142^{a} *(8)*	0.301^{ab} *(7)*	0.0903^{ef} *(1)*	53.6^{b} *(9)*	20417^{c} *(5)*	4.94^{d} *(10)*	55.0^{b} *(5)*	0.00393^{cd} *(9)*	7.33^{abc} *(7)*
	Ducati	1.40^{de} *(2)*	1350^{a} *(7)*	0.282^{abc} *(6)*	0.0970^{e} *(3)*	65.1^{b} *(5)*	24300^{a} *(5)*	8.40^{cd} *(8)*	67.2^{a} *(5)*	0.0102^{cd} *(10)*	8.76^{a} *(23)*
	Kühl	1.11^{g} *(3)*	1197^{a} *(7)*	0.301^{ab} *(5)*	0.0785^{f} *(6)*	42.1^{b} *(2)*	21367^{bc} *(5)*	5.47^{d} *(3)*	57.7^{b} *(3)*	0.00328^{d} *(9)*	8.33^{ab} *(5)*
	No brand	1.45^{d} *(4)*	1233^{a} *(8)*	0.313^{a} *(19)*	0.162^{b} *(3)*	1602^{a} *(27)*	23783^{ab} *(2)*	8.75^{cd} *(6)*	37.8^{f} *(3)*	0.270^{a} *(13)*	7.57^{abc} *(7)*
ROASTED GROUND	Do Ponto	1.79^{c} *(2)*	1408^{a} *(7)*	0.282^{abc} *(5)*	0.185^{a} *(3)*	99.5^{b} *(8)*	22467^{abc} *(4)*	14.7^{b} *(3)*	49.2^{cd} *(3)*	0.0178^{cd} *(11)*	6.67^{bc} *(6)*
	Morro Grande	1.34^{ef} *(3)*	1205^{a} *(6)*	0.221^{c} *(3)*	0.118^{cd} *(7)*	113^{b} *(16)*	22700^{abc} *(4)*	12.5^{bc} *(5)*	40.8^{ef} *(4)*	0.0218^{cd} *(5)*	7.55^{abc} *(3)*
	Melitta	1.90^{b} *(2)*	1322^{a} *(4)*	0.248^{bc} *(2)*	0.125^{c} *(1)*	64.7^{b} *(4)*	21617^{bc} *(3)*	11.6^{bc} *(4)*	36.8^{f} *(2)*	0.0109^{cd} *(4)*	6.58^{bc} *(3)*
	União	1.13^{g} *(2)*	1317^{a} *(2)*	0.221^{c} *(2)*	0.0912^{ef} *(1)*	137^{b} *(17)*	22317^{abc} *(1)*	15.8^{ab} *(4)*	38.7^{f} *(2)*	0.0334^{bc} *(10)*	7.42^{abc} *(3)*
	Pelé	2.58^{a} *(1)*	1387^{a} *(5)*	0.229^{c} *(6)*	0.130^{c} *(2)*	250^{b} *(11)*	21967^{abc} *(3)*	12.2^{bc} *(3)*	52.5^{bc} *(2)*	0.0516^{b} *(9)*	6.49^{bc} *(1)*
	Pilão	1.25^{f} *(1)*	1253^{a} *(5)*	0.269^{abc} *(1)*	0.111^{d} *(4)*	69.2^{b} *(4)*	22433^{abc} *(1)*	21.6^{a} *(4)*	44.1^{de} *(1)*	0.0114^{cd} *(6)*	7.80^{ab} *(3)*
	Ituano	0.825^{h} *(2)*	1151^{a} *(15)*	0.086^{d} *(3)*	0.0840^{ef} *(5)*	54.6^{b} *(10)*	20800^{c} *(1)*	3.89^{cd} *(8)*	57.0^{b} *(2)*	0.00693^{cd} *(17)*	5.67^{c} *(4)*

Means with the same letter are not significantly different by Tukey's test (p>0,05).

4 Acknowledgements

The authors are in debt with FAPESP, which provided the financial support.

5 References

1. Krivan, V., Barth, P. and Morales, A.F. Multielement analysis of green coffee and its possible use for determination of origin. *Mikrochim. Acta*. **110** (1993) pp. 217-236.
2. Orlic, J. Makjanic, V. and Valkovic, V. Optimization of XRFS for the analysis of toxic elements and heavy metals in coffee products. *J. Radioanal. Nucl. Chem.* **102** (1986) pp.203-210.
3. Martín, M.J., Pablos, F. and González, A.G. Characterization of green coffee varieties according to their metal content. *Anal. Chim. Acta*. **358** (1998) pp.177-83.
4. Fernandes, E.A.N. Scandium as tracer in the sugar and alcohol agroindustry. *J. Radioanal. Nucl. Chem.* **168**-1 (1993) pp 41-6.

KINETICS OF β-[^{14}C]CAROTENE IN A HUMAN SUBJECT USING ACCELERATOR MASS SPECTROMETRY

Stephen R. Dueker[1], Yumei Lin[1], Jennifer R. Follett[1], Andrew J. Clifford[1], Bruce A. Buchholz [2], John S. Vogel [1,2]

Department of Nutrition[1], Univ of Cal, Davis, CA; and Center for Accelerator Mass Spectrometry[2], Lawrence Livermore National Laboratory, Livermore, CA.

β-Carotene is a tetraterpenoid distributed widely throughout the plant kingdom. It is a member of a group of pigments referred to as carotenoids that have the distinction of serving as metabolic precursors to vitamin A in humans and many animals [1,2]. We used Accelerator Mass Spectrometry (AMS) [3] to determine the metabolic behavior of a physiologic oral dose of β-[^{14}C]carotene (200 nanoCuries; 0.57 μmol) in a healthy human subject. Serial blood specimens were collected for 210-d and complete urine and feces were collected for 17 and 10-d, respectively. Balance data indicated that the dose was 42% bioavailable. The absorbed β-carotene was lost slowly via urine in accord with the slow body turnover of β-carotene and vitamin A [4]. HPLC fractionation of plasma taken at early time points (0-24-h) showed the label was distributed between β-carotene and retinyl esters (vitamin A) derived from intestinal metabolism.

MATERIALS AND METHODS: β-[^{14}C]carotene was prepared by growing spinach in an atmospherically sealed labeling chamber pulsed with $^{14}CO_2$ [5,6,7]. The β-carotene was extracted and final radiometric and chemical purity was determined to be greater than 98% by RP-HPLC (S.A. 0.35 mCi/mmol). The β-carotene was dispersed in olive oil and a 200 nanoCi aliquot was transferred to a gelatin capsule equivalent to 0.57 μmol (0.306 mg) for ingestion. Pre-dose urine and feces were collected a day before dosing. Pre-dose blood was drawn just before the dosing. Blood samples were drawn every 30-min for 10-h postadministration and at longer intervals thereafter for 210-d. Complete 6-h urine collections (4) were made during the first 24-h period. Complete 24-h collections were continued from day 2 through day 17. Complete fecal collections were made through day 10. Plasma was separated from blood cells by centrifugation. Fecal collections were dispersed with 0.5 M KOH; aliquots were removed for AMS and carbon analysis. Plasma samples were analyzed neat and following HPLC separation of major metabolites. Samples were combusted and reduced to graphite [8] for AMS measurement [9,10]. Natural ^{14}C was subtracted from measured isotope ratios.

RESULTS AND DISCUSSION: AMS measurements were done to ±3% precision (instrument precision) as measured by the standard deviation of 3 or more measurements of the ^{14}C concentration [3,10]. The assay precision was determined from the amount of scatter in the results obtained from repeat analyses of a homogenous sample and is the primary determinant of detection limit. Accordingly, absolute sensitivity per sample based on double the uncertainty in the assay precision was 18, 32, and 17 amol β-[^{14}C]carotene per plasma, urine, and fecal specimen, repectively. Collected HPLC fractions had detection limits of 3 amol β-[^{14}C]carotene/fraction.

The concentrations versus time course of the label in plasma analyzed neat (A) and following HPLC fractionation (B) are shown in figure 1. The neat concentration profile displayed a 5.5 hour delay before appearance of label (A). The 5.5 hour lag is in accord with the relatively slow absorption kinetics of lipophilic compounds via the lymphatic system. Slow gastric emptying may also contribute to this lag. The initial peak occuring between 5.5 and 10 hours reflects the appearance of label into circulation aboard short-lived chylomicron particles of intestinal origin. These particles and their lipophilic contents are rapidly removed from circulation by the liver ($t_{1/2}$ ~11 m) which explains the bell-shaped (guassian) profile. Subsequent resecretion of labeled compounds from the liver aboard longer-lived lipoproteins and specific carrier proteins results in a broad secondary peak of higher concentration. Sufficient signal was present for determinations

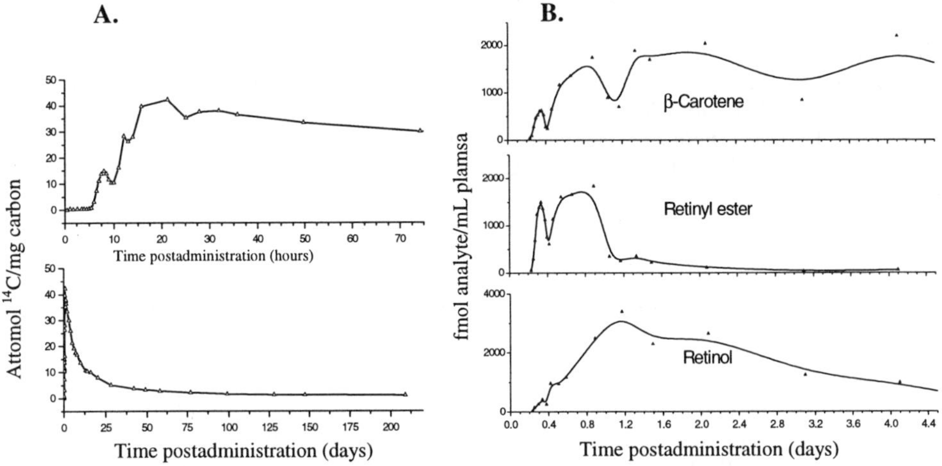

Fig. 1. Profiles of concentration versus time course of label in plasma analyzed neat (A) and following HPLC fraction of select metabolites (B).

209 days postadministration (A, bottom panel). Fractionation of labeled metabolites by HPLC followed by AMS analysis revealed the concentration time course of β-carotene and its metabolites, retinyl esters and retinol (B). It is apparent that the initial absorptive peak starting at 5.5 h is similar in shape to that of the neat profile and is a composite of the β-carotene and retinyl ester concentration profiles, suggesting these compounds share a common absorption process. Plasma retinol rises linearly (bottom panel) for ~ 24-h

postdosing reflecting the postabsorptive hepatic biotransformation of retinyl esters into retinol and the subsequent release into circulation with specific retinol binding proteins.

Cumulative urinary and fecal output as a percentage of dose were determined. Absolute bioavailability determinations using this approach are quantitative and facile. Based upon the recovery of 58% of the dose after 48-h in the feces, dose bioavailability was 42%. Urinary excretion rates were low: 5% of the administered dose and 2% of the absorbed dose was excreted in the urine after 17-d. The slow elimination rate is attributed to the low aqueous solubility of the compounds and long biological half-lives (estimated from the final elimination portions of the concentration vs. time curves, β-carotene and retinol have half-lives of ~40 and 210-d, respectively).

CONCLUSION: AMS is an excellent bioanalytical tool for in vivo human nutrient studies using long-lived radioisotopes such as ^{14}C [11]. AMS sensitivity (attomolar) enables radiocarbon tracer studies to be conducted at low microSievert radiation exposures using physiologic-sized doses. The methodology will facilitate determination of fundamental kinetic parameters for nutrients and bioactive phytochemicals.

References

1. Olson, JA. (1989) Provitamin A function of carotenoids: the conversion of beta-carotene into vitamin A. *J Nutr 119*:105-8
2. Olson, JA. (1981) Formation and Function of Vitamin A. In, Biosynthesis of Isoprenoid Compounds, J. W. Porter & S. L. Spurgeon, eds. (New York: John Wiley & Sons), pp. 371-412
3. Vogel, JS; Turteltaub, KW; Finkel, R; Nelson, DE. (1995) Accelerator mass spectrometry: Isotope quantification at attomole sensitivity. *Anal Chem* 67(11):353A - 359A.
4. Parker, RS. (1996) Absorption, metabolism, and transport of carotenoids. *Faseb J,* 10: 542-51
5. Segall, HJ; Brown, CH; Paige, DFJ. (1983) *J Lab Compd Rad*, 20: 671-89.
6. Lamé, MW; Wilson, DW; Segall, HJ. (1996) *J Lab Compd Rad,* 38, No. 12:1053-60.
7. Dueker, SR; Clifford, AJ; Lame, MW; Segall, HJ; Buchholz, BA; Vogel JS. (1998) Production of carbon-14 labeled phytonutrients in spinach for human nutrition studies employing accelerator mass spectrometry detection. *Faseb J*, 12:a210
8. Vogel, JS. (1992) Rapid production of graphite without contamination for biomedical AMS. *Radiocarbon,* 34:344-350.
9. Creek, MR; Frantz, CE; Fultz, E; Haack, K; Redwine, K; Shen, N; Turteltaub, KW; Vogel, JS. (1994) 14C AMS quantification of biomolecular interactions using microbore and plate separations. *Nuc Inst & Meth,* B92:454-458.
10. Vogel, J.S., and K.W. Turteltaub. (1992) Biomolecular tracing through accelerator mass spectrometry. *Trens Anal Chem*, 11:142-49
11. Vogel, JS; Turteltaub, KW. (1998) Accelerator mass spectrometry as a bioanalytical tool in nutrition research. In, Advances in Experimental Medicine and Biology, edited by A. Clifford & H-G Müller. (New York: Plenum) 397-410

AN APPROACH TO EVALUATE SOIL ORGANIC MATTER AS A SCAVENGER OF HEAVY METALS USING A RADIOTRACER TECHNIQUE

RYOKO FUJIYOSHI, TOSHIHIDE KAWAI, MASASHI NAKAYAMA, HIROYUKI HIRASHIMA AND SADASHI SAWAMURA

Division of Quantum Energy Engineering, Graduate School of Engineering,

Hokkaido University, Sapporo 060-8628, Japan

E-mail: ryoko@ccms1.hucc.hokudai.ac.jp

Sorption of Zn(II) on several top soil samples was investigated to evaluate the sorption capacity of those soils, and also to elucidate possible role of soil organic matter as a Zn(II) scavenger using a radiotracer technique with Zn-65 as a sorbate. Sorption isotherms of Zn(II) were fitted to the Langmuir equation to give the amounts of Zn(II) at the maximum sorption(A_m), corresponding to 100 percent surface coverage by the ions. Comparing A_m(Zn) values of the original soil samples with those after heating at 450°C suggested that the contribution of soil organic matter to Zn(II) scavenging was different depending on the soil characteristics. Among them, the A_m(Zn) values decrease with increasing humification of the soils with similar plant habitats. Competitive sorption occurred in some samples, which was elucidated by using a double tracer technique with Mn-54 as a second tracer.

1 Introduction

It is well known that the fate of minor or trace amount of pollutants like heavy metals and radionuclides released into the aquatic environment depends strongly on their chemical forms which also change with time under "in situ" conditions including temperature, pH, redox potential and various solid and dissolved species etc.. In the soil-solution system, extensive studies have been carried out to elucidate interactions between these pollutants and soil surfaces. Most of the results obtained so far have been deduced from model calculation, model experiments and separation of pertinent species by sequential extraction as an example[1,5,6,8]. However, it is often difficult to estimate the ability of soil organic matter as a scavenger of heavy metals, and also difficult to evaluate competitive sorption among them.

The present authors have investigated the sorptive behavior of heavy metals to various geological samples such as marine sediments, lacustrine sediments and soils using radiotracer technique with Zn-65 as a sorbate[2,3, 4]. A sorption isotherm gives an amount at the maximum sorption(A_m) of zinc(II) which depends on the properties of soil samples. In the present study, possible role of the soil organic matter as a scavenger of Zn(II) was estimated by comparing the Am(Zn) of individual samples with those heated at 450°C. Competitive sorption to these soils was also investigated using Mn-54 as a second tracer.

2 Method

Top soils were collected at several locations in Hokkaido(Japan)[7]. Some of the soil

Properties such as water content, organic content, soil pH, cation exchange capacity, surface area and x-ray diffraction analysis were investigated according to the procedures described previously[2,3,4]. Sorption experiments were carried out on all the samples using Zn-65 and/or Mn-54 as a tracer[4].

3 Results and discussion

Isotherms of the Zn(II) sorption fit well to the Langmuir equation which lead to get A_m(Zn) values for individual soil samples. Possible role of the soil organic matter as a scavenger of Zn(II) was estimated by comparing the A_m(Zn) of a sample with that heated at 450°C[9].

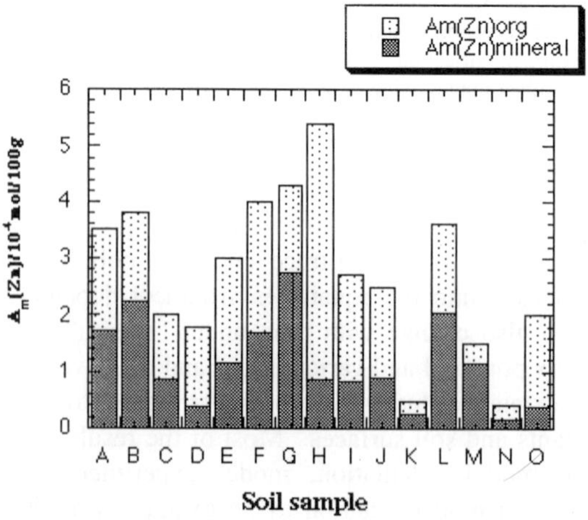

Fig. 1 Sum of the A_m(Zn) values of organic and mineral components of the soil samples

Considering that most of the soil organic matter is decomposed at 450°C, the A_m(Zn) values after heating could be regarded as the sorptive capacities of the non-organogenic mineral components(mostly oxides, clays and silicate minerals) of the soils.

Figure 1 shows the result of $A_m(Zn)$ values for individual samples, where $A_m(Zn)_{org}$ and $A_m(Zn)_{mineral}$ denote to $A_m(Zn)$ values estimated for the organic and mineral components of a sample, respectively. In any samples, soil organic matter appears to provide sorptive sites for Zn(II) ions, although the effect is different with different soils. For example, in the samples(H and O), soil organic matter may play an important role to sorb Zn(II), whereas in the sample M, there exists a certain mineral component which scavenges Zn(II) predominantly. Proton exchange process is likely to occur on the soil surfaces in the case of zinc(II) as a sorbate. Considering $A_m(Zn)$ as an indicator of the zinc(II) uptake to soils, a certain relationship may exist between the $A_m(Zn)$ and degree of humification of the soils with similar plant habitats.

Further investigation is going with soil core samples containing organic components in a varying degradation stage.

References

1. Bolton K. A., Sjoberg S. and Evans L. J., Proton binding and cadmium complexation constants for a soil humic acid using a quasi-particle model. Soil Sci. Soc. Am. J., **60** (1996) pp. 1064-1072.
2. Fujiyoshi R., Gomei T. and Katayama M., Sorption of Zn(II) on marine sediments by a sequential extraction-radiotracer technique. Appl. Radiat. Isot., **47**(2) (1996) pp.165-169.
3. Fujiyoshi R., Gomei T. and Sawamura S., Sorptive behavior of Zn(II) on a lake sediment by sequential extraction-radiotracer technique. Talanta, **44** (1997), pp. 1055-1061.
4. Fujiyoshi R., Hirashima H. and Sawamura S., Investigation of the sorption of Zn(II) on surface soils using a radiotracer technique. Appl. Radiat. Isot., **49**(1-2) (1998) pp.1-4.
5. Lee S-Z., Allen H. E., Huang C. P., Sparks D, L., Sanders P. L. and Peijnenburg W. J. G. M., Predicting soil-water partition coefficients for cadmium. Environ. Sci. Technol., **30** (1996) pp.3418-3424.
6. Rigol A., Vidal M. and Raupet G. J., Capillary zone electrophoresis to studythe humic fraction in organic soils and its relationship with radiocaesium mobility. J. Radioanal. Nucl. Chem., Articles, **208**(2) (1996) pp. 617-630.
7. Sakaguchi Y. Peats. In: Geology of the Peatland(in Japanese), Tokyo daigaku shuppankai, (1980):Tokyo, Japan.
8. Spark K. M., Wells J. D. and Johnson B. B., Sorption of heavy metals by mineral-humic acid substrates. Aust. J. Soil Res., **35** (1997) pp.113-122.
9. Wong C. S., Chin Y. P. and Gschwend P. M. ,Sorption of radon-222 to natural sediments. Geochim. Cosmochim. Acta **56** (1992) pp. 3923-3932.

Session I3

AMS and Radiocarbon Dating Techniques

ISOTOPIC (^{14}C) AND CHEMICAL COMPOSITION OF ATMOSPHERIC VOLATILE ORGANIC COMPOUND FRACTIONS – PRECURSORS TO OZONE FORMATION*

GEORGE A. KLOUDA**
**National Institute of Standards and Technology, Gaithersburg, MD 20899 USA
E-mail: george.klouda@nist.gov

CHARLES W. LEWIS‡, REINHOLD A. RASMUSSEN†, DAVID C. STILES††, WILLIAM D. ELLENSON†† AND JULIE L. MAROLF**
‡U.S. Environmental Protection Agency, Research Triangle Park, NC 27711 USA
†Oregon Graduate Institute, Beaverton, OR 97006 USA
††ManTech Environmental Technology, Inc., Research Triangle Park, NC 27709 USA

Atmospheric volatile organic compounds (VOCs) are an important factor in the production of ozone near ground level [3]. Many hydrocarbons originate from auto exhaust. However, a number of VOCs, *e.g.*, isoprene, are known to be natural in origin. To develop reliable models for understanding how ozone levels might be controlled, it has been recognized that a key variable is the fractional amount of fossil *vs.* non-fossil (biogenic, uncontrollable) VOCs [1]. These two source categories are uniquely discriminated from each other based on the ^{14}C abundance of VOC fractions. Combining isotopic measurements with chemical information will yield greater confidence in source estimates. Over the last 8 years, NIST, OGI and the U.S. EPA have explored sampling methods and analytical procedures to collect enough carbon from atmospheric non-methane VOC fractions to measure the ^{14}C composition [4,9]. In doing so, this work also provided an opportunity to enhance the sensitivity of GC-FID and GC-MS for chemical identification and quantification. This paper will briefly describe the methods studied and present preliminary ^{14}C and chemical results from Azusa, CA, Houston, TX, and Nashville, TN, USA.

1 Introduction

Many regions throughout the United States have difficulty complying with the National Ambient Air Quality Standard (NAAQS) for ozone mixing ratio, 80 nmol mol^{-1} averaged over 8 h. Atmospheric non-methane VOCs play an important role in ozone formation when combined with nitrogen oxides and sunlight [1]. Radiocarbon (^{14}C) measurements of VOC fractions are valuable in identifying the origin of the compounds that make up this material [4,6,7,10]. ^{14}C is absent in fossil fuels (*e.g.*, auto emissions), yet present in living materials (*e.g.*, vegetative and charbroiling emissions) at

*Contribution of the National Institute of Standards and Technology; not subject to copyright.

measurable levels, $^{14}C/^{12}C \approx 1.2 \times 10^{-12}$. By measuring the ^{14}C content of atmospheric VOCs, the portion of biogenic and fossil VOCs can be accurately estimated. Consideration of carbon isotope measurements of the VOC fraction, its chemical composition, and meteorological information is expected to lead to more accurate estimates of the controllable sources of VOC emissions. With a better understanding of sources and their contributions, more effective strategies can be implemented to control ozone levels and comply with the NAAQS.

2 Methods

2.1 Sampling and Chemical Characterization

Air samples were collected either by compressing 0.1 m^3 of air into summa-polished canisters (32 L) or pumping 0.3 m^3 of air into liquid nitrogen cooled canisters. The cryogenic method provided a means to recover more VOC-carbon for ^{14}C measurement. Capillary gas chromatography-flame ionization (GC-FID) and -mass spectrometry (GC-MS) detection methods were used to determine the chemical composition of each air sample [8]. Samples were also analyzed for their VOC-C mixing ratio by measuring the total FID response without chromatography -data needed to evaluate the recovery of VOC-C separated from air.

2.2 VOC Separation and Preparation for ^{14}C Measurement

Each compressed air sample was processed through a column of LiOH monohydrate to reduce the CO_2 level to nominally 50 nmol mol^{-1} while reconstituting the air sample in a separate evacuated canister at –196 °C [9]. The next step was to pass the air through a series of cryo-traps on a vacuum manifold to condense VOCs and residual CO_2 at –196 °C while all other components of air were exhausted [4]. Concentrated VOCs were then oxidized to CO_2 over hot platinized alumina pellets at 850 °C. Once distilled from –78 °C to remove H_2O, the CO_2 was quantified by manometry in a calibrated volume. Finally, sample CO_2 was reduced to graphite [11] for ^{14}C accelerator mass spectrometry measurement [2].

3 Model Validation: Ancillary Samples and Reference Materials

Samples of whole gasoline and local vegetation, representative of fossil fuel- and biomass sources of VOC, respectively, and atmospheric CO_2 were collected and analyzed for ^{14}C to determine the fraction of modern carbon *(f_M)*. f_M is defined as the $^{14}C/^{12}C$ ratio of the sample divided by 0.95 times the $^{14}C/^{12}C$ ratio of SRM 4990b Oxalic Acid, *i.e.*, modern carbon *($f_M = 1.0$)*. The lack of ^{14}C in whole gasoline *($f_M = 0.0048 \pm 0.0026^{\dagger}$)* and the consistency of the ^{14}C content of living vegetation *($f_M = 1.049 \pm 0.018$)* with atmospheric CO_2 separated from whole air *($f_M = 1.018 \pm 0.042$)*, on average, provided the basis of a mass weighted *(ϕ)* two-source model, $f_M\ (meas.) = \phi \bullet f_M\ (biogenic) + (1-\phi) \bullet f_M\ (fossil)$. Normalizing $f_M\ (meas.)$ to the current atmospheric CO_2 level allows one to calculate the portion of biogenic and fossil VOC-C.

To validate the VOC sample preparation process, the following VOC mixtures were treated and measured for their chemical and isotopic composition: 1) whole gasoline vapor in air and in N_2 [4], 2) a distillate (vapor) of lime oil in air, and 3) zero air. The mass of dead VOC-C processed *(x)* and the corresponding ^{14}C content *(y)* of these materials were used to determine the carbon mass and isotopic composition of the process blank. A non-linear regression of a decay hyperbolic function, $y = ab/(b+x)$, was fit to the data resulting in a *VOC process blank (b)* $= 4.5 \pm 0.6\ \mu gC$ with an f_M $(a) = 1.07 \pm 0.03$. These estimates were consistent with a *chemical blank* $= 3.3 \pm 0.7\ \mu gC$ determined by mass balance assuming contemporary carbon contamination.

4 Results and Discussion

Air samples identified in Figure 1 were collected during the summer in and around three U.S. cities. In Azusa, CA (1997), air was compressed into canisters on several days during the following periods: *(i)* 0600-0900 h; A2, A8, A11, A14, A15, *(ii)* 1300-1600 h; A1, A3, A12, A16, and *(iii)* 1700-2000 h; A13. Three air samples were cryo-collected in Nashville, TN (1995), nominally from 0730-1130 h at a site 24 km southeast of the city center in a rural area, and combined into two samples, N1 and N3. A third composite sample from Nashville 1994 (N) was comprised of 12 32-L compressed air samples collected atop the city center Polk Bldg. representative of 1200 to 1800 h. In Houston, TX (1994), samples were collected at three sites: (1) a northern suburban/rural site, H3 (a.m.) and H2 and H4 (p.m.), (2) a city site (Clinton) in the ship-channel area oil-refinery complex, H7 (p.m.), and (3) the Sam Houston National Forest 80 km north of Houston, H8 (p.m.).

---

† All uncertainties are reported as standard uncertainties, *i.e.*, one standard deviation.

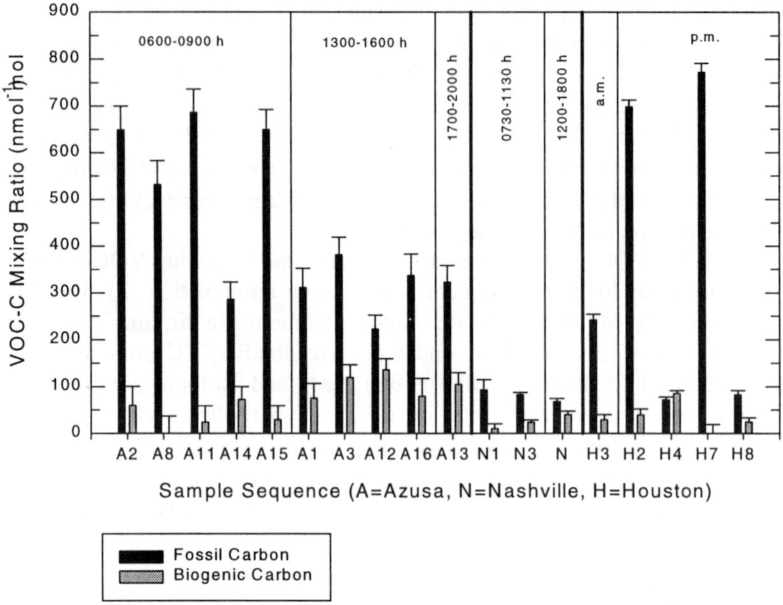

Figure 1. Atmospheric volatile organic compound (VOC) carbon mixing ratios from fossil and biogenic sources are plotted with error bars representing one standard deviation.

Figure 1 is an ordered plot of VOC -biogenic and -fossil carbon mixing ratios and their uncertainties. Also, note that the samples collected during a similar time- of-day are grouped from left (a.m.) to right (p.m.) within study areas. In Azusa, total VOC-C mixing ratios range from 400 nmol mol^{-1} to 700 nmol mol^{-1} with fossil VOC-C dominant in the early morning while biogenic emissions increase significantly in the afternoon. This is consistent with high pollution events driven more by fossil fuel than by non-fossil related emissions.

In Nashville, observed mixing ratios are considerably lower than Azusa and Houston. The city center (N) shows a higher biogenic fraction than at the rural site (N1, N3), counter to what one might expect. However, a VOC mixing ratio as low as 100 nmol mol^{-1}, with 37 % ± 6 % biogenic fraction, may indicate an intrusion of clean background air or some other source of living carbon. The three Nashville results are quite similar to each other. On the other hand, Houston results show major extremes in the VOC mixing ratio. The ship channel site (H7) is entirely void of ^{14}C in contrast to the National Forest sample (H8) that shows a surprisingly low but significant biogenic fraction, 23 % ± 8 %. The largest percentage of biogenic VOCs observed, 55 % ± 4 % (H4), is from the Houston suburban/rural site in the afternoon.

5 Conclusions

For the regions studied, the results suggest that biogenic sources are not the major contributor to atmospheric VOCs. When all ^{14}C data are combined, there is a general trend of decreasing fraction of biogenic VOCs with increasing VOC-C mixing ratio. Uncertainties are the greatest for samples with the lowest mixing ratio and the least amount of air processed. Under these circumstances, *e.g.*, when VOC levels might be representative of clean background air, ^{14}C measurements obtained from the methods reported here might be inadequate for modeling purposes. An alternative field method now under investigation is designed to separate oxygenates from hydrocarbons in a sample of at least 2.5 m^3 of air for individual ^{14}C and ^{13}C analysis [5]. This new method is expected to yield biogenic estimates of these two VOC classes with a higher level of confidence than the method presented here.

6 Acknowledgements and Disclaimer

This work was funded in part by the U.S. Environmental Protection Agency Interagency Agreement No. DW13937923 with the National Institute of Standards and Technology, and the Texas Natural Resources Conservation Commission (TNRCC). The paper has been reviewed in accordance with the U.S. Environmental Protection Agency's peer and administrative review policies and approved for presentation and publication. Mention of trade names or commercial products does not constitute endorsement or recommendation for use. The authors wish to thank S.L. Dattner of TNRCC for field support.

References

1. Chameides, W.L., Lindsay, R.W., Richardson, J. and Kiang, C.S., The role of biogenic hydrocarbons in urban photochemical smog: Atlanta as a case study. *Science* **241** (1988) pp. 1473-1475.

2. Donahue, D.J., Jull, A.J.T. and Toolin, L.J., Radiocarbon measurements at the University of Arizona AMS facility. *Nucl. Intrumen. Methods Phys. Res.* **B52** (1990) pp. 224-228.
3. Haagen-Smit, A.J., Chemistry and Physiology of Los Angeles Smog. *Indust. Eng. Chem.* **44** (1952) pp. 1342-1346.
4. Klouda, G.A., Lewis, C.W., Rasmussen, R.A., Rhoderick, G.C., Sams, R.L., Stevens, R.K., Currie, L.A., Donahue, D.J., Jull, A.J.T. and Seila, R.L., Radiocarbon measurements of atmospheric volatile organic compounds: quantifying the biogenic contribution. *Environ. Sci. & Technol.* **30(4)** (1996) pp. 1098-1105.
5. Klouda, G.A., Rasmussen, R.A., Stiles, D.C., Lewis, C.W., Ellenson, W.D., Gunawardena, R. and Mohan, J.F., A sampling method to enhance the sensitivity of chemical and carbon-isotope measurements of atmospheric volatile organic compounds. To be submitted to *J. Geophys. Res.* (1999).
6. Larsen, B., Brussol, C., Kotzias, D., Veltkamp, T., Zwaagstra, O. and Slanina, J., Sample preparation for radiocarbon (^{14}C) measurements of carbonyl compounds in the atmosphere: quantifying the biogenic contribution. *Atmos. Environ.* **32(9)** (1998) pp.1485-1492.
7. Lewis, C.W., Klouda, G.A., Marolf, J.L., Stiles, D.C., Kronmiller, K.G., Oliver, K.D. and Adams, J.R., The SCOS97 radiocarbon experiment. In proceeding of the A&WMA/EPA Symposium, *Measurement of Toxic and Related Air Pollutants, September 1-3, 1998, Cary, NC, pp. 511-519.*
8. Lewis, C.W., Stevens, R.K., Rasmussen, R.A., Cardelino, C.A. and Pierce, T.E., Biogenic fraction of ambient VOC: comparison of radiocarbon, chromatographic, and emissions inventory estimates for Atlanta GA. *J. Air Waste Manage. Assoc.* **49** (1999) pp. 299-307.
9. Rasmussen, R.A., Lewis, C.W., Stevens, R.K., Ellenson, W.D. and Dattner, S.L., Removing CO_2 from atmospheric samples for radiocarbon measurements of volatile organic compounds. *Environ. Sci. & Technol.* **30(4)** (1996) pp. 1092-1097.
10. Tanner, R.L., Zielinska, B., Uberna, E. and Harshfield, G., Concentrations of carbonyl compounds and the carbon isotopy of formaldehyde at a coastal site in Nova Scotia during the NARE summer intensive. *J. Geophys. Res.* **101(D22)** (1996) pp. 28,961-28,970.
11. Verkouteren, R.M., Kleindinst, D.B. and Currie, L.A., Iron-manganese system for preparation of radiocarbon AMS targets: characterization of procedural chemical-isotopic blanks and fractionation. *Radiocarbon* **39(3)** (1997) pp. 269-283.

THE HIGH PRECISION DATING OF LIULIHE XIZHOU TOMBS

Zhang Xue-lian Qiu Shi-hua Cai Lian-zhen
Institute of Archaeology, Chinese Academy of Social Sciences,
27 Wangfujing Street, Beijing, 100710, P. R. China

On the base of high-precision measurement, this article uses high-precision wiggle-matching to research Liulihe-tomb date of Xia-Shang-Zhou Chronology Project , the precision being enhanced and calibration error being reduced.

The high-precision wiggle-matching is a method which uses a series of samples with sequence and age gap between two samples as dating samples, Bayesian paradigm as statistical framework and the high-precision radiocarbon calibration curve since 1986 as calibrating basis to do data fitting. In this way, the calibration error is reduced efficiently.

Because of bone's complicated component and its large quantity needed in conventional method, the sample-preparing of a series of bone samples becomes a difficult problem. In this article, it is researched also. By selecting better bone samples to make gelatin, satisfactory dating result is gained.

The research results are shown that the calibration error of M1193 chamber wood of Liulihe-site is reduced from over 100 to 30 years. By dating the bone samples of tombs of 3 phases 6 sections of Liulihe-site, the dates of history period from earlier Xizhou to later Xizhou dynasty of Liulihe-site are acquired for the first time and the date of each phase is also given, which provides a basis for researching into Xizhou period of Xia-Shang-Zhou Chronology Project. That "wiggle-matching" must combine with archaeology is also a characteristic.

1 Introduction

In May 1996, the Xia-Shang-Zhou Chronology Project of China started. The Xia-Shang-Zhou Chronology Project is a big program, a combined operation which involves four disciplines of history, archaeology, astronomy and 14-C dating technique. The project aims at establishing Xia-Shang-Zhou and its relative chronology table, clarifying the date problems of Xia-Shang-Zhou before BC 841. The task of 14-C dating technique is to date a series of samples which have definite archaeology phase and chronology significance.

Since Libby established 14-C dating method, it has passed 50 years. And it is also over 30 years since it was introduced into China. Through the development of these years, sample-preparing technique is improved and dating precision is enhanced, clarifying many date problems for archaeology researching. But the Xia-Shang-Zhou dating is different from those. The three dynasties of Xia, Shang and Zhou are less than 2000 years in all, and there are only several hundreds of years between every two dynasties. If we date in old way, the calendar transferring error could be bigger, sometimes up to one hundred years, which could not satisfy the need of clarifying Xia-Shang-Zhou dates.

On the base of high precision measurement, this paper uses the method of high precision wiggle-matching [1,2] to research into the dates of the Western Zhou (XiZhou) Liulihe-tomb of the Xia-Shang-Zhou Chronology Project, the precision being enhanced and calibration error being reduced.

The high precision wiggle-matching is a method which uses a series of samples with sequence and age gap between every two samples as dating samples, Bayesian paradigm as statistical framework and the high precision radiocarbon calibration curve since 1986 as calibrating basis to do date fitting. In this way, the calibration error is reduced efficiently.

Because of bone's complicated component and its large quantity needed in conventional method, the sample-preparing of a series of bone samples becomes a difficult problem. In this article, it is studied also. By selecting better bone samples to make gelatin, satisfactory dating result is gained.

2 Situation of Liulihe Site

Liulihe Site is one of the important sites in Xia-Shang-Zhou chronology.

By history literature and archaeological researching, after Wu King′s (Zhou Wu King) sending punitive expedition against Shang King, the Shang Dynasty was overthrown and the Western Zhou Dynasty was established. Then the Western Zhou began its system of enfeoffment. Liulihe Site is one of the earlist enfeoffments. It was enfeoffed as Yan State (•) . And it existed to the Chunqiu Dynasty. So, studying the site conduces to determination of the line of demarcation between the Shang and the Western Zhou Dynasty, and can provide a basis for the history researching of the Western Zhou Dynasty.

Liulihe Site is located in Fangshan County, Beijing. It covers about 4 square kilometers, including Dongjalin ruins of city, Huangtupo tombs and Liulidian commonalty residential quarters.

In 1960's, a general survey for the site was started. And it has been excavated from 1970's to 1990's. During about 20 years' excavating, the ruins of city, tombs and sites have been found. Among these tombs, there is such large tomb as Tomb 1193, unearthing bronzes with inscription and chamber wood well-preserved. Over 200 tombs are divided into 3 phases of early, medium and late or are divided in details into 6 sections (It means that each phase is more divided into 2 sections of early and late). Human skeletons are found in almost every tomb. Sites are divided into 2 phases of early and late. And there are samples of char excavated from stratigraphic sequence.

3 Method and Operation

The method includes taking a series of samples unearthed, preparing them and measuring. Then using Bayesian statistics and the high precision 14-C•dendrochronology curve since 1986 do the date fitting for calibration.

Samples taken are the bones from the tombs of 3 phases and 6 sections and the chamber wood from Tomb 1193.

3.1 Sample-Preparing

Bone

The procedure is selecting series of bones well-preserved in accordance with "phase" and "section" and preparing them into gelatin:

Immersing the bones in 1N HCI , making them into collagen; hydrolyzing into gelatin at 90•, PH=3. And then drying.

Wood

The chamber wood of Tomb 1193 is well-preserved and amounts to 156 rings. From the bark of the tree we began taking samples. We take 10 rings for every sample, and 10 samples are taken in all. Then the samples are cut into pieces, treated with acid-alkali-acid, washed and dried. Then gelatin and dried wood pieces are combusted into dioxide, converted to acetylene, then to benzene, and measured. Radiocarbon was determined by liquid scintillation counting.

3.2 Measurement

The equipment for measuring is DYS-92 low background liquid scintillation counter made in China. Cycle measuring is taken. The counting time for every sample is 2000 minutes in all . The material of modern is Chinese sucrose char.

The results are as follows.

(They have been calibrated by •13-C)

Wood:

Table 1. 14-C result of Tomb 1193 chamber wood

No.	Ring	Weight of benzene	14-C year (5568,1950)
5834B	47-56	4.7612	2905±37
5833B	37-46	4.7182	2860±37
5832B	27-36	6.9585	2872±32
5831B	17-26	6.6107	2854±33
5830B	7-16	5.6823	2821±35
5829B	1-6	6.4259	2848±33

Bone:

Table 2. 14-C result of Liulihe-tomb bone

Phase(section)	No.	Weight of benzene	14-year (5568,1950)

1 phase early	5802	5.3933	2884±35
	5800	6.2815	2874±33
1 phase late	5807	6.3133	2847±31
	5806	6.2046	2845±32
	5808	6.2633	2840±20
	5804	6.3754	2825±31
2 phase early	5809	6.2151	2835±32
	5812	3.2570	2828±44
2 phase late	5805	6.4181	2761±31
	5811	5.2447	2746±35
3 phase	5822	1.9996	2707±37
	5826	4.7697	2621±32
	5803	7.3677	2535±31

3.3 Transferring from 14-C year to calendar year and results

We use Oxford software•Oxcal to do transferring from 14-C year to calendar year. This software runs under 3.1 windows.

The results of transferring are as follows.

Table 3. The result of transferring

Sample	Phase(section)	Calendar year
Wood		1016BC(1.00)985BC
Bone	1 Phase early	1039BC(1.00)1018BC
	1 Phase late	1018BC(1.00)974BC
	2 Phase early	976BC(0.04)974BC
		958BC(0.96)923BC
	2 Phase late	926BC(0.67)892BC
		875BC(0.33)852BC
	3 Phase	858BC(1.00)808BC

4 Discussion

Wood

By the relics from Tomb 1193 and their inscription, some archaeologists consider that the Tomb 1193 may be a tomb of Yan Marquis in the early Western Zhou Dynasty. From 14-C dating, the chamber wood's date is 985BC•1015BC. It is can be seen that the 14-C dating is in accord with above archaeology researching.

Several years ago, some researcher did the same work with the same chamber wood. But the transferring error is over 100 years for she didn't apply the method of high precision wiggle-matching. From the result shown, it can be seen that the error is reduced to 30 years.

Bone

Selecting the bones well-preserved and preparing them gelatin for dating acquires satisfactory result, which becomes another example of bone conventional dating. Bone dating is paid more attention for its not tending to be so old as char easily in age, but it becomes a difficult problem to preparing sample for its component's being contaminated of easily. So, many dating researchers have done a great deal of work on it for a long time, which provides a good reference for us [3.4] Combing these literature with our experiences, we find that by selecting the bones well-preserved and preparing them into gelatin, the problem can be solved better.

During doing date fitting ,we got the exact phases of these tombs by consulting the archaeologists who have been working in Liulihe Site, and then acquired the result with less transferring error. Definite archaeology phases conduce to limiting the transferring error. The more detailed the archaeology information, the less the transferring error.

From the result shown, the dates of history period from the early Western Zhou to the late Western Zhou can be seen. And it also shows the dates of every phase. These dates are not known before. While it is very important to archaeology researching.

5 Conclusions

The research results are shown that the calibration error of Tomb 1193 chamber wood of Liulihe Site is reduced from over 100 years to 30 years. By dating the bone samples of tombs of 3 phases and 6 sections of Liulihe Site, we acquire the dates of history period from the early Western Zhou to the late Western Zhou Dynasty of Liulihe Site for the first time. And the dates of each phase are also given, which provides a basis for researching into the Western Zhou period of Xia-Shang-Zhou Chronology Project. That "wiggle-matching" must combine with archaeology is also a characteristic.

6 Acknowledgment

We are very grateful to the Xia-Shang-Zhou Chronology Project for financial aid

Reference

1. Yasushi, K.; Kalin, R. M. & Long, A.,1994, High-Precision " Wiggle-Matching" in Radiocarbon Dating. Journal of Archaeological Science 21, 475-479

2. Christen, J. A. & Litton, C. D., 1995, A Bayesian Approach to Wiggle-Matching. Journal of Archaeological Science 22, 719-725
3. Hedges & G. J. Van Kinken, 1992, A Review of Current Approaches in the Pretreatment of Bone for Radiocarbon Dating by AMS, Radiocarbon, Vol. 34, No. 3, 279-291
4. Long, A., Wilson, A. T. & Hare P. E., 1989, AMS Radiocarbon Dating of Bones at Arizona, Radiocarbon Vol. 31 No. 3, 231-238

A LOW ENERGY TABLE-TOP APPROACH TO AMS

R.B. SCHUBANK
Ultra Woodwind Pads, 10501 Hollymount Dr., Richmond, BC Canada V7E 4Z3

A design for a low energy table-top approach to AMS of 14C and other select nuclides is described. The design consists of an ion source capable of producing high intensity +3 ion beams with low emittance at 5-10keV/nucleon, pair of E-B analyzers, metal foil/vapour chargex to produce negative ions, another pair of E-B analyzers, and a low-energy ion detector with optional passive absorber. Capital costs, operating costs, and running times may be an order of magnitude less than conventional 3MV AMS.

1 Introduction

The conventional approach to radiocarbon AMS (chemical pre-processing, negative ion source, mass analysis, acceleration in a multi-MV tandem, stripping to +3, mass analysis, detection) has been well documented [1-4] and is used in several labs worldwide. However, an opposite approach, (using +3 ions, mass analysis, charge-exchange to negative ions, mass analysis) may prove superior in terms of higher sample throughputs and lower capital and operating costs. The opposite approach, described in greater detail below was perhaps first hinted at by K.Purser in AMS-1 [3] in 1978, and more vaguely by G.E.Kocharov et al. [5] in 1990, yet there has been no follow-up in the literature since then (except for TRIUMF-LS17 proposal in 1994 and related patents.)

2 Description of the proposed facilty

(The facility herein described emphasizes 14C, but obviously may be applied to ^{3}H, ^{26}Al, ^{41}Ca, ^{55}Fe, ^{129}I, and perhaps other, selected isotopes. Bold section numbers refer to diagram numbers in **Figure-1**.)

1) Chemically pre-processed samples are injected as solids, liquids or gas either directly into the plasma or into the Argon carrier gas. The latter mode allows for external volatilization of liquids, including colloidal suspensions, using a suitable nozzle to form an aerosol, and of solids using thermal, RF, laser or other techniques. "Memory effects" may be reduced by maintaining the argon discharge between sample insertions while simultaneously purging; this may not be necessary for solid-samples directly inserted (e.g., via carousel).
The preferred sample-forms best suited to precision radiocarbon dating (where $^{14}C/^{12}C$ ratios are less than 1ppt) would still be either pure CO_2 or graphite, to

reduce initial levels of hydrogen an nitrogen. Modern to supra-modern samples may not even require chemical pre-processing!

2) A medium-power Electron Cyclotron Resonance source (ECR) is used to generate a beam of +2 or +3 ions at medium currents (few mA) and up to 10keV/nucleon energy, and moderate emittance (<100pi-mm-mrad). The intense discharge also helps to reduce molecular-ion mass isobars for $q<+3$, and completely eliminate them for $q>=+3$. Alternatively, other sources could be used, such as the new generation of PIG's, Bernas, even Middleton sputter sources simply modified to use I^- with no electron-suppression, instead of Cs^+, etc.

3,4) Standard vacuum baffles and Electrostatic lenses (Einzel or quadrupole triplet) are used to isolate the relatively high pressure in the region of the ECR from the high vacuum of the remaining system and to transfer the ions of interest from the source into the remaining system. Focus is at infinity (not critical)

5,8) Medium resolution energy and Momentum analyzers are used to filter-out and monitor ion species of extraneous mass. The simple and inexpensive configuration shown in Figure-1 also provides dispersion matching to enhance mass separation efficacy. Resolving Power needed is only $3\times$ (q=3) $\times$ (M=14) =126 (FWHM) to adequately separate ^{14}C from the tails of ^{12}C and ^{13}C.

6, 16, 19) Slit systems

7,11,13,17) Drift lengths (dimensions not critical, except to match analyzers)

9) Slit system, including positionable Faraday cups to monitor mass 12-16 beams for normalization and removable FC system to pass only central (mass 14) beam

10,14) Electrostatic quadrupole-triplets to re-shape the beam to desired profiles.

12) A thin metal foil or gas charge-exchange canal or supersonic vapor jet [6] is used to convert ions to a negative charge state with up to 10% efficiency [5,7], as well as provide additional dissociation of any molecular fragments. Divergence increases by 10-15 mrad after charge-exchange [8], so a point-focus is chosen to minimize the increase in beam emittance.

23,24) Optional accelerating voltage at charge-ex could decrease the beam emittance and relative energy spread, and reduce ion source accelerating voltage.

15,18) Medium resolution energy and momentum analyzers are used to analyze the negative ions. Dispersion matching improves both collection efficiency and final mass separation, since the beam energy spread is several keV.

20) A simple ion detector is used to count the final individual ions and measure their energy (E). Counts are normalized to the integrated currents of $^{12,13}C$ beams in faraday cups at 9). A Faraday cup must be inserted to measure more intense beams as needed. The efficiency correction factors are determined at regular intervals by using sample "standards" of known ^{14}C concentration as well as periodically re-tuning both magnetic analyzers for transmission of ^{12}C.

21) A reference scale length of 30cm is shown, which corresponds to the preferred dimensions of the energy and momentum analyzers optimized for ^{14}C.

22) Central beam trajectory; Beam envelopes calculated to 2nd odrer.

3 References

1. K.H. Purser, R.B. Liebert, A.E. Litherland, R.P. Beukens, H.E. Gove, C.L. Bennett, M.R. Clover and E. Sondheim, Rev. Phys. Appl. 12 (1977) 1487
2. D.E. Nelson, R.G. Korteling and W.R. Stott, Science 198 (1977) 507
3. Proc. First Conference on Radiocarbon Dating with accelerators, ed., H.E. Gove, Univ. of Rochester, USA (1978)
4. Proc. Fifth Int. Symp. on Accelerator Mass Spectroscopy, Nucl. Instr. and Meth. B52 (1990) 211- 634; Other AMS conference proceedings appear in NIM.
5. ``The possibilities of cosmogenic isotope investigation by mean of mass-spectrometrical methods '', G.E. Kocharov, V.T. Kogan, A.N. Konstantinov and A.K. Pavlov, NIM B52 (1990) p384
6. Low-energy fusion cross sections of d + d and d + ^{3}He reactions '', A. Krasss, H.W. Becker, H.P.Trautvetter, C.Rolfs, & K.Brand, Nucl. Phys. A465 (1987) p150
7. `Equilibrium Charge State Distributions'', K.Shima, T. Mikumo and H.Tawara, Atomic and Nuclear Data Tables 34 (1986) p357
8. ``Angular Distributions of ions scattered in thin carbon foils'', G. Hoegberg and H. Norden, Nucl. Instr. and Meth. 90 (1970) p283

Figure-1 : Sketch of proposed typical layout. See text for descriptions.

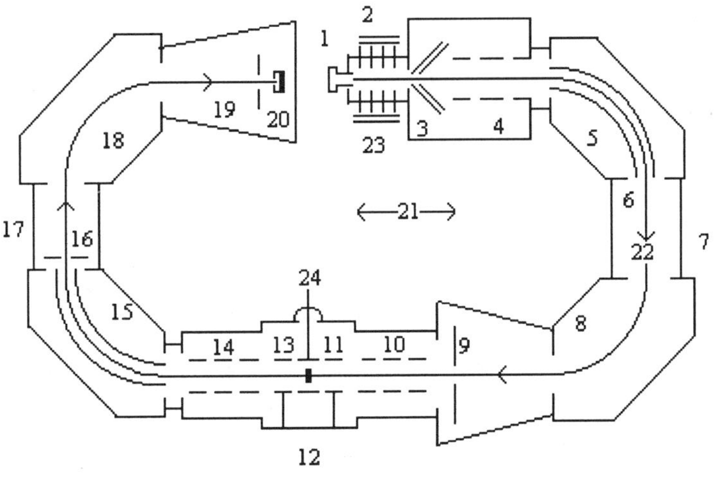

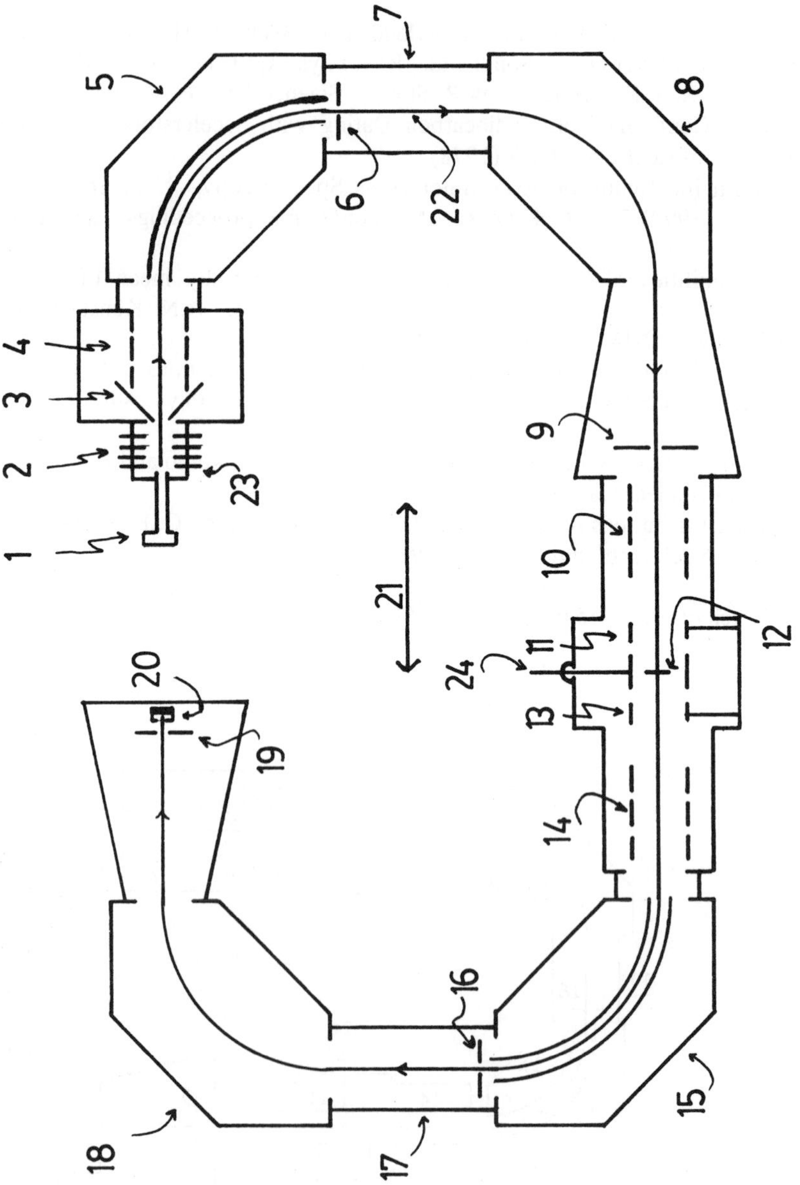

FIGURE–1

CARBON-14 LABELLING STUDIES AT ULTRA-HIGH SENSITIVITY IN AGRICULTURAL AND HORTICULTURAL RESEARCH

P. B. ROBERTS AND K.KIM

Institute of Geological & Nuclear Sciences, Lower Hutt, New Zealand
E-mail: p.roberts@gns.cri.nz

An example will be provided of the application to agricultural/horticultural research of a method to detect carbon-14 labelled biomolecules at ultra-trace levels. This great sensitivity is based on the use of accelerator mass spectrometry (AMS) to detect and count the carbon-14 atoms. The labelled animal neurotoxin, ^{14}C-paxilline, has been produced biosynthetically at low specific activity. Using conventional decay counting it was not possible to detect paxilline or its metabolites in the brain tissue of mice sacrificed at the time they displayed neurological symptoms. AMS results have been obtained that include measurements of nanogram amounts of paxilline in brain tissue and also its distribution within the major segments of the brain. Our results suggest that animal and plant biologists, biochemists and toxicologists should consider what new information can be obtained by combining tracer studies with the power of AMS detection. In particular, it opens up new possibilities such as labelling studies in large animal or plant systems, determination of receptor binding sites, studies at true dietary or environmental levels and use of carbon-14 labelled compounds in field trials.

1 Introduction

A series of papers by Vogel, Turteltaub and co-workers [2, and references therein] has demonstrated the usefulness for biomedical research of the detection of ^{14}C atoms by accelerator mass spectrometry (AMS). This paper describes an example of the use of AMS in animal health research. Paxilline is an intermediate in the biosynthesis of lolitrem neurotoxins that cause the disorder rye grass staggers in livestock grazing on pastures infected with certain fungi [1]. Paxilline itself produces similar symptoms and the ^{14}C-labelled compound has been produced biosynthetically at low specific activity. Using AMS, the presence and distribution of paxilline in the brain tissue of test mice has been measured for the first time.

2 Methods

The preparation of paxilline and its ^{14}C-labelled form (specific activity 8.5kBq, 0.23μCi or 511,000 dpm per mg) will be described elsewhere (paper in preparation). A paxilline dose of 8mg/kg (i.p.) and a total activity of 11,000 dpm (obtained by using a mix (1:9) of labelled and unlabelled paxilline) was given to female Swiss mice (approximately 25g). When the mice showed a peak neurological (tremorgenic) response after 45 minutes, they were sacrificed and the brain removed. In one experiment the whole brain was homogenized; in the other, the main brain segments were dissected out and then homogenised.

Preparation of samples for AMS measurement followed the methods described by Vogel [2]. Enough tissue to yield about 1mg of carbon was placed in a quartz tube and dried in a Jouan Rotary Evaporator. The tube was evacuated and sealed after the addition of excess copper oxide and baked at 900C for 2 hours. The resulting CO_2 was then reduced to graphite. After attaching the tube via a breakseal to a vacuum line, water in the tube was condensed with a dry ice-isopropanol trap. A second tube on the vacuum line contained cobalt as catalyst and TiH_2/ZnO to generate H_2 *in situ*. When this tube had been evacuated, the breakseal was snapped and CO_2 distilled into the second tube using liquid nitrogen to cool its tip. Non-condensable gases were pumped away and the tube sealed and baked at 500C for 4 hours.

The resulting graphite powder was pressed into an aluminium target holder and the ratio of ^{14}C atoms to total carbon atoms was determined using the radiocarbon AMS measurement system based on the 6MV Tandem Van de Graaff accelerator at the Institute [3]. The data were generated initially in the form used for radiocarbon dating as 'percent modern carbon'. This was converted to an activity or concentration of tracer using the equivalence that modern carbon is 6.11fCi, 0.0136 dpm or 98amoles per mg total carbon and the measurement that the total organic carbon was 51% of the dry weight of the whole brain.

3 Results and Discussion

Using liquid scintillation decay counting, no activity could be detected in the brain tissue. Table 1 shows the ^{14}C activity obtained using AMS detection and expressed as the mean and s.e.m of 4 replicates from the whole brain of a control (no paxilline) mouse and a mouse receiving labelled paxilline. It also shows the concentration of paxilline equivalent (total paxilline plus metabolites) calculated from the mean values for each sample, with the natural abundance of ^{14}C in the control being expressed as paxilline equivalent. The activity and concentration detected in other brain segments is also given in Table 1. As only small amounts of some of these tissues were available, the results are the average of two replicates.

The results show, for the first time, the presence of paxilline or its metabolites in the brain of a tremoring animal and that detection would have been possible with a paxilline dosage containing ten times less activity (about 1000 dpm, 20Bq or less than 1nCi). AMS has made possible the detection in tissues of an animal biotoxin that can only be produced at low specific activity and only a small proportion of which reaches the target organ. The results also indicate that it would be feasible to study ^{14}C-paxilline in sheep or cattle using reasonable amounts of compound. The activity present in the spinal cord was probably significantly lower than the average in the whole brain. In other major brain segments activity was close to that in the whole brain. This experiment therefore provided no information on the site of

Table 1. ^{14}C activity and paxilline plus metabolite concentration in mouse brain.

Brain Segment	dpm/mg dry tissue	pg/mg dry tissue
Whole brain (control)	0.0075 (0.00004)	15
Whole brain	0.5101 (0.0076)	985
Cerebellum	0.4970	959
Medulla	0.4635	893
Mid-brain	0.5289	1021
Hippocampus	0.5627	1088
Cerebrum	0.5880	1137
Spinal cord	0.3736	719

binding of paxilline to any specific receptors in the brain. However, the sensitivity of AMS and the small tissue amounts required clearly make possible measurements of tracer within localised sites in a tissue.

The data emphasise the power of AMS to study animal toxins of low specific activity, to make large animal studies using tracers feasible and to investigate the localisation of activity within tissues.

4 Acknowledgements

We thank Sarah Finch of AgResearch, New Zealand, for permission to use unpublished data, and staff of our Institute for assistance with AMS measurements.

References

1. Miles C.O., Wilkins A.L., Gallagher R.T., Hawkes A.D., Munday S.C. and Towers N.R., Synthesis and tremorgenicity of paxitrols and lolitriol: possible biosynthetic precursors of lolitrem B, *J.Agric. Food Chem.*, **40** (1992) pp. 234-238.
2. Vogel J.S., Turteltaub K.W., Finkel R. and Nelson D.E., Accelerator mass spectrometry: isotope quantification at attomole sensitivity, *Anal. Chem.* **67** (1995) pp. 353A-359A.
3. Zondervan A. and Sparks R.J., Development plans for the AMS facility at the Institute of Geological and Nuclear Sciences, New Zealand, *Nuclear Instr. and Methods B* **123** (1997) pp. 79-83.

RADIOCARBON DATING AT SITES OF EAST JAVA AND CENTRAL JAVA PROVINCES AS SUPPORTING DATA FOR THE TEMPLES CHRONOLOGY IN INDONESIA

B. Wasito, W. Faisal, A. Taftazani, K. T. Basuki
Yogyakarta Nuclear Research Center
and
Siswanto
Directorate of Archaeology, Education and Culture Department

The absolute and relative dating using radiocarbon dating and relative methods of Kidal, Sumberawan, Tegowangi and Surawana temples of East Java Province have been carried out. These dating results were then compared with the dating of Pringapus and Gondosuli temples in Central Java Province. The dating results of East Java Provinces were 1053±50 AD, 1645.9±81 AD, 1373.7±84.7 AD and 1065±51 AD for Kidal, Sumberawan, Tegowangi and Surawana temples respectively. While the datings from Central Java Province were 429.5±63.8 AD and 473.2±40.4 AD for Pringapus and Gondosuli temples respectively. From macroscale point of view, both the absolute and relative datings have a similarity and interpretation results. These results indicated that the influence of Hindus-Buddhist cultures in the Central Java seems to be older than that of in East Java.

INTRODUCTION

The aim of the work is to predict the age of temples in the Central and East Java then is to connect the chronological culture of those areas. Dating is an important part in the field of archaeology. There are two ordinary dating techniques used to measure age of temples in the archaeological field, i.e. relative and absolute dating methods. The dating results of both the methods, however, can be used to set up the culture chronology, especially in Java.

Relative dating method is applied by the archaeological scientist due to its simple interpretation, i.e.availability data of object samples, such as shape of temples, ceramics, ornaments and inscriptions. Then by relating those samples with the history of the previous known temple, relative dating of the new temples can be obtained. Such method may have problems in interpreting dating of the samples themselves, particularly, when the samples were movable objects [1,3,4]. The absolute dating method can be used by the archaeological and other scientists. Reinfrew and Bahn [2] explained that absolute radiocarbon dating gave more accurate dating than that of relative dating. However, the obtained data are also related with the fluctuation of carbon contained, so that the calculations have to be corrected using the tree-ring correction method. In the past, radiocarbon dating analysis were done by sending the samples to USA, France or Australia [1]. At the moment, the software and hardware of the radiocarbon dating analyzer has already been available in Indonesia. This research work is a joint research of Archaeological Directorate and BATAN in Yogyakarta.

In order to set up the chronological culture in the Central and East Java, the dating of several temples were studied. The chronological spreading of the temples in the Central Java Province is from the north side of the island [1]. This classification also represents the ages of the temples erection [1]. In conjunction with this classification, the temples in the East Java have not been clearly classified, because shape of temples in the East Java almost having a similar architecture to the temples in the history of Majapahit Kingdom. Therefore, they assumed the ages of temples in the East Java are younger than that of in the Central Java.

In order to confirm and cross checking of the logical culture that have been set up by relative dating method, radiocarbon dating technique were used to actualize the dating of the temples, so that the dating of the temples as well as the set up chronological culture became more accurate.

EXPERIMENTAL WORKS

Experimental work was divided into two parts, i.e. samples collection and radiocarbon dating analysis.

1. Samples collection

Samples collection consist of three main activities

a. observation, planning and mapping the sites for excavation. b. excavation. c. classification of the samples.

2. Radiocarbon dating analysis

a. Materials/ Chemicals.

Charcoal/ samples, oxalic acid (NBS 4990 C); lithium (LCA, USA); dry ice, CO_2 and O_2 gases (Aneka Gas); liquid nitrogen (YNRC); benzene and H_3PO_4 (PA, Merck); PPO and POPOP (Packard); chromium alumina (Kal-Chemie); glass wool or quartz wool (Heraeous); ascarite II (Thomas Scientific); HCl, H_2SO_4, $KMnO_4$, P_2O_5, NaOH and CuO (PA, Fisher) and B&W, colour and slide film.

b. Equipments

Benzene Synthesizer (Task.Inc.USA), Liquid Scintillation Counter (Packard 2000 CA/LL), Mass Spectrometer (VG ISOGAS SIRA 9) and excavation equipments (Archaeological Directorate)

c. Procedure

The first step is washing the samples using tap water then aquadest. After that they are soaked in HCl 0.2 N, washed with aquadest, soaked in NaOH 0.2 N, washed with aquadest and the last they are washed using HCl 0.2 N. Then they are dried in oven at 125°C. These samples are synthesized into benzen. The analysis are continued by counting ^{14}C using a Low Level Liquid Scintillation Counter. The obtained activities are compared with the standard, so that the dating of the sample can be found [3,4].

RESULT AND DISCUSSION

The result of absolute dating of the observed temples using radiocarbon dating (RCD) method are presented as numbers of year (Table 2). However, as examples, the firing and counting data from Tegowangi and Sumberawan are given in Table 1.. The resulted absolute dating of all the observed temples are presented in Table 2.

Table 1. Firing and Counting Result of samples and Standard (oxalate).[1,3]

No.	Type of sample	Weight sample g	CO_2 Press. (inHg)	Weight of benzene (g)	Count of benzen cpm	$d^{13}C$ (°/oo)
1.	Tegowangi charcoal	2.6	13	0.7958	6.87	- 25.07
2.	Sumberawan charcoal	3	14	0.8729	7.66	- 26.70
3.	Oxalic acid NBS (Std)	20	45	2.6477	27.85	- 17.45

Table 2. Comparison of Absolute and Relative Dating.[1,3]

No	TEMPLE SITE	SAMPLE TYPE	DATING BP	Confidence level 68.3 %	Confidence level 95.4 %	Relative Dating AD
1	Kidal	Charcoal	897 ± 50.25	1038 - 1172	1022 - 1230	1248
2	Surowono	Charcoal	885 ± 51.46	1044 - 1214	1038 - 1262	± 1400
3	Sumberawa	Charcoal	304.1 ± 81.04	1482 - 1659	1432 - 1658	1359
4	Tegowangi	Charcoal	576.27 ± 84.72	1294 - 1416	1267 -1460	± 1400
5	Pringapus +	Charcoal	1520 ± 63.78	450 - 598	416 - 644	before 850
6	Gondosuli +	Charcoal	1476 ± 40.39	550 - 624	452 - 650	832

+ Temples in the Central of Java

The results of relative and absolute dating show that both have similar interpretations. The temples in the Central of Java are older than that of East Java. That mean, the flow of culture comes from Central Java to East Java. With regard to the culture, although the temples are different, the people at that time built places for praying almost on the same time.

SUMMARY

The radiocarbon dating is one of important technique for setting up chronological culture. This experiment can also be used to get more information on dating of the unknown temples. The chronological of Javanese culture happens from Central Java to East Java Provinces.

ACKNOWLEDGMENT

The authors would like to acknowledge to Mr. Sukarman A., Dr. N.R.Stevenson and Ms. A.W.Fong, who have helped in financial support for this presentation, the Indonesian Government, all families and friends for the confidential support.

LITERATURE

1. SISWANTO, **1996**, *"Excavation at the Sites of Sumberawan, Malang and Tegowangi Kediri East Java"*, Report of Archaeology, Archaeology Directorate Yogyakarta (in Indonesian).
2. RENFREW C. AND BAHN P., **1991**, *Archaeology Theories, Methods and Practice*. Thames and Hudson Ltd. USA.
3. Faisal, W., Arumbinang, H., Taftazani, A., Sumiyatno, Suhardi, Siswanto and Djiono, **1997**, *"Comparison of Relative and Absolute Carbon Datings at Pringapus dan Gondosuli"*, Prosiding PPI.
4. FAISAL, W., Taftazani, A., and Wasito, B., **1997**, *"Problems and Development of Radiocarbon Dating method"*, Achaeological Bulletin , XVII, No. 1, May.

Session J

Poster Session

^{11}C-METHYLATIONS USING TETRABUTYLAMMONIUM FLUORIDE AS THE BASE

M.J. ADAM, S. JIVAN, J.M. HUSER, AND J. LU

TRIUMF University of British Columbia, Vancouver, Canada, V6T 2A3

E-mail: adam@triumf.ca

Carbon-11 Methylation reactions on functional groups such as phenols and amides usually require a base when using ^{11}C-methyl iodide. Our standard procedure for labelling compounds such as Raclopride is to add a few microliters of concentrated sodium hydroxide to the dissolved precursor. However, this method requires between 1-2 mg of precursor for optimum radiochemical yield. Other groups have successfully used Alumina/KF as the base for some compounds but this method isn't universal and requires filtration prior to purification. However, the Alumina/KF method does allow one to use much less precursor than the hydroxide method, which is thought to be due to binding to the surface of the Alumina. To overcome the problem of having to filter the Alumina/KF we decided to investigate the use of Tetrabutylammonium Fluoride (TBAF) as an organic soluble base for these methylation reactions. We report here the use of TBAF as a base in the methylation labelling of four radiopharmaceuticals; Raclopride, Dihydrotetrabenazine, Methylphenidate and PK11195 and compare these results with the other methods.

1 Introduction

Precursors required for many ^{11}C-radiopharmaceuticals are either in short supply or are very difficult to make and some of the intermediates needed to synthesize them, such as tetrabenazine, are no longer available. Therefore, methods that reduce the amount of precursor in these labelling reactions to sub-milligram levels are important.

We have been working on methods to reduce the amount of precursors used in the ^{11}C-methylations of Raclopride, SCH23390, Dihydrotetrabenazine, Methylphenidate and PK 11195 with ^{11}C-methyliodide. Recently, Schmitz et. al. (1) have reported excellent results with the use of Al_2O_3/KF to carry out S- N- and O-[^{11}C]methylations of nor precursors. The reported benefits of using this base were faster reaction times, cleaner reactions and a simplified purification. Since the precursor is apparently bound to the surface during these methylations we speculated that one might be able to reduce the amount of precursor used as compared with a base such as sodium hydroxide. We also wanted to compare and test the organic soluble base tetrabutylammonium fluoride.

Our standard procedure for labelling compounds such as Raclopride was to add a few microliters of concentrated sodium hydroxide to the dissolved precursor. However, with this method one must use between 1-2 mg of precursor for optimum radiochemical yields. In this current study we report ^{11}C-methylation yields using both Alumina/KF and the organic soluble base, tetrabutylammonium fluoride (TBAF) in reactions using sub-milligram amounts of precursor (Table 1).

Table 1: Labelling Conditions

Compound	precursor amount	solvent	base	reaction temp (oC)	%yield (decay corrected)	Specific Activity EOS (Ci/mmol)
Dihydrotetrabenazine	200µg	CH3CN	KF/Al2O3	90	38@	
	200µg	CH3CN	TBAF(0.6mg)	90	48	3,000-5,000
Methylphenidate	200µg	CH3CN	KF/Al2O3	85	38*	
	200µg	CH3CN	TBAF(0.6mg)	85	48*	
	350µg	CH3CN	KF/Al2O3	85	46@	
	350µg	CH3CN	TBAF(0.6mg)	85	48	1500-2500
PK11195	2mg	DMSO	5N NaOH(20µl)	85	3	
	2mg	DMSO	10mg NaOH	80	2	
	2mg	DMF	3mg TBAF	80	1	
	2mg	DMSO	KF/Al2O3	100	no yield	
	2mg	DMSO	TBAF(3mg)	100	27#	2000
Raclopride	200µg	CH3CN	KF/Al2O3	85	no yield	
	200µg	DMF	KF/Al2O3	85	no yield	
	200µg	acetone	KF/Al2O3	85	no yield	
	200µg	CH3CN	TBAF(0.6mg)	85	no yield	
	1.7mg	DMSO	5N NaOH(8µl)	80	35@	7,000-10,000

*Worked up with IX method (2) #final product in 10%ethanol @currently used routinely

2 Experimental

Briefly, the procedure is carried out as follows: The nor precursor (200µg) is dissolved in 450µl of DMF or acetonitrile. KF/Al_2O_3 (10mg, Aldrich) or TBAF is added to the solution before bubbling the ^{11}C-methyliodide through the solution. The methyl iodide is bubbled through the solution at 0°C. After addition, the vial is heated for 5 minutes. The crude mixture is filtered through a 13mm HV Millipore millex filter unit and the resulting solution was worked up on preparative HPLC or on SepPaks. Radiochemical yields in table 1 are decay corrected to the amount of methyl iodide trapped in the precursor solution.

3 Results and Discussion

One of the main drawbacks to the Alumina/KF method is that filters tend to clog up when filtering on-line before injection into HPLC. Forunately, tetrabutylammonium fluoride (TBAF) overcomes this problem since it is an organic soluble base. Surprisingly, as shown in Table 1, the radiochemical yields for

PK11195, dihydrotetrabenazine and methyl phenidate resulting from the use of TBAF were at least as good as those obtained with Alumina/KF even at the low level (200 μg) of precursor. More importantly TBAF was the only base that worked well with PK11195 and gave much better results that reported for the hydroxide method (3). Unfortunately, Raclopride labelling is only achieved with the hydroxide method.

4 Conclusion

In conclusion we have shown that a significant reduction in precursor amounts can be achieved in some ^{11}C-methylation reactions, by using Alumina/KF or tetrabutylammomiun fluoride as the base. Also, it was demonstrated that TBAF works as well or better than the Alumina/KF method.

5 References

1. Schmitz F., Plenevaux A., Del Fiore G., Lemaire C., Aerts J., and Luxen A., *J. Lab. Compd. Radiopharm.* 37, 749-751 (1995).
2. Adam M.J., Lu J., Huser J., and Ding Y.-S. XII th Int. Symp. Radiopharm. Chem. Uppsala, Sweden, 1997.
3. Pike V.W., Halldin C., Crouzel, et .al. Nucl. Med. Biol. 20, 503-525, (1993).

6 Acknowledgements

The authors wish to thank the Medical Research Council of Canada for support and Dr. Michael Channing of the NIH PET facility for helpful discussions.

AN UNIQUE TARGET FOR PET RADIOISOTOPE PRODUCTION WITH THE SUPERCONDUCTING ION CYCLOTRON RESONANCE ACCELERATOR

BHASKAR MUKHERJEE

Safety Division, Australian Nuclear Science and Technology Organisation, PMB 1, Menai, NSW 2234, AUSTRALIA, E-mail: mukherjee@ieee.org

TERRY L. GRIMM, RICHARD C. YORK AND CHRIS T. RAMSELL

National Superconducting Cyclotron Laboratory, Michigan State University, East Lansing, MI 48824, USA, E-mail: grimm@nscl.msu.edu

The compact, lightweight, superconducting Ion Cyclotron Resonance Accelerator (ICRA) developed at the National Superconducting Cyclotron Laboratory (NSCL) of Michigan State University has been successfully tested to accelerate a proton beam to a kinetic energy of 50 keV. In the ICRA a proton beam orbits around a DC magnetic field produced by a superconducting solenoid and drifts axially while being accelerated azimuthally by a RF cavity. A 10 MeV version of the ICRA has been designed and evaluated. This paper highlights the operating principle of an annular ($H_2{}^{18}O$) target to produce the common PET radioisotope ^{18}F via the $^{18}O(p,n)^{18}F$ reaction. This unique target is designed for an optimal utilisation of the axially drifting accelerated proton beam from the ICRA. The important operational aspects of the annular target including the isotopic yield, characteristics of the parasitic neutron and gamma radiation field and the vital radiological shielding aspects of this novel radioisotope generating machine will be presented.

1 Introduction

Hospital based Medical Cyclotrons for the production of short-lived positron emitting radioisotopes are being increasingly used worldwide [1]. High reliability, lightweight, low-footprint area and efficient radiological shielding are the essential criteria of a modern Medical Cyclotron. This paper summarises the design concept of a novel target for the production of PET radioisotopes using the ultra compact lightweight Superconducting Ion Cyclotron Resonance Accelerator (ICRA).

2 Principle of ICRA Target

Protons from the ICRA spread out from the accelerating cavity in an expanding helical (tornado like) trajectory [2, 3]. The proton energy rises and the proton current drops with the increasing radial distance from the z-axis (Figure 1). The proposed target is essentially an annular groove (cross section: 0.44 cm^2) cut around a solid circular aluminium target body acting as a beam stop. A 50 μm thick nickel window encloses the 41 ml annular target cavity (Figure 1). The isotopic yield (Y) of ^{18}F produced via the $^{18}O(p,n)\,^{18}F$ reaction is described by the charged particle activation equation [4].

$$Y\,[GBq] = \sigma_{av} \times I \times a^{-1} \times N \times S(t) \times 6.24 \times 10^{-24}$$

Where, σ_{av}, I, a, N and S(t) are the average cross section [mb] of the $^{18}O(p,n)^{18}F$ reaction [5], the proton current [μA], the entrance area of the target [cm^2], the number of ^{18}O atoms in the target ($H_2{}^{18}O$) volume and the saturation factor respectively. The saturation factor is $S(t) = 1-\exp(-0.693 \times t/T_{1/2})$, where t and $T_{1/2}$ are the target bombardment time (min) and half life of ^{18}F (110 min) respectively. The target cavity will contain 40 ml of ~95% enriched $H_2{}^{18}O$ (water). An isotopic yield (EOB) of 128 GBq (^{18}F) for a proton current of 275 μA with proton energies ranging from 5-10 MeV and a bombardment time of 60 minutes has been calculated.

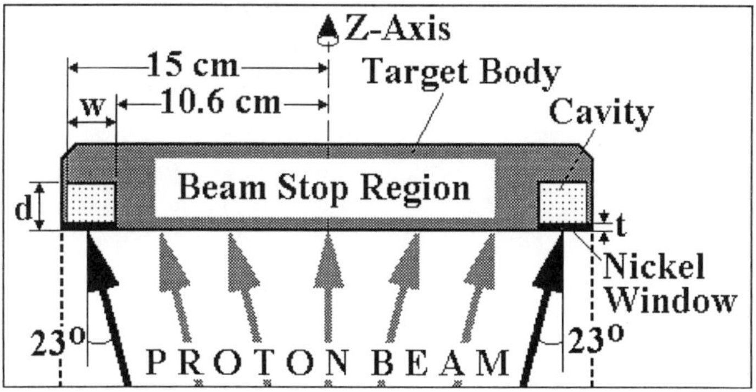

Figure 1: Schematic cross section diagram showing the principle of the annular PET radioisotope production target for ICRA.

3 Summary and Conclusion

Present model calculations support the feasibility of the production of a substantial activity of ^{18}F using the proposed target. The large target volume (40 ml) suggests the design of a ring of multiple targets for simultaneous production of several radioisotopes. It is planned to select a suitable material for the "Beam Stop Region", with the goal being to minimize the production of secondary neutrons [6]. The cylindrical shape of the target (Figure 1) facilitates an effective utilisation of a compact neutron/gamma biological shield made of a composite material [7].

References

1. Directory of cyclotrons used for radionuclide production in Member States, IAEA-TECDOC-1007, 1998.
2. T L Grimm et al. In *Proceedings of the 1997 Particle Accelerator Conference*, Vancouver, 1997.
3. C T Ramsell et al. In *Proceedings of the 1999 Particle Accelerator Conference*, NY NY, 1999.
4. B Mukherjee. Diplomarbeit (unpublished), Technische Universität Berlin, 1975.
5. T J Ruth et al. *Act.* **26**(1979) pp 21-24.
6. J K Blair, P D Miller and B W Wieland. *Appl. Radiat. Isot.* **32**(1981) pp 389-395.
7. T Kurosawa et al. *Radioisotopes* 45(1996) pp 675-684.

YIELD OF THE ^{123}Te(p,p′)^{123m}Te REACTION AT 15 MEV

VLADIMIR JAKOVLEV AND LEONID SOLIN

Laboratory of Nuclear Reactions and Nuclear Medicine, V. G. Khlopin Radium Institute, Second Murinsky Avenue 28, 194021 St. Petersburg, Russia

E-mail: jakovlev@atom.nw.ru, solin@mail.convey.ru

The continuously growing amount of isomer ^{123m}Te ($T_{1/2}$ = 119.7 days) drastically increases activity of target material used for production of iodine-123 with the ^{123}Te(p,n)^{123}I reaction. The yield of isomer ^{123m}Te formed via reaction ^{123}Te(p,p′) for a few tellurium targets at the proton energy 15 MeV has been measured. The results are discussed.

1 Introduction

Tellurium targets used for the production of ^{123}I with the reaction ^{123}Te(p,n) are usually prepared by smelting of the tellurium dioxide on a platinum plate. The target activity after irradiation and extraction of ^{123}I belongs mainly to platinum backing and isomer ^{123m}Te ($T_{1/2}$ = 119.7 days) formed via reaction ^{123}Te(p,p′). Activation of the platinum plate is mostly conditioned by the activity of ^{196}Au ($T_{1/2}$ = 6.18 days). This part of the target activity can be excluded by cooling for about two weeks or by change of the platinum plate. In case of isomer ^{123m}Te there is no any known way to remove it from the tellurium-123 target but a rather long cooling of the target. The goal of the present work is the measurement of the yield of ^{123m}Te at the proton energy 15 MeV that is usual at the production of iodine-123 in the Laboratory of Nuclear Reactions and Nuclear Medicine of the V. G. Khlopin Radium Institute.

2 Experimental

Two 95% enriched tellurium-123 dioxide targets were each of 320 mg/cm^2 thickness on 0.3 mm thick platinum backing. One target was freshly made and the other was used for production of ^{123}I for two years. The targets were irradiated two and four times, respectively, by the external beam of the V.G. Khlopin Radium Institute (St. Petersburg, Russia) MGC-20 cyclotron. Proton beam energy was 15.5 MeV. Because of energy loss in a 0.1 mm thick aluminium window and helium-cooling gap the proton energy decreased to 14.8 MeV. Farther protons were slowing down in the tellurium dioxide down to the energy 8 MeV on platinum plate surface. Each irradiation was carried for about 16 hours with the beam current of 5.5 μA.

The main gamma line of ^{123m}Te is of the same energy of 159 keV as of the ^{123}I one. Therefore gamma measurements of the ^{123m}Te activity were carried out two weeks after the end of bombardment when the rest of ^{123}I remaining after it's extraction has completely decayed.

3 Results and discussion

The yield of ^{123m}Te isomer measured with gamma spectrometry was 14.18 ± 0.35 and 15.30 ± 1.55 kBq/(μAh) for the new and for old targets, respectively. Value of the ^{123m}Te yield averaged with weights was equal to 14.24 ± 0.34 kBq/(μAh).

Dose of the tellurium target after irradiation with average proton beam current integral of 100 μAh was estimated as 160 μSv/h. For ten months of weekly irradiations the dose will increase up to 3.3 mSv/h.

The estimated values are in good agreement with results of the dose monitoring of the tellurium targets carried out in the Laboratory of Nuclear Reaction and Nuclear Medicine of the V. G. Khlopin Radium Institute.

REGENERATION OF TELLURIUM DIOXIDE IN IODINE-123 PRODUCTION

B. K. KUDELIN, E. A. GROMOVA, L.V. GAVRILINA AND L.M.SOLIN

V.G.Khlopin Radium Institute, 194021 St.-Petersburg, Russia

There is a regular necessity of the target material regeneration during iodine-123 production from tellurium isotopes. It was found that some impurities are able to trap iodine-123 and prevent its complete distillation from irradiated tellurium dioxide target. The regeneration procedure usually is connected with significant losses (up to 15%) of valuable enriched tellurium isotopes. A new regeneration (purification) procedure of tellurium and tellurium dioxide in presence of organic and inorganic impurities is described. The enriched isotope losses are less than 5%.

1 Introduction

A dry distillation or sublimation method is usually used for iodine separation in ^{123}I production by irradiation of targets with tellurium dioxide. During the target heating TeO_2 is partially evaporated and remained on walls of sublimation apparatus. Tellurium dioxide losses in the process of iodine sublimation can reach 10 mg (5%) from one target [1]. Thorough selection of sublimation process parameters allows us to reduce the losses to 0.3-0.6 mg (0.2-0.3%), but periodically it is necessary to clean the sublimator and return TeO_2 into production.

Usually a regeneration procedure includes dissolution of tellurium dioxide in hydrochloric acid and reduction of tellurium by hydrazine hydrate. But this procedure does not provide necessary purity of the obtaining material and leads to near 15% loss of TeO_2 during regeneration.

The aim of our investigation was to develop an improved regeneration and purification procedure.

2 Methods and results

A procedure developed is based on TeO_2 volatility at high temperature. A quartz apparatus was constructed for vacuum evaporation of tellurium and its dioxide. Starting material in a crucible is placed at the bottom of a tube 15 cm long and 3 cm diameter, which is capped by cooling finger so that the distance between the crucible and the finger is near 1 cm. The tube is evacuated by vacuum pump to $4 \bullet 10^{-3}$ mm Hg and heated in electric oven to 780-800 °C. The finger is cooled by water flow. 100-150 mg of TeO_2 is evaporated during 30-40 minutes.

Whole scheme of regeneration procedure is shown in Fig.1. At the first stage starting material (TeO_2 with impurities) is heated in the described apparatus. In our experiments samples of pure TeO_2 in such conditions were evaporated wholly and condensed on the cooling finger surface. In presence of impurities only a part of TeO_2 was evaporated and in the crucible a nonvolatile residue was found. The residue contained tellurium and its mass depended on impurities quantity.

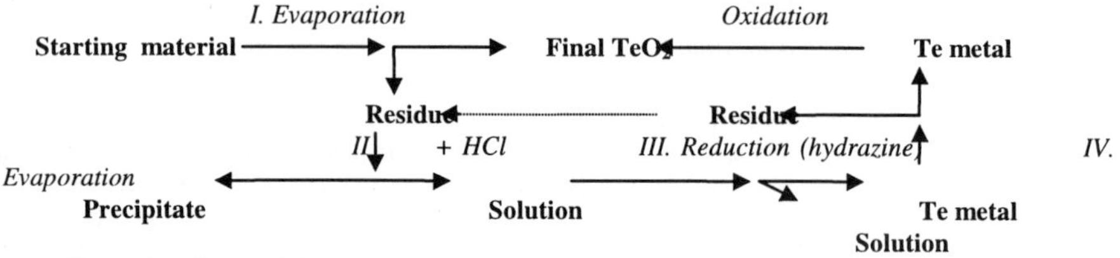

Figure 1. Scheme of the regeneration process

We found that targets with contaminated TeO_2 have a diminished iodine-123 yield. It was assumed that some impurities were able to trap iodine and prevent its complete recovering from irradiated target.

Chemical spectral analysis data for main impurities obtained at different stages of the regeneration process are shown in Table . TeO_2 from a target with low ^{123}I yield (70-80% of expected) was used as starting material for this experiment.

Table. Chemical spectral analysis data. Relative error 15%.

Chemical element	Metal content, mass %		
	Precipitate after stage II	Dried solution after Te reduction on stage III	Final TeO_2
Te	7.0	2.1	>76
Al	35.0	15.5	0.01
Fe	5.4	22.5	0.05
Mg	1.5	6.8	0.04
Si	10.2	1.2	1.20
Ca	1.2	5.7	0.12
Cu	0.6	6.3	0.06

A lot of impurities were found (columns 2 and 3) and their influence on iodine-123 yield is the subject of our further investigations. The sample of TeO_2 obtained had very good iodine-123 yield after irradiation.

The procedure of regeneration described allows to obtain high quality TeO_2 suitable for iodine-123 production. Tellurium losses during regeneration are 3-5%.

References

1. Oberdorfer F., Helius F. and Maier-Borst W., Experiences in the routine production of ^{123}I via the $^{124}Te(p,2n)^{123}I$ reaction with a low energy cyclotron, *J. Radioanal. Chem.* **65** (1981) pp. 51-56.

MEDICAL AND ECOLOGICAL ISOTOPES PRODUCTION AT THE MGC-20 CYCLOTRON

L.M.Solin, E.A.Gromova, V.A.Jakovlev, V.A.Kalinin, Ju.A.Protasenko
V.G.Khlopin Radium Institute, St.Petersburg, Russia

We have developed the cheapest isotope production methods for medicine and ecology using the MGC-20 cyclotron. The most popular medical nuclide iodine-123 we produce with high radionuclidic purity via ^{123}Te(p,n) ^{123}I reaction at 15 MeV proton beam energy. Every week radiopharmaceuticals: sodium iodide, iodine-123 and o-iodohippurate, iodine-123 are delivered to 20 hospitals of St.Petersburg. Moreover we have developed methods of gallium-67 and indium-111 production also via p,n reaction with high radionuclidic purity. We found the optimal solutions for target materials, regimes of irradiation, chemical procedure for extraction of produced nuclides and for target material recovery. We have developed method for gallium-67 citrate production. We have made probe runs to estimate our possibilities in technetium-95m and xenon-127 productions. We have produced technetium-95m in the ^{93}Nb(α,2n) reaction and xenon-127 in the ^{127}I(p,n) reaction. We have produced plutonium-236, 237 and neptunium-235 as traces for environmental investigations.

The Radium Institute cyclotron is a typical cyclotron of MGC-20 type. There are four cyclotrons of such type in Russia and several ones abroad: in Finland, in Hungary, in North Korea and in Egypt (under construction). Wide application of the MGC-20 cyclotrons is obliged to the following reasons: four types of accelerated particles - p, d, α and ^{3}He, broad range of the beam energies, and the suitable beam currents. The Radium institute cyclotron MGC-20 is used since 1989 mainly for radioisotope production. The proton beam with the energy of about 15 MeV proved itself to be the most suitable for the production of the most popular medical nuclides such as iodine-123, gallium-67 and indium-111.

Iodine-123. We consider the reaction ^{123}Te(p,n)^{123}I to be suitable for MGC-20 cyclotron. The main reason is simplicity of extraction of iodine-123 from raw material in the form suitable for labeling different compounds. We use enriched tellurium-123 (in dioxide form) deposited on the platinum background as a target. The target is bombarded by proton beam with the energy of about 15 MeV. After irradiation the target is heated and escaped iodine is absorbed NaOH solution. For the ^{123}Te(p,n)^{123}I reaction the radionuclidic purity is determined by admixture in target of ^{124}Te that is to be the origin of ^{124}I via (p,n) reaction. It was shown that the highest radionuclidic purity 99.976% may be obtained for the target containing 0.06% of ^{124}Te.

Gallium-67. The $^{67}Zn(p,n)^{67}Ga$ nuclear reaction proved itself to be the most suitable for gallium-67 production in the case of the MGC-20 cyclotron application. Enriched target material (Zn-67 ~ 90%; Zn-68 ~ 5%) is irradiated by proton beam current 10-15 μA at the energy of 15 MeV. It turned out that powder tablets of oxide ZnO are the best version of the target both for the gallium-67 extraction and for recovery of Zn-67. The irradiated powder is solved in HCl. Separation of the manufactured nuclide is performed using the method of cation exchange. The product's radionuclidic purity depends on the admixture in the starting material of isotope ^{68}Zn and on the yield of the $^{67}Zn(p,2n)$ reaction. The daughter products of all the isotopes except ^{68}Ga (9.4 hours half-life and with 511, 1039 and 2752 keV gamma ray energies) are short-lived enough to decay in a few hours after the end of the bombardment. At the end of bombardment contribution its activity achieved a value of about 20 %. However, after cooling during 2 days the product's radionuclidic purity is over 99.2 %.

Indium-111. Manufacture of ^{111}In is performed by irradiation of the starting material Cd enriched by isotope ^{111}Cd (95%) with a proton beam using the $^{111}Cd(p,n)$ ^{111}In reaction. The beam energy is about 15 MeV, beam current is 10 to 15 μA. Separation of the manufactured nuclide is performed using the method of extraction chromatography similar to that for gallium-67. The daughter products of all the isotopes except ^{110m}In are short-lived enough and thus decayed in a few hours after the end of the irradiation. The only product that can compete with ^{111}In is ^{110m}In. ^{110m}In is a product of the $^{111}Cd(p,2n)$ reaction. ^{110m}In has a half-life of 4.9 hours and energies of main γ-lines 657.7, 707.4, 884.7 937.5 and 997.2 keV. At the end of irradiation contribution its activity achieves the value of about 20 %. However, the product radionuclidic purity in 24 hours is over 99.9 %.

Technetium-95m. Technetium-95m is also used in medicine. It may be produced by means of the MGC-20 cyclotron by bombardment of a Nb foil with ^{4}He nuclei. We have carried out a proper run with the alpha beam of 20 MeV energy. After target cooling during two weeks and radiochemical procedures we have obtained technetium-95m of a high radionuclidic purity.

Xenon-127. We have used $^{127}I(p,n)$ reaction for ^{127}Xe production. As a result we had the yield about 0.55 MBk/μA h at the energy of protons 15,5 MeV.

Isotopes for ecology. The environmental investigations on movement of transuranium elements are held in some countries by traces method using nuclides plutonium-236, 237 and neptunium-235 as traces. ^{237}Pu and ^{236}Pu are produced at the MGC-20 together via reactions $^{237}Np(p,n)$ and $^{237}Np(p,2n)$ consequently at proton beam energy about 16 MeV. ^{235}Np is produced via reaction $^{235}U(p,n)$ at the same proton energy. In both reactions we use the powder of metal oxide as a target. For manufacturing nuclides we employ the method of extraction chromatography. High quality of final products was determined by utilizing high quality raw material for the target and appropriate technological procedure.

DISTRIBUTION AND INVENTORY OF $^{239+240}$PU AND ^{137}CS IN MARINE SEDIMENTS FROM THE JAPANESE COAST

MASATOSHI YAMADA AND YUTAKA NAGAYA

*Geochemical Research Section, Research Center for Radioecology National Inst*itute of Radiological Sciences,*Isozaki 3609, Hitachinaka, Ibaraki 311-1202, JAPAN*

E-mail: m_yamada@nirs.go.jp

Artificial radionuclides ^{239}Pu, ^{240}Pu and ^{137}Cs have been mainly delivered to the environment by global fallout derived from atmospheric nuclear weapons tests. The purposes of our investigation were to determine the concentrations of $^{239+240}$Pu and ^{137}Cs in sediments from the Japanese coast and to discuss the behavior of artificial radionuclides by measuring the distributions of $^{239+240}$Pu/^{137}Cs activity ratios and inventories of these nuclides in marine sediments. The sediment samples were collected from the Japanese coast during the *R/V Tansei Maru* cruise. At Stn. 8, KT-91-3 (35-04.8°N, 139-16.2°E, water depth 1329 m), the $^{239+240}$Pu concentration increases gradually with depth to 11.2 mBq/g at 12-14 cm depth, at which the maximum is identified, and then decreases gradually on going deeper. The profiles of $^{239+240}$Pu and ^{137}Cs in sediments from Stn. 8, KT-91-3 show nearly the same variation with depth. The inventories in the sediment column at Stn. 8, KT-91-3 are estimated to be 569 MBq/km^2 for ^{137}Cs and 434 MBq/km^2 for $^{239+240}$Pu. The $^{239+240}$Pu inventory is approximately 10 times greater than that of the estimated direct fallout input at the same latitudes of 30-40°N. These results may suggest that other input sources such as land wash by erosion and river runoff contribute to the $^{239+240}$Pu inventory in marine sediments for Stn. 8, KT-91-3.

1 Introduction

Artificial radionuclides ^{239}Pu ($T_{1/2}$= 24400 years), ^{240}Pu ($T_{1/2}$= 6580 years) and ^{137}Cs ($T_{1/2}$= 30 years) have been mainly delivered to the environment by global fallout as a result of release from atmospheric nuclear weapons testing, which took place mostly during 1950s and early 1960s[1]. It is important to understand the degree of artificial radioactive contamination and the behavior of these nuclides in the marine environment. The purposes of our investigation were to determine the concentrations of $^{239+240}$Pu and ^{137}Cs in sediments from the Japanese coast and to discuss the behavior of artificial radionuclides by measuring the distributions of $^{239+240}$Pu/^{137}Cs activity ratios and inventories of these nuclides in marine sediments.

2 Experimental

The sediment samples were collected from the Japanese coast during the *R/V Tansei Maru* cruise. The samples were collected with a box corer, which was used to get relatively undisturbed surface sediments, and analyzed for their $^{239+240}Pu$ and ^{137}Cs concentrations. The analytical procedure for $^{239+240}Pu$ and ^{137}Cs was the same as that described previously[2].

3 Results and Discussion

The vertical profile of $^{239+240}Pu$ is shown in Figure 1 for Stn. 8, KT-91-3 (35-04.8• N, 139-16.2°E, water depth 1329 m). The $^{239+240}Pu$ concentration increases gradually with depth to 11.2 mBq/g at 12-14 cm depth, at which the maximum is identified, and then

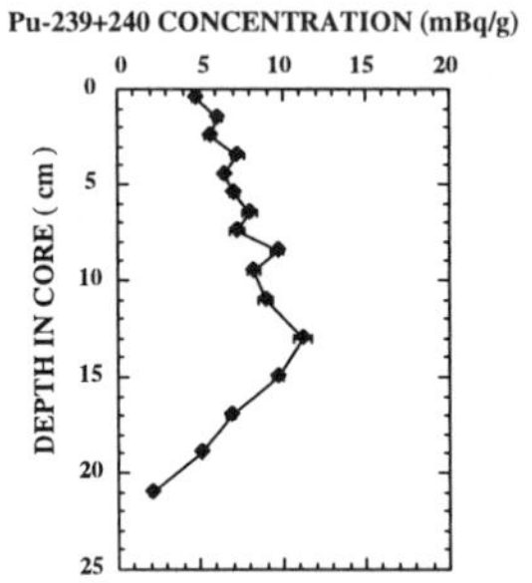

Figure1: Vertical Profile of $^{239+240}Pu$ at Stn. 8, KT-91-3.

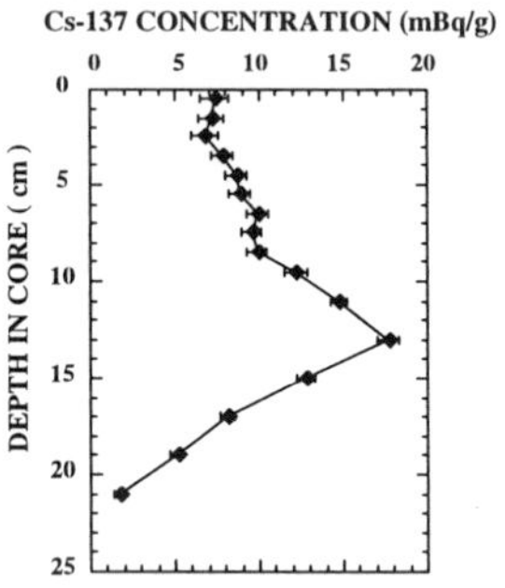

Figure 2: Vertical profile of ^{137}Cs at Stn. 8, KT-91-3.

decreases gradually on going deeper. The profiles of $^{239+240}Pu$ and ^{137}Cs in sediments from Stn. 8, KT-91-3 show nearly the same variation with depth. The ^{137}Cs concentration is 7.4 mBq/g in the surface layer and increases with depth to a maximum value (17.7 mBq/g) at 12-14 cm depth and then decreases as shown in Figure 2. The inventories in the sediment column at Stn. 8, KT-91-3 are estimated to be 569 MBq/km^2 for ^{137}Cs and 434 MBq/km^2 for $^{239+240}Pu$. The $^{239+240}Pu$ inventory is approximately 10 times greater than that of the estimated direct fallout input at the same latitudes of 30-40°N[1]. These results may suggest that other input sources such as land wash by erosion and river runoff contribute to the $^{239+240}Pu$ inventory in marine sediments for Stn. 8, KT-91-3.

References

1. UNSCEAR., Ionizing Radiation: Sources and Biological Effects, United Nations, 1982
2. Nakamura K. and Nagaya Y., Distribution of ^{137}Cs and $^{239+240}Pu$ in sediment of the Seto Inland Sea, *J. Radioanal. Nucl. Chem.* **138** (1990) 153-164

DETERMINATION OF ^{239,240}Pu AND ^{137}Cs CONCENTRATIONS IN MARINE ORGANISMS FROM THE SEAS AROUND JAPAN

TATSUO AONO, MASATOSHI YAMADA AND SHIGEKI HIRANO

Research Center for Radioecology, National Institute of Radiological Sciences
Isozaki 3609, Hitachinaka, Ibaraki 311-1202,
JAPAN
E-mail: t_aono@nirs.go.jp

To investigate artificial radioactivity levels in Japan, several marine biological samples were collected annually in the seas around Japan from 1992 to 1998, and their ^{239,240}Pu and ^{137}Cs activity concentrations were measured. The activity concentrations of ^{239,240}Pu and ^{137}Cs in marine organisms ranged from less than the detection limit to 20.8 mBq/kg-raw and from 11 to 141 mBq/kg-raw, respectively. Bivalves and algae showed higher ^{239,240}Pu activity concentrations than fishes. On the other hand, ^{137}Cs activity concentrations in fishes were similar to those in algae. As a result, the concentration factors of ^{137}Cs in fishes and algae ranged from 20 to 100 and from 27 to 120, respectively. The variation of ^{137}Cs and ^{239,240}Pu activity concentrations in organisms seemed attributable to their marine environment.

1 Introduction

Artificial radionuclides such as ^{239,240}Pu and ^{137}Cs have entered terrestrial and marine environments through radioactive fallout, and liquid and solid-state radioactive wastes. This study monitors radioactive pollution in the marine environment, and measures the activity concentrations of ^{239,240}Pu and ^{137}Cs in marine organisms from the fishing grounds along the Japanese coastlines. The concentration factors (CFs) and radioactivity concentration mechanisms in marine organisms are also discussed.

2 Sampling and analysis

Fish, shellfish and algae samples were collected annually from fishing grounds along the Japanese coastlines from 1992 to 1998. Radiochemical analysis procedures used for the separation and purification of plutonium and cesium were based on established techniques[1]. Each purified fraction of the isotopes was measured for its plutonium content with an alpha spectrometer and for its cesium content with a beta counter.

3 Results and discussion

3.1 ^{239,240}Pu and ^{137}Cs in fishes and shellfish

Activity concentrations of ^{239,240}Pu were not detected in some parts of fish samples, and of these samples with detected activities, the activity concentrations of

$^{239,240}Pu$ in fish samples ranged from less than 0.04 mBq/kg-raw in the muscle part of *Zeus japonicus* and *Limanda yokohamae* sampled from Ibaraki(1997) to 1.42 mBq/kg-raw in the gill part of *Limanda yokohamae* sampled from Ibaraki(1997). On the other hand, shellfish samples showed remarkably higher activity concentrations of $^{239,240}Pu$ than fish samples, and were ranged from 1.72 mBq/kg-raw in *Voluthar pa ampullance perryi* to 20.8 mBq/kg-raw in *Spisula sachalinensis*, both from Ibaraki(1993). The activity concentrations of ^{137}Cs in fishes and shellfish samples ranged from 50 mBq/kg-raw in the viscera part of *Paralichthys olivivaceus* sampled from Fukuoka(1994) to 287 mBq/kg-raw in the muscle part of *Seriola quinqueradiata* sampled from Ibaraki(1992), and from 39 mBq/kg-raw in 1994 to 120 mBq/kg-raw in edible tissue of *Spisula sachalinensis* from Ibaraki(1997).

3.2 $^{239,240}Pu$ and ^{137}Cs in algae (Sargassum thunbergii)

As *Sargassum thunbergii* is distributed widely in the seas around Japan, the distributions and annual variations of radionuclide activity concentrations can be compared with the same indicator organism. The activity concentrations of $^{239,240}Pu$ and ^{137}Cs in *Sargassum thunbergii* samples ranged from 4.84 mBq/kg-raw from Hokkaido(1994) to 39.51 mBq/kg-raw from Aomori(1995) and, from 107 mBq/kg-raw from Ibaraki(1997) to 360 mBq/kg-raw from Nagasaki(1994), respectively. The activity concentrations of $^{239,240}Pu$ and ^{137}Cs showed great variation at the same locations within one year.

3.3 Concentration factors and activity ratios in marine organisms

Based on the activity concentration of ^{137}Cs being about 3 mBq/kg in sea water, CFs of ^{137}Cs in fishes, shellfish and algae ranged from 20 to 100, 10 to 50, and 27 to 120, respectively. The different activity concentrations of ^{137}Cs in fishes and shellfish reflected differences in diet and environmental factors such as water temperature and salinity[2,3]. The activity ratios of $^{239,240}Pu$ relative to ^{137}Cs in algae ranged from 0.03 to 0.25, and the average of the ratios was 0.12. As CFs of $^{239,240}Pu$ were higher than those of ^{137}Cs, the accumulation of plutonium in algae was greater than that for cesium. The effect of the particulate matter from surface area of sediments could not be disregarded. It seemed that the variation of $^{239,240}Pu$ and ^{137}Cs activity concentrations in organisms was attributable to their marine environment.

References

1. Nagaya, Y. and Nakamura, K. $^{239,240}Pu$ and ^{137}Cs concentrations in some marine biota, Nippon Suisan Gakkaishi **53**(1987) pp.873-879.
2. Kasamastu, F., Ishikawa, Y., Kawabe, K. and Kawamura, H. Variation of ^{137}Cs

concentrations in marine fishes, JCAC **34**(1999) pp.30-36(in Japanese).
3. Suzuki, Y., Nakahara, M. and Nakamura, R. Roles of food and sea water in the accumulation of radionuclides by marine fish, Bull. Japan. Soc. Fish **45**(1979) pp.1409-1416.

PRODUCTION OF NEPTUNIUM-235 TRACER FOR LABORATORY USE

JORMA AALTONEN
Laboratory of Radiochemistry, Department of Chemistry, University of Helsinki P.O. Box 55, FIN-00014 Helsinki, Finland
E-mail: jorma.o.aaltonen@helsinki.fi

PETER DENDOOVEN
Department of Physics, University of Jyväskylä P.O. Box 35, FIN-40351 Jyväskylä, Finland
E-mail: dendooven@phys.jyu.fi

EVGENIA GROMOVA AND VLADIMIR JAKOVLEV
Laboratory of Nuclear Reactions and Nuclear Medicine, V. G. Khlopin Radium Institute, Second Murinsky Avenue 28, 194021 St. Petersburg, Russia
E-mail: jakovlev@atom.nw.ru

The production of ^{235}Np by the reaction $^{238}U(p,4n)^{235}Np$ using the Department of Physics, University of Jyväskylä K-130 cyclotron was carried out. Neptunium of nanogramme quantities was isolated from the irradiated uranium targets by chemical methods including ion-exchange, extraction chromatography with TRU-resin (Eichrom Industries Inc., USA) and with metilthreeoctilamonium nitrate (MTOA). High decontamination of neptunium from uranium, plutonium and fission products was achieved.

1 Introduction

The X-ray-emitter ^{235}Np ($T_{1/2}$ = 396 d) is mainly used as a tracer for the control of ^{237}Np release into the environment from the nuclear fuel cycle because ^{235}Np is not formed in the nuclear reactors in a significant amount and it has suitable decay characteristics. The goal of the present work has been to produce high quality radionuclide tracer of ^{235}Np.

2 Experimental

The production of ^{235}Np has been carried out by means of the reaction $^{238}U(p,4n)^{235}Np$ which is the most productive at the proton energy lower than 40 MeV [1]. Two 0.17 mm thick natural uranium foils were irradiated separately in the external beam of the Department of Physics, University of Jyväskylä K-130 cyclotron. The proton beam energy was 30 MeV and the collected beam charge was about 24 μAh.
The chemical separation of neptunium from the irradiated uranium targets and radiochemical analysis took place in the Laboratory of Radiochemistry at the University of Helsinki. Separation methods were based on ion-exchange, extraction

chromatography with MTOA-resin [2,3] and TRU-resin [4]. The gamma and alpha spectrometry data handling took place in the V.G. Khlopin Radium Institute.

3 Results and discussion

The method of separation and purification of neptunium developed in this work led to a high amount and quality of ^{235}Np tracer. The activity of ^{235}Np in the 1 ml final solution was 277 kBq at 42 days after EOB. The gamma activity of fission products was less than 0.01% and the alpha activity of other nuclides less than 0.001% of that of ^{235}Np.

4 Acknowledgements

The authors express their gratitude to the staff of the University of Jyväskylä Accelerator Laboratory for assistance during the irradiations and to MAP Medical Technologies Oy (Jyväskylä, Finland) for transportation of the irradiated targets. This work was supported by the Academy of Finland.

References

1. Bellido L.F., Robinson V.J. and Sims H.E., Production of ^{235}Np and ^{236}Pu for environmental applications, *Radiochimica Acta* **64** (1994) pp. 11-14.
2. Alexandruk V.M., Babaev A.S., Demjanova T.A. and Stepanov A.V., Luminescence determination of neptunium in soils of the 30 kilometers area of the Cherbobyl nuclear power station, *Radiochimia* **5** (1991) pp. 203-208.
3. Markov V.K., Usolkin A.N. and Ternovskiy A.I., Using nitrates of the quaternary amonium bases for extraction-chromatographic separation of neptunium, *Radiochimia* **6** (1979) pp. 862-867.
4. Horwitz E.P., Chiarizia R., Ditz M.L., Diamond H. and Nelson D.M., Separation and preconcentration of actinides from acidic media by extraction chromatography, *Analytical Chimica Acta* **281** (1993) pp. 361-372.

A NOVEL BORON-ISOTOPE SPECIFIC RESIN HAVING TRIETHANOLAMINE GROUPS

AKINARI SONODA, YOJI MAKITA, KENTA OOI, AND TAKAHIRO HIROTSU

Shikoku National Industrial Research Institute, Hayashi, Takamatsu 761-0395, JAPAN
E-mail: sonoda@sniri.go.jp

HIRONORI MAEDA

Kumamoto Institute of Technology, Ikeda, Kumamoto 860-0082, JAPAN

We designed and synthesized a novel boron-isotope specific resin having triethanolamine groups. The resin was made from poly(glycidylmethacrylate) crosslinked with divinybenzene. The triethanolamine type functional groups were introduced by additions of diethanolamine. We have evaluated the fractionation of boron isotopes in the isotopic exchange reactions with 3-coordinated and 4-coordinated boron complexes in experimental and theoretical approach. The isotope separation factor of the resin was obtained by batch-wise test. The isotopic ratio of resin phase was calculated from the concentrations and the ratios of the initial and final solution phase. In this case, the isotope separation factor was **1.040** that is the highest value for boron-isotope specific adsorbents. The values have been also estimated from the ratio of the reduced partition function ratios of boric acid and triethanolamine borate, which are calculated from the normal vibration modes using two kinds of ab initio molecular orbital methods, Gaussian and Dmol programs.

1 Introduction

The boron-isotope separation factors with anion exchange resins (*S*, defined as Eq.1) are restricted to the upper limit of 1.0186 at 300K [1].

$$S = (^{10}B/^{11}B)_{resin} / (^{10}B/^{11}B)_{solution} \quad \text{(Eq.1)}$$

Isotope fractionation between two chemical species is principally due to a difference in intermolecular forces acting on the atom between the species. Accordingly, we have evaluated the fractionations of boron isotopes in the isotopic exchange reactions with 3-coordinated and 4-coordinated boron complexes. We selected boric acid for a 3-coordinated boron species because of the simplicity and the stability in aqueous systems. For a 4-coordinated boron species, we considered triethanolamine borate having a tetrahedral structure around the boron atom.

In this study, we designed and synthesized a boron-isotope specific resin having triethanolamine groups (Fig.1). We evaluated the boron-isotope selectivity of the triethanolamine type functional group in experimental and theoretical approach.

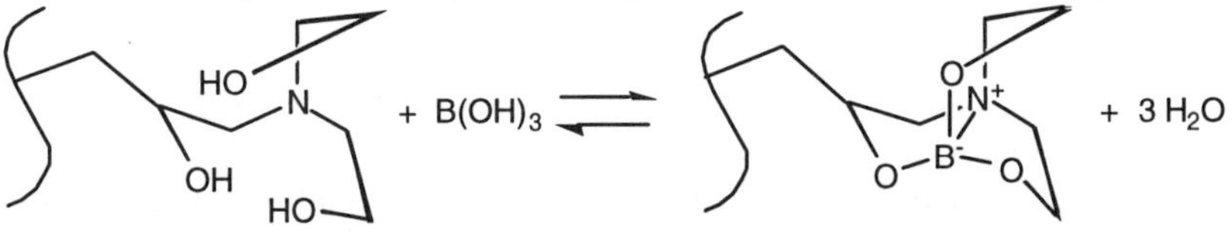

Figure 1. A novel boron-isotope specific resin having triethanolamine groups and the complex.

2 Experimental

The boron-isotope specific resin was made from beads of poly(glycidylmeth-acrylate) crosslinked with 7.5% DVB, which were synthesized by suspension polymerization. The triethanolamine type functional groups were introduced by additions of diethanolamine in dioxane at 100°C for 4 h.

The ***S*** values of the resin were obtained by shaking the resin in an aqueous boric acid solution at constant temperatures in equilibrium. The boron-isotope ratios of the solution phase were determined by NTIMS. The boron concentrations of the solutions were determined by ICP-AES. The ***S*** values were determined from the boron-isotopic ratios and the boron concentrations of the solution phase.

A theoretical ***S*** value was also estimated from a ratio of the reduced partition function ratios (***K***) of boric acid (**A**) and triethanolamine borate (**B**) as in Eq.2,

$$A_{light} + B_{heavy} = A_{heavy} + B_{light}, \quad K = f_{\mathbf{A}}{}^{\mathbf{r}} / f_{\mathbf{B}}{}^{\mathbf{r}}, \qquad \text{(Eq.2)}$$

where $f^{\mathbf{r}}$ is the reduced partition function ratio [2]. The value of $f^{\mathbf{r}}$ depends upon only the vibrations, and is defined as in Eq.3,

$$f^{\mathbf{r}} = \Pi_i[\, u_i \exp(-u_i/2)\{1-\exp(-u_i')\} \;/\; u_i' \exp(-u_i'/2)\{1-\exp(-u_i)\}] \qquad \text{(Eq.3)}$$

where $u_i = h\nu_i / kT$, ν_i is the ith vibrational frequency, and the primed symbols refer to the lighter isotope. The normal vibration analyses were carried out two kinds of ab initio calculations using the Gaussian and the Dmol programs.

3 Results and Discussion

The amount of triethanolamine type functional groups on the resin was about 3 mmol/g-resin from elemental analysis. The boron uptake of the adsorbents was dependent on pH of solution. The capacities of polymethacrylate-base resins were greater than that of polystyrene-base resins with same functional groups. This is probably due to a difference in a hydrophilicity of polymer. The ***S*** value of a polymethacrylate-base resin with triethanolamine groups was **1.040** that is the highest among the reported boron-isotope specific adsorbents. Calculated ***K*** values between boric acid and triethanolamine borate were about 1.03. These values are closed to the experimental results. These results are showing that it is possible to estimate maximum isotope separation factor using ab initio MO calculation.

References

1. Kakihana H., Ion exchange, Isotope Exchange and Isotope Separation, *J. Chromatogr.* **102** (1974) pp.47-55.
2. Bigeleisen J. and Mayer M. G., Calculation of Equilibrium Constants for Isotopic Exchange Reactions, *J. Chem. Phys.* **15** (1947) pp. 261-267

MODELLING ATMOSPHERIC PB-212 CONCENTRATION

ERNEST C.Y. HO AND DAVID F. MEASDAY

Department of Physics and Astronomy, University of British Columbia, 6224 Agricultural Road, Vancouver, BC, Canada. V6T 1Z1.

Abstract: The implementation of the Comprehensive Test Ban Treaty is resulting in the construction of a world wide system of 80 monitoring stations that will be able to detect the air-borne radioactivity not only from atomic bombs but also from other anthropogenic and natural sources, thus guarding the health of all people. A prototype monitoring station has been operating since April 1996 in Vancouver, BC, Canada. This station provides daily reports of natural radioactivity, including Be-7 and products of Rn-220 (thoron). Data for Pb-212 concentrations have been analysed over a six month period. The concentration is reduced by rainfall or high wind velocity and it also depends on the wind direction. We present a relatively simple model which is easy to use and which offers predictive powers that can be applied to other similar environmental situations.

1 Monitoring Station

The monitoring station designated CA002 is situated on the campus of the University of British Columbia at 49^0 15' N, 123^0 15' W. The site is a plateau about 50 metres above sea level. The samples are collected on a fibreglass filter, measuring 60 x 60 cm^2, which is centred 1.4 metres above ground level. The air flow rate is about 970 cubic metres per hour with a linear face velocity of 1.1m/s and as the sampling time is about 24 hours, the typical sample is 23000m^3 of air. The filter is changed once a day and, during the time when the data for this research was taken, it was placed immediately in front of the germanium detectors.

The first 4 hours of data collection was placed in a separate file and the data collection restarted. Since all Rn-222 products have about a half-hour half-life, they all decay away during this time. The second data acquisition period started immediately after this and lasted about 20 hours until the next sample was taken. The spectra are transmitted to the PIDC for automatic analysis. The line of 239keV, produced by the gamma from Pb-212, was selected for analysis.

2 Data Analysis

The fitting equation was taken to be:

$$C = \frac{\sum_{i=1}^{9} k_i W_i (1 + a\overline{V}_i)}{(1 + aV)(100 + k_{10} R_0 + k_{11} R_1 + k_{12} R_2 + k_{13} R_3)}$$

where C is the predicted Pb-212 concentration in gamma rays per hour per cubic metre of air. ks are the parameters that are to be fitted. (k_1 to k_9 represent wind directions[1]; k_{10} to k_{13} represent rainfall) R_0 to R_3 are the rainfall in mm on the day, the day before, 2 and 3 days before data collection. W_i are the "wind factors" which record the hourly wind direction during the data collection period. V is the wind velocity of the day and a is the parameter for wind velocity. The $(1+a\overline{V}_i)$ is the normalising factor so as to make the k factors (for different a parameters) can be directly compared to each other.[2] The fitting results are tabulated on Table 1.[3]

3 Discussion

1. The $\chi^2 / D.O.F$ fit is big, which means the fitting is poor; thus although the general trend is followed, more parameters are needed to model the data.
2. The stability of the k_i parameters for the variation of the wind speed correction from a=0 to a=0.2 means different wind directions do bring in a consistently different amounts of Pb-212. Referring to Table 1, it is clear that winds blowing from the sea (i.e. westerly wind[4]) bring in a comparatively less Pb-212 content.
3. From the simulation results, one notes that there is a general underestimation of summer data and a bit overestimation of winter data. Maybe the soil temperature plays a role in the Pb-212 content (see Porstendorfer) or one may include more rain parameters or increasing the database from half a year to a full year.

4 Selected References

1. Porstendorfer, J. Properties and behaviour of radon and thoron and their decay products in the air. J. Aerosol Science 23: 219-263; 1994.
2. Stocki, T. J; Measday D. F. Protecting the World from nuclear fallout through detection. Bulletin of the Canadian Protection Association 19: 8-10; 1998.

---

[1] k_1 represents wind blowing from 045^0 k_2 represents wind direction 090^0 and so on, k_9 represents still air.

[2] Bar denotes average; the wind velocity for each direction is averaged over the whole six month period.

[3] Owing to the page limit requirement, the authors are not possible to outline the complete data analysis procedure. The results on Table 1 (except 0.07R) is actually a fit keeping k_{10} to k_{13} fixed. The values of k_{10} to k_{13} were determined previously using a cruder model without the a parameter and a more primitive definition of W_i. The values of k_{10} to k_{13} were determined to be 9.1, 3.7, 1.4, 2.0 respectively, showing a generally decreasing but positive importance of rainfall with time in dampening Pb-212 content. 0.07R is actually a refit (seeing χ^2 reaching a minimum on a=0.07) releasing all rain parameters and adding two more (k_{14} and k_{15}).

[4] Please refer to any maps of Vancouver, BC, Canada for a geographic description of the area.

$a =$	0	0.07	0.07R	0.1	0.2	$\overline{V}$, km/h
k_1	432	279	271	252	206	6.82
k_2	248	252	249	251	247	12.29
k_3	355	344	392	333	307	13.76
k_4	268	274	240	268	245	13.87
k_5	255	188	175	176	156	13.11
k_6	147	189	168	194	203	13.84
k_7	70	92	127	91	85	17.58
k_8	311	272	268	268	263	7.21
k_9	685	596	641	577	540	0
$(\chi^2/N)^{1/2}$	55.11	52.16	49.9	52.27	53.06	

Table 1: χ^2 fitting results, N is the number of data that are used for fitting; here, N is 172. For 0.07R, k_{10}=7.2, k_{11}=3.0, k_{12}=1.7, k_{13}=1.4, k_{14}=0.9, k_{15}=1.6.

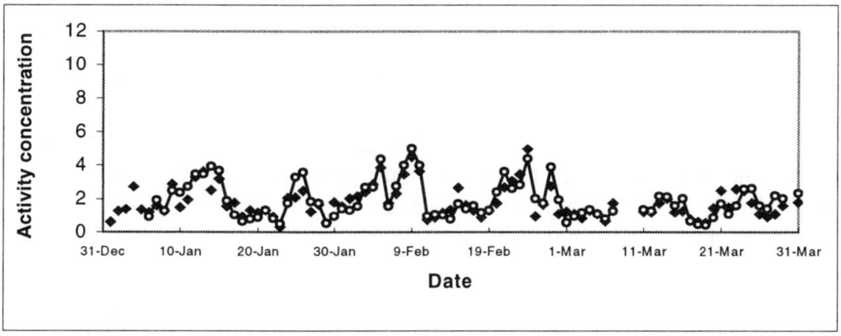

Figures 1 and 2: Comparison between simulation results and actual data for January to June 1997 for a=0.07R (the best fit). Full diamonds are actual data while open circles are simulation results.

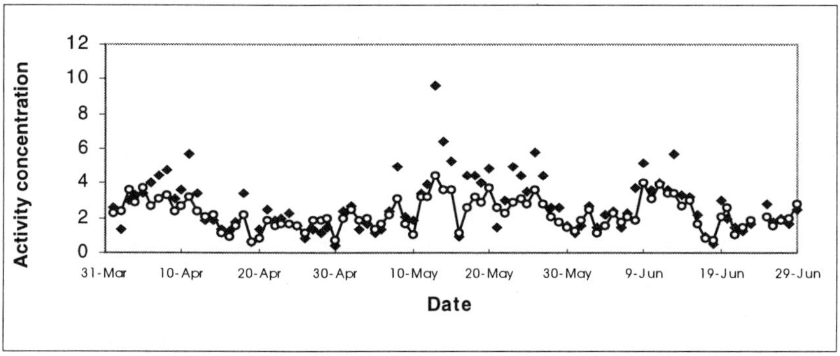

RECOVERY OF GE-68 FROM TARGET COOLING WATER

K. L. KOLSKY, D. H. HUNTER, AND L. F. MAUSNER
Medical Department, Brookhaven National Laboratory, Upton, NY 11973 USA

V. T. HAMILTON, D. J. JAMRISKA, AND D. R. PHILLIPS
Nuclear and Radiochemistry Group, Los Alamos National Laboratory, Los Alamo, NM 87545 USA

During the 1998 production cycle at the Brookhaven Linac Isotope Producer (BLIP), two Ge-68 targets developed unrelated failures that resulted in the release of over 16 GBq of Ge-68 to the target cooling water. A method was developed to recover this valuable isotope by pumping the entire cooling water volume (~ 1100 L) over an Alumina-N column. Ge-68 was then removed from the column by distillation into a water trap and the activity was concentrated by solvent extraction. Work supported by USDOE under Contract #DE-AC02-98CH10886 (BNL) and Contract # W-7405-ENG-36 (LANL).

Germanium-68 is widely used in the nuclear medicine community to calibrate PET cameras. It decays ($t_{1/2}$ = 271 d) by electron capture to ^{68}Ga ($t_{1/2}$ = 68.1 m) which then decays by positron emission (90.5%). Germanium-68 has been produced at Brookhaven National Laboratory (BNL) for many years using the ^{nat}Ga(p,xn) reactions by irradiating ^{nat}Ga targets with 45 MeV protons. Targets are ^{nat}Ga metal (~ 140 g) encapsulated in niobium (~ 7.6 cm diameter x 0.8 cm thick) shells with window thickness ~ 0.25 mm. These targets are cooled during irradiation by having cooling water flow past the target faces. The cooling water is a closed system (1100 L) with no active purification. Rather, the cooling water is replaced with fresh deionized water about every two months. During the 1998 production cycle, two separate targets developed leaks near the end of their irradiation times. One target developed a leak from a crack that appeared in the downstream window, the other target developed a leak from a crack that developed along the electron beam welded joint. Both leaks became quickly evident from routine cooling water sample analysis that showed the presence of ^{65}Zn, ^{69}Ge, and ^{68}Ge. In both cases, approximately 50% of the target mass was lost to the cooling water along with 80% of the ^{68}Ge activity (~ 16 GBq). Due to its monetary value, at the end of the production cycle, a method was developed to recover and purify this ^{68}Ge activity from the cooling water.

The recovery and purification process consisted of two separate phases. The first phase involved developing a remotized process to recover ^{68}Ge from 1100 L of water also containing ~ 7.8 GBq of ^{65}Zn, 92.5 GBq ^{7}Be, and lesser amounts of ^{56,58}Co and ^{54}Mn (< 10 MBq each). From the literature, it is well known that germanium and gallium are easily adsorbed from an aqueous medium on the inorganic adsorbent alumina, Al_2O_3. Experiments were performed to investigate retention of ^{68}Ge as a function of flow rate and large elution volumes. Results showed that even with very high flow rates, > 600 mL/min, ^{68}Ge was quantitatively retained by Alumina resin. The prototype column was designed in collaboration with Suppelco from a Porozorb™ cartridge: 500 mL (13.10 x 8 cm) containing activated neutral Alumina (60-325 mesh). The removable cap of the column was attached with a screw thread and rubber gasket and the column inlet and outlet were ¼ inch male pipe fittings. The BLIP cooling water was first pumped from the tank, using a submersible pump, to a 55-gallon drum. The water was then pumped over the column (~ 500 mL/minute) using a peristaltic pump. The eluant was collected in 55-gallon drums and sampled for breakthrough. Under these conditions, the entire volume of cooling water and rinses were pumped across the column in 60 hours and the column retained approximately 93% of ^{68}Ge, 75% of ^{65}Zn, 50% of ^{54}Mn, and 96% of ^{7}Be and ^{56,58}Co originally present in the cooling water.

The second phase of this project was to remove the ^{68}Ge activity from the resin and the radioimpurities that were also retained on the column. Preliminary tests showed ^{68}Ge could be distilled from Alumina resin in 6 N HCl. The resin (500 mL) was first transferred in its entirety to a 1 L distillation pot. To this were added 800 mL 8 N HCl. The distillation pot was then connected to a distillation system: condenser connected to a 250 mL receiving flask with a drip tube that extended to nearly the bottom of the vessel. To this flask were added 15 mL H_2O such that the water level submerged the drip tube. The receiving flask was immersed in an ice bath to reduce volatility losses. The distillation was allowed to proceed until 125 mL of distillate were collected. No activity was recovered in the distillate presumably due to dissolution of Alumina in the acid distillation pot during heating which lowered the acidity (^{68}Ge will not distill at < 5 N HCl). The resin was then distilled in batches. Each batch consisted of ~175 mL resin contacted with 800 mL 8N HCl. Assay of the combined distillate fractions (650 mL total) showed ~ 56% recovery. No radioimpurities were detected though it appeared that some dissolved aluminum co-distilled. The resins were assayed after distillations and showed only traces of ^{68}Ge (< 5% loss). The low recovery yield is not fully understood. The most likely cause for loss is volatilization.

After distillation, the ^{68}Ge was concentrated and separated from dissolved aluminum by extracting the activity into toluene. The total volume of distillate was chilled and then divided into three fractions (~ 200 mL each). The first fraction was added to 80 mL chilled toluene. An additional 500 mL of 12 N HCl were added to bring the aqueous phase acidity to 10 N HCl. After the extraction, the aqueous phase was removed and the second fraction was added to the toluene, the aqueous phase was again made 10 N HCl and extracted. These steps were repeated for the third fraction as well. The toluene was back extracted with 3 portions of 0.5 N HCl (8, 7, and 5 mL). The recovery yield (in 20 mL 0.5 N HCl) from extractions was 88%. The overall ^{68}Ge recovery for the entire procedure was approximately 50% with no detectable quantities of radioimpurities.

XENON-127 REVISITED

L.F. MAUSNER, H. SCHNAKENBERG, K.L. KOLSKY, S.C. SRIVASTAVA

Medical Department, Brookhaven National Laboratory
Upton, NY 11973, USA
E-mail: mausner@bnl.gov

Xe-127 [$t_{1/2}$ =36.3 d, Eγ=202.9keV (68%), 172.1keV (25%)] was originally developed at BNL in 1973 for lung ventilation imaging. It offers the advantages of less radiation dose to the patient, and better resolution images than Xe-133. Due to various reasons demand declined substantially in the early 1990s and BNL production ceased in 1993. However, a recent surge of interest in using Xe-127 for absolute cerebral brain blood flow measurement and imaging has encouraged us to provide this isotope once again. New features of its production include thinner targets, which allow higher incident beam current, and with two thin diaphragms at the periphery for needle puncture. We now extract the xenon gas by puncturing the target can while maintaining a gas seal, dissolving the salt with flowing water, and then purge the water of xenon with a helium flow. This contrasts with the original method , which melted the target CsCl salt and flowed helium across the molten surface. The aqueous system is more efficient and safer for hot cell use. Distribution of Xe-127 to the nuclear medicine community is anticipated to begin in March of 2000. Work supported by USDOE under Contract #DE-AC02-98CH10886.

1 Introduction

Xe-127 was originally developed at BNL in 1973 for lung ventilation imaging. It nuclear properties are summarized below. It offers the advantages of less radiation dose to the patient, and better resolution images than Xe-133. Due to various reasons demand declined substantially in the early 1990s and BNL production ceased in 1993. However, a recent surge of interest in using Xe-127 for absolute cerebral brain blood flow measurement and imaging has encouraged us to provide this isotope once again.

2 Methods

We produce xenon-127 by the (p,2p5n) reaction on ^{133}Cs at proton energies of approximately 190 MeV. The target is a pressed pellet of CsCl, 0.44 cm thick, encapsulated in a stainless steel can. This target is reduced in thickness from 0.81 cm in order to better cool the target and allow higher incident beam current. It also features two thin diaphragms at the periphery, used for needle puncture. The target contains internal channels connecting the puncture locations with the main target cavity.

We now extract xenon gas by puncturing the target can while maintaining a gas seal, dissolving the salt with flowing water, and purging the water of xenon with a helium flow. This contrasts with the original method, which melted the target CsCl salt and flowed helium across the molten surface. The aqueous system is more efficient and safer for hot cell use. A schematic of the entire gas extraction , purification, trapping, and dispensing apparatus is shown in Figure 1. Water is pumped from a reservoir vessel through the target and back to a reservoir. Simultaneously, helium gas is bubbled through the reservoir to sweep the Xe-127 out of the salt solution and into the gas train.

The gas flowing (25 cc/min) out of the reservoir passes through a gas purifier which removes water vapor, oxygen and carbon dioxide. It then flows through a "u" tube filled with silver beads heated to 400°C. This trap reacts with and removes residual radioiodine species. Then the gas mixture passes through a trap with large volume and surface area, cooled to liquid nitrogen temperature. Xenon-127 freezes out and the helium carrier gas passes through a back up cold trap and is then vented to hot cell atmosphere.

After collecting all Xe-127 from the target into the LN coil, the upstream section of the apparatus is valved off and the Xe-127 is cryogenically transferred to an evacuated steel storage vessel. Dispensing to individual users is also done by cryogenically transferring from the storage vessel to glass ampoules through needle valves. The quantity of radioactivity transferred is measured by a small diode detector for each ampoule and the storage vessel. Residual Xe-127 in the manifold is frozen back into the storage cylinder for reuse. Finally ampoules are remotely flame sealed and drop off the manifold. These ampoules are shipped to users and are crushed to release the Xe-127.

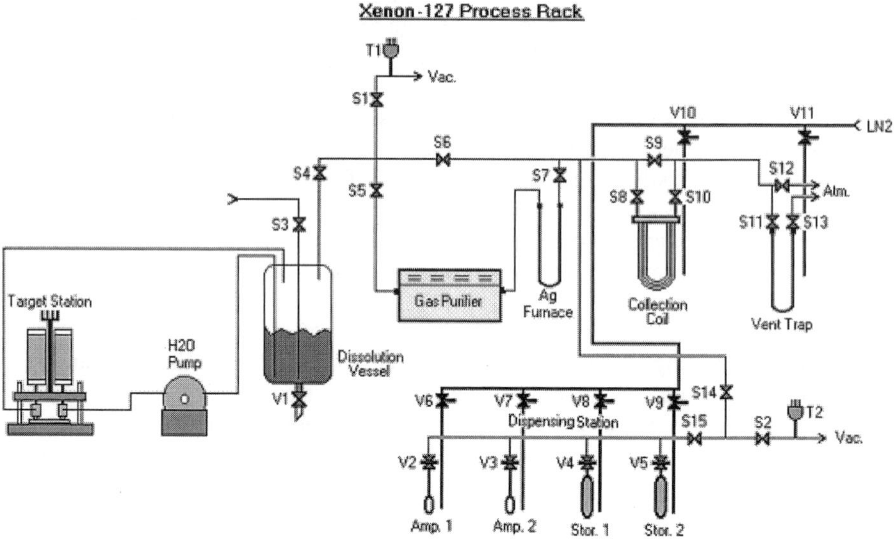

Figure 1. Schematic of Xe-127 Processing Apparatus

A CONVENIENT METHOD TO PREPARE [^{14}C]PNU-171829, A MATRIX METALLOPROTEINASE INHIBITOR.

E. FONTANA[1], A.PIGNATTI[1] AND F. ABRATE[2]

Dept. of Drug Metabolism Research[1], P&U Preclinical DevelopmentDept. of Chemistry[2], P&U Discovery Research Oncology Pharmacia & Upjohn S.p.A Viale Pasteur 10, 20014 Nerviano (MI), Italy

Overexpression of matrix metalloproteinase (MMP) enzymes, a family of zinc-dependent enzymes that degrades all major components of extracellular matrix, has been linked to a variety of deseases including cancer. The development of MMPs inhibitors as potential adjuvant of cytotoxic agents or radiation therapy to limit tumor growth and progression, is an active area of the on-going pharmaceutical research [1]. (2S)-2-Cyclohexyl-2-{(3S)-3-[2-(hydroxyamino)-2-oxoethyl]-3-isobutyl-2-oxopyrrolidinyl}-N-methylethanamide, PNU-171829, is a lactam hydroxamate that showed an interesting profile both "in vitro" and "in vivo". In order to perform further "in vitro" and " in vivo" studies with this new MMPs inhibitor, a radiolabelled form was required. Previous "in vitro" metabolism studies indicated that the methyl group attached to the nitrogen of the amido group could reasonably be expected to remain present in the possible PNU-171829 metabolites. A convenient six-steps sequence starting from the commercially available carbon-14 labelled methylamine hydrochloride (**1**) was developed to introduce carbon-14 in the above indicated position. According to this procedure [^{14}C]PNU-171829 was obtained >98% radiochemically pure with an overall radiochemical yield of about 16% from **1**. The synthetic pathway will be presented and discussed.

1 Results and Discussion

The synthetic route followed to prepare [^{14}C]PNU-171829 is depicted in Fig.1. In the original large scale synthesis of PNU-171829, more than a six-fold molar excess of methylamine was used in the first step. In our case, these conditions would result in a dramatic reduction of the radiochemical yield. Moreover, the use of volatile radiolabelled materials in large excess would not be appropriate due to radioprotection problems as well as to the production of undesirable radioactive waste. A number of different synthetic methods could be applied to prepare the carbon-14 labelled amide **3**. Among these, the use of O-(benzotriazol-1-yl)-N,N,N',N'-tetramethyluronium tetrafluoroborate (TBTU) seemed to suit well our purposes for the acceptable [^{14}C]methylamine:compound **2** molar ratio (1:1) and the easy and safe reaction work-up.

The reaction of [^{14}C]methylamine.HCl **1** and compound **2** (1:1 molar ratio) in acetonitrile in the presence of TBTU and 4-methylmorpholine (NMM) afforded, after stirring overnight at room temperature, the amide **3** >97% radiochemically pure, with a radiochemical yield of about 85%. The removal of the Boc protective group was carried out with trifluoroacetic acid in anhydrous methylene chloride at 0°C. The crude intermediate **6** was obtained by condensation of the amine **4** with the lactol the presence of $NaCNBH_3$ in methanol at room temperature followed by heating with silicagel in dimethylformamide at 150°C for 24 hours. After purification by TLC on silicagel plates (Merck 60 F254; 1 mm thick; ethyl acetate:n-hexane 1:1 by volume) the compound **6** was recovered >97% radiochemically pure. The oxidative cleavage of the double bond with potassium permanganate/sodium periodate in tert-butyl alcohol (tBuOH) followed by treatment with O-benzylhydroxylamine.HCl ($BnONH_2$) in the presence of carbonyldiimidazole (CDI) and NMM gave the O-benzylhydroxamate **8**. After purification by flash-chromatography (silicagel Merck 60; 230-400 mesh; ethyl acetate:n-hexane 3:1 by volume), the intermediate **8** was hydrogenated over 10% $Pd/BaSO_4$ catalyst in methanol. The desired [^{14}C]PNU-171829 was obtained >98% radiochemically pure with an overall radiochemical yield of about 16% from **1**.

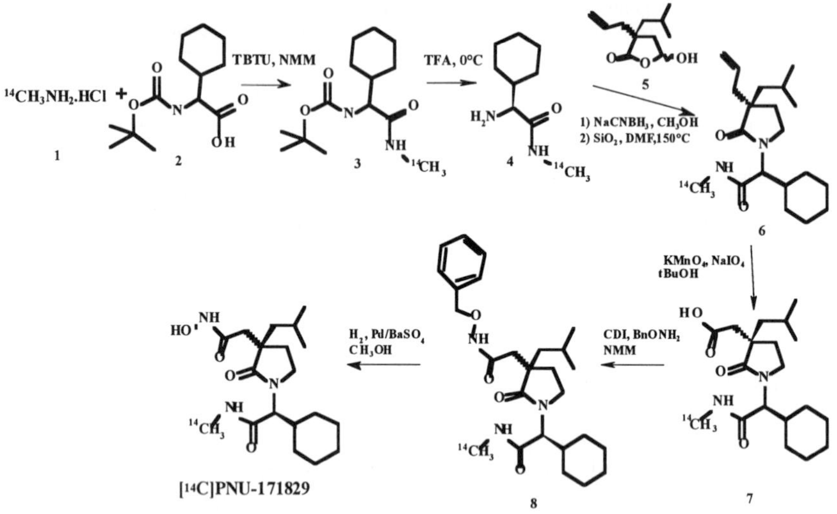

Figure 1. Scheme of the preparation of [^{14}C]PNU-171829

Reference

1. Levy D.E. and . Ezrin A.M, Matrix Metalloproteinase Inhibitor Drugs. *Emerging Drugs* **2** (1997) pp. 205-230.

SYNTHESIS OF ISOTOPICALLY LABELED 1-[(2,4-DI-1-PYRROLIDINYL-9*H*-PYRIMIDO[4,5-B]INDOL-9-YL)-ACETYL)PYRROLIDINE MONOHYDROCHLORIDE, PNU-142731A, AN ORALLY ACTIVE ANTIASTHMA AGENT

JOHN A. EASTER

Pharmacia & Upjohn, 301 Henrietta St., Kalamazoo, MI 49007, USA
E-mail: john.a.easter@am.pnu.com

1-[(2,4-Di-1-pyrrolidinyl-9H-pyrimido[4,5-b]indol-9-yl)-acetyl]pyrrolidine monohydrochloride, PNU-142731A, is under investigation as a potential orally active antioxidant for the treatment of asthma. Stable isotope and radioisotope forms of this compound were needed for drug adsorption, distribution, excretion, and metabolism studies. The synthesis of carbon-14 PNU-142731A, labeled at the C-2 position of the pyrimidine ring was accomplished in 6 steps. [(2,6-Di-1-pyrrolidinyl-4-[2-^{14}C]pyrimidinyl)amino]acetyl pyrrolidine, [^{14}C]PNU-145325 was synthesized in 4 steps from 300 mCi (nominally 55 mCi/mmol) of carbon-14 labeled urea and unlabeled diethyl malonate. The synthesis was completed by the cyclization of [^{14}C]PNU-145325 with 2-bromocyclohexanone, followed by dehydrohalogenation to afford 31 mCi (10.3% overall yield) of [^{14}C]PNU-142731A as the monohydrochloride salt. The specific activity was 112.02 μCi/mg (51 mCi/mmol) with a radiochemical purity of 99% as determined by HPLC and TLC. In a similar manner, [^{13}C, $^{15}N_2$]urea and [1,3-^{13}C]diethyl malonate were used to prepare 379 mg of [$^{13}C_3$, $^{15}N_2$]PNU-142731A with excellent purity and 99% atom incorporation.

1 Introduction

Clinical links to oxygen radicals include stroke, myocardial infarction, arrhythmia's, Alzheimer's, Parkinson's, MS, ALS, and asthma. It has been reported that the novel pyrrolo[2,3-d]pyrimidine is a potent antioxidant and inhibitor of lipid peroxidation. [1] The Reactive Oxygen Project Team at Pharmacia & Upjohn has identified several leads that have exhibited potential as orally active antioxidants for the treatment of asthma and neurodegenerative disorders. The latest in this series, PNU-142731A is currently undergoing Phase I clinical trial. To support the development of this compound, carbon-14 labeled PNU-142731A was synthesized.

2 Discussion

Carbon-14 labeled PNU-142731A was synthesized using the chemistry shown in Scheme 1. Following a literature preparation for unlabeled barbituric acid, 300 mCi of carbon-14 labeled urea was converted to [^{14}C]barbituric acid, ***3***. [2] The

barbituric acid which upon treatment with phenylphosphon
ic dichloride gave 2,4,6-trichloro-[^{14}C]pyrimidine, ***4*** in 74% yield. Treatment of ***4*** glycine ethyl ester hydrochloride in the presence of diisopropylethylamine gave a mixture of products from either the displacement of the chlorine at C-2 or C-6 of the pyrimidine ring in a ratio of 1:2 (v:v). Subsequent column chromatography using 25% ethyl acetate:hexane resulted in the isolation of 164 mCi of ***5***, a 50% yield. Reaction of ***5*** with excess pyrrolidine at reflux gave [^{14}C]PNU-145352, ***6*** in 63% yield. In the presence of the non-nucleophilic base diisopropylethylamine, ***6*** underwent cyclization with 2-bromocyclohexanone to give ***7*** in 89% yield. The synthesis of [^{14}C]PNU-142731A was completed by the dehalogenation of ***7*** using 10% palladium on carbon in decahydronaphthalene with heating to afford 31 mCi (10.3% overall yield) of ***8***, [^{14}C]PNU-1427361A. The specific activity was 112.02 µCi/mg
(51 mCi/mmol), and a radiochemical purity of 99% as determined by HPLC and TLC. Stable isotope labeled PNU-142731A was also synthesized using a parallel route.

Scheme 1

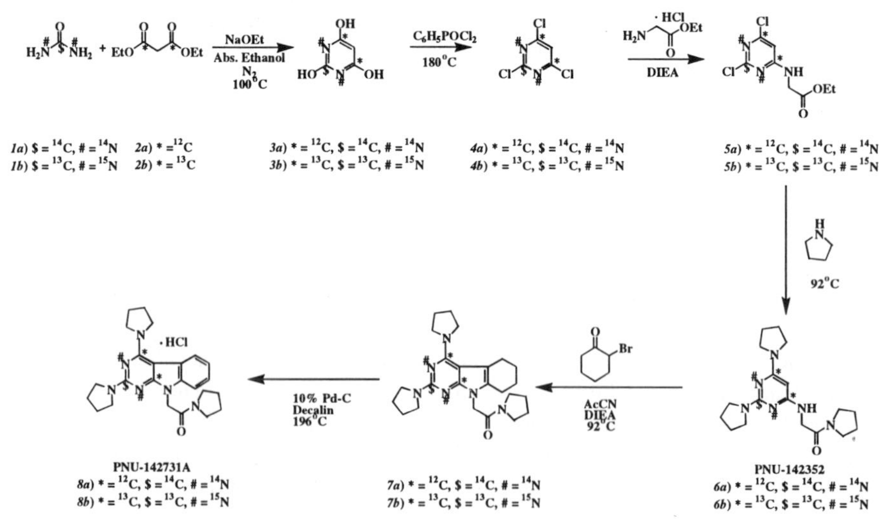

References

1. Bundy GL, Gremban RS, Palmer JR, Mizsak SA and Han F, J. Org. Chem., **63** (21), 7542-7546, 1998.
2. Dickey JB and Gray AR, Preparation of barbituric acid, Org. Syn. Coll. Vol. II, 1943:60.

SYNTHESIS OF SEVERAL ANALOGUES OF A NEW CLASS OF SYNTHETIC ANTIBACTERIALS, OXAZOLIDINONES, RADIOLABELED WITH CARBON-14, AND TRITIUM AND STABLELABELED WITH CARBON-13 AND DEUTERIUM

JOHN A. EASTER, EDWARD H. CHEW, AND RICHARD S.P. HSI

Pharmacia & Upjohn, 301 Henrietta St., Kalamazoo, MI 49007, USA
E-mail: john.a.easter@am.pnu.com

The emergence of strains of bacteria, which are resistant to current drug therapy, is becoming a problem of increasing significance. The development of new antibacterials with potent activity versus resistant organisms is urgently needed, especially in light of the discovery of vancomycin resistant *Staphylococcus aureus*. The oxazolidinones are a new class of totally synthetic antibacterial agents with activity against several important human pathogens including multiple resistant *Staphylococcus aureus* and *Staphylococcus epidermidis*, the streptococci, and the enterococci. In an effort to support drug developmental studies, such as adsorption, metabolism, distribution, and excretion studies. Several analogues, PNU-100480, PNU-100592 (eperezolid), and PNU-100766 (linezolid) have been radiolabeled with Carbon-14 and stablelabeled with Carbon-13 and Deuterium. In support of on-going discovery activities into understanding this class of compounds, several analogues have also been radiolabeled with tritium. The radiosynthesis of the above compounds will be discussed.

1 Introduction

The Oxazolidinone Project Team at Pharmacia & Upjohn identified two first generation analogs PNU-100592 (eperezolid) and PNU-100766 (linezolid), as candidate drugs for development for the treatment of human infections caused by a broad spectrum of Gram-positive bacteria including methicillin-resistant *Staphylococcus aureus (MRSA) and Staphylococcus epidermidis* (MRSE), penicillin and cephalosporin-resistant *Streptococus pneumoniae,* and vancomycin-resistant enterococci. [1] These compounds represent a novel class of orally active, totally synthetic antibacterial agents, which act as bacterial protein synthesis inhibitors. This inhibition occurs at an early stage of the initiation phase. In an effort to support the development of these compounds, the radiolabeled and stable isotope syntheses were undertaken.

2 Discussion

PNU-100592 and PNU-100766 were first synthesized using carbon-14 labeled acetyl chloride to afford the label at the acetamide position.

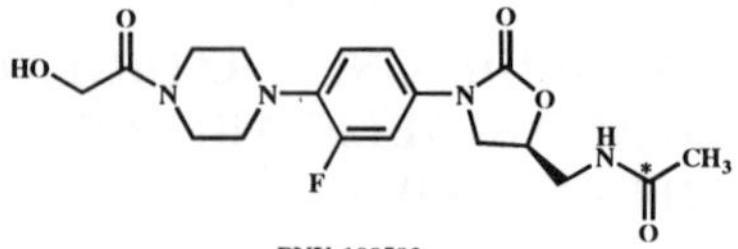

PNU-100592

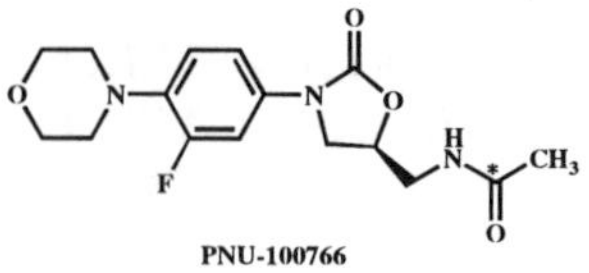

PNU-100766

PNU-100766 was also labeled with carbon-14 in the oxazolidinone ring using a seven step synthesis starting with carbon-14 phosgene.

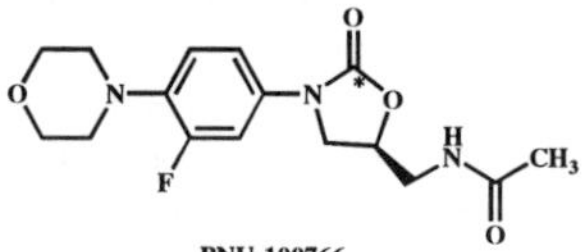

PNU-100766

There is an on-going effort within Drug Discovery to obtain a better understanding of this new class of antibacterials. In an attempt to support some of these efforts with radiolabeled drugs, we have labeled three analogs, PNU-93936, PNU-100592, and PNU-141059 with tritium. The method involves the activation of sodium acetate with *O*-benzotriazol-1-yl-*N,N,N',N'*-tetramethyluronium hexfluorophosphate (HBTU), followed by reaction with the corresponding amine. Workup of the reaction involves removal of the solvents, followed by either prep-TLC or HPLC purification to the desired tritium labeled compound.

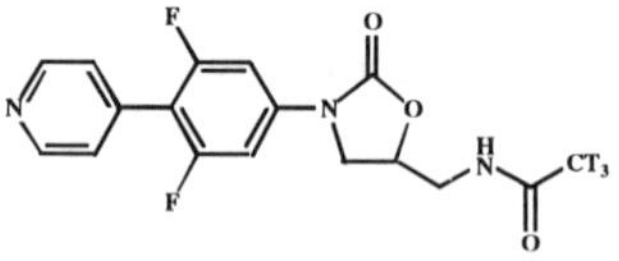

PNU-93936
Sp. Act. 25.6Ci/mmol

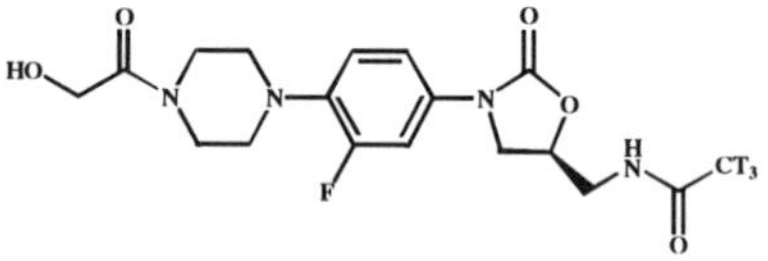

PNU-100592
Sp. Act. 31 Ci/mmol

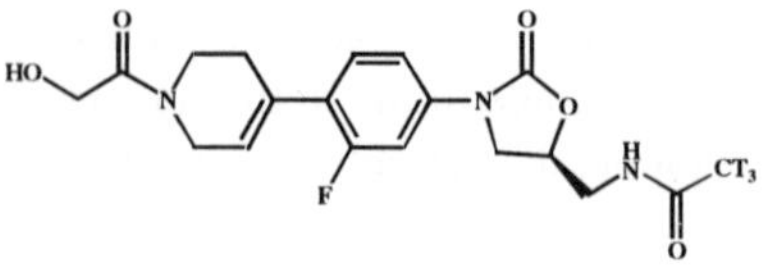

PNU-141059
Sp. Act. 32 Ci/mmol

T = Tritium
* = Carbon-14

USING NATURAL RADIONUCLIDES TO TRACE SOURCES OF POLLUTANTS IN THE MANSFELD MINING DISTRICT, CENTRAL GERMANY

PETER C. SCHRECK

UFZ Centre for Environmental Research Leipzig-Halle Ltd., Department of Hydrogeology, Th. Lieser-Str. 4, D-06120 Halle/Saale, Germany
E-mail: schreck@hdg.ufz.de

KLAUS FREYER & HANNS-CHRISTIAN TREUTLER

UFZ Centre for Environmental Research Leipzig-Halle Ltd., Department of Analytical Chemistry, Permoser Str. 15, D-04318 Leipzig Germany
E-mail: freyer@ana.ufz.de

In metal-mining districts containing outcropping ore bodies, mining waste heaps and dumped residues from ore-smelting, it is often difficult to distinguish between the various sources of environmental pollution due to their very similar geochemical composition. One example is the copper-mining district of Mansfeld in Central Germany, where uraniferous copper shale has been mined and processed for more than 800 years. The abundance of key radionuclides of the ^{238}U decay series in sediments and soils is a powerful tool for distinguishing between geogenic and anthropogenic metal sources, as well as for detecting pollutant migration pathways.

1 Introduction

Copper mining in Central Germany dates back to the early 13th century. In the industrial age, large-scale processing was introduced in two major smelters. By 1990, when mining and smelting was abandoned, 2.2 million tons of copper and about 20 other metals of economic importance including zinc, lead, cadmium, tin, molybdenum, silver, rhenium and selenium had been recovered. Volatilizable metals were extracted from the flue dust scrubbed from the furnace top gas [1,2].

Nowadays we find about 50 million tons of slag, waste rock and about 300,000 tons of flue dust ("Theisen sludge") around the former smelters [1]. The waste is subject to erosion and metal enrichments in soils, in stream and lake sediments, in the groundwater and in the surface water are common. Copper shale ore crops out at many locations in the mining district and is supposed to contribute to the metal load of rivers and sediments. However, the efficacy of rehabilitation and restoration measures in the mining district depends on the knowledge of pollutant pathways and dissemination mechanisms.

2 Methods

Sediment samples were collected from rivers, lakes and from mining galleries in the mining district. Sample preparation included drying at 60°C, screening (<250μm) and homogenizing. Activity measurements were carried out by low-level γ-spectroscopy using coaxial Low-Energy HPGe detectors, n-type, with an active volume of 39 cm^3 and 0.5 mm Be-window.

3 Results

Smelting of uranium-bearing copper shale leads to radioactive disequilibrium in the ^{238}U decay series. The abundance of key radionuclides of the ^{238}U decay series in sediments and soils can be used for distinguishing between geogenic (copper shale) and anthropogenic (slag, flue dust) metal sources, as well as for detecting pollutant pathways.

Most of the investigated stream sediment samples are enriched by metals from different metal sources. Nevertheless clear differences exist between the different source materials and all samples could be classified properly:

- Sediment samples with low ^{238}U in radioactive equilibrium with ^{226}Ra and ^{210}Pb indicate copper shale as the major source of metals.
- Samples with elevated ^{238}U and ^{226}Ra but low ^{210}Pb content are dominated by slag.
- Excess enrichment of ^{210}Pb but low levels of ^{226}Ra and ^{238}U denote flue dust as the major polluter.

On the edges of the Mansfeld mining district, many samples are dominated by copper shale and more to the central part slag is the major metal carrier. Downstream the flue dust deposits till to the Lakes of Mansfeld samples are polluted by flue dust. Sediments from mining galleries exhibit elevated shares in ^{210}Pb, indicating an illegal underground disposal of flue dust.

4 References

1. Schreck P., Environmental impact of uncontrolled waste disposal in mining and industrial areas in Central Germany. *Environ. Geol.* **35** (1998) pp. 66–72
2. Weiss H., Morency M., Freyer K., Bourne J., Fontaine D., Ghaleb B., Mineau R., Möder M., Morgenstern P., Popp P., Preda M., Treutler H.C. and Wennrich R., Physical and chemical characterization of a complexly contaminated scrubber dust - a by-product of copper smelting in Sachsen-Anhalt, Germany. *Sci. Total Environ.* **203** (1997), pp. 65–78

Synthesis of [^{14}C] SDZ 267-489

GRAZYNA CISZEWSKA, LAWRENCE JONES and TIMOTHY DUELFER

DMPK, Isotope Lab., Novartis Pharmaceuticals Corporation, East Hanover, NJ 07936, USA.

E-mail: grazyna.ciszewska@pharma.novartis.com

SDZ 267-489 ((R)-3-methyl-5-phenyl-5-[14C]-5H-thiazolo[2,3-b]quinazoline) a new dyslipidemia and atherosclerosis candidate, labelled with [^{14}C] was required for pharmacokinetic ADME studies. The carbon at 5 position of the 5H-thiazolo[2,3-b]quinazoline moiety was chosen as the site for the label.

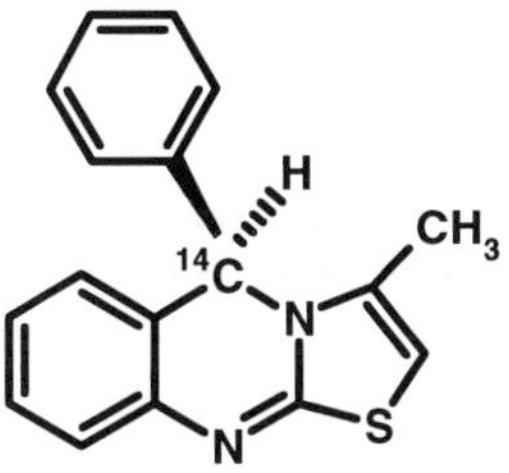

SDZ 267 - 489

The synthesis consisted of seven steps. Compound (7), SDZ 267-489 was obtained in 5.4% overall radiochemical yield. The label was introduced in the first step. 2-Iodobenzamine (1) was reacted via Rosenmund reaction with Cu^{14}CN (2.2: 1 equivalent ratio) to obtain labelled 2-aminobenzonitile (2). This compound, as well as all other intermediates in the synthesis, was identified by radio-TLC versus cold standard and by mass spectroscopy. Compound 2 was then converted to the labelled intermediate 3 via Grignard reaction with phenyl magnesium bromide in the presence of the catalytic amount of cuprous bromide followed by acid hydrolysis. Labelled 2-aminobenzophenone (3) was reduced with sodium borohydride and then cyclized in the presence of ammonium thiocyanate to give the thioquinazoline derivative (4). Hetero-annulation of compound 4 with chloroacetone produced the racemic drug substance (5). Compound 5 was then resolved with D-tartaric acid via the D-tartrate salt (6) which, in the presence of sodium bicarbonate, afforded the desired R-enantiomer (7).

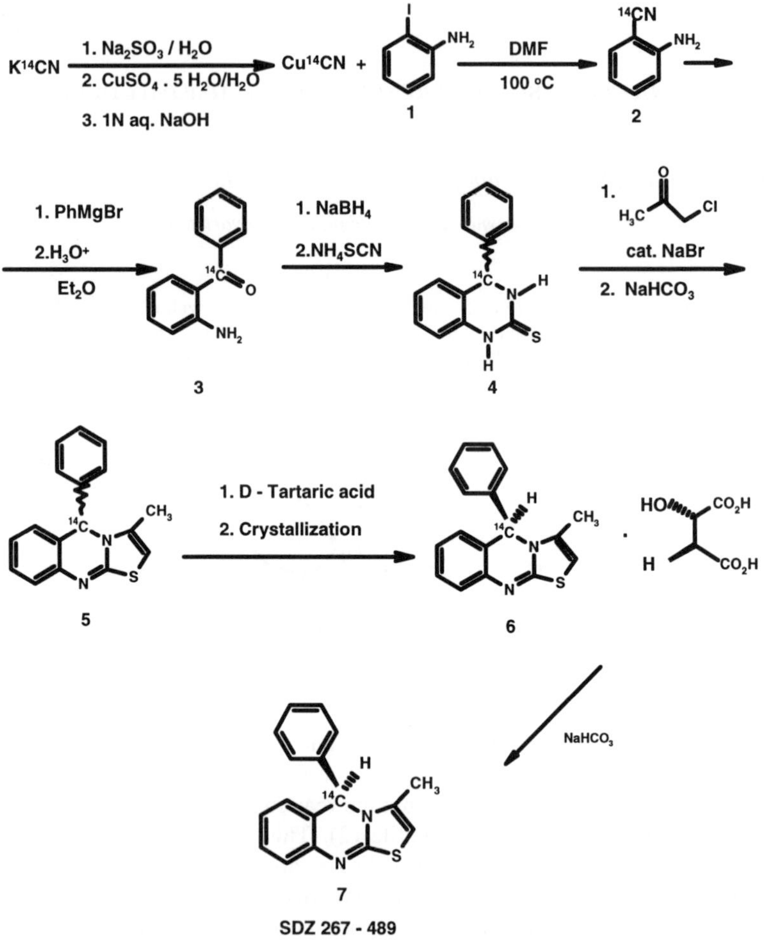

Full analysis of [^{14}C] SDZ 267-489 by our analytical group revealed that greater than 98% of the radioactivity was associated with the peak of interest. The retention time of the sample corresponded with that of the reference standard. Optical purity indicated that greater than 98% of the compound 7 was associated with the correct enantiomer.

EVALUATION OF ANTIBODY BIOTINYLATION REAGENTS FOR APPLICATION TO TARGETED RADIOTHERAPY

D.K. HAMLIN, D.S. WILBUR, P.M. PATHARE, D.S. SUBRAMANIAM, M. CORCORAN AND O.W. PRESS
Departments of Radiation Oncology, Medicine, and Biological Structure
University of Washington, Seattle, WA 98195
USA
dswilbur@u.washington.edu

Five biotinylation reagents have been prepared, and their conjugation yields, streptavidin/avidin binding, and stability towards the enzymatic action of biotinidase have been studied. Along with the new reagents, the commonly employed biotinylation reagent, biotin-caproate NHS ester (biotin-LC NHS) was studied. Very little difference was noted for conjugation yields and binding with streptavidin or avidin between the new biotinylation reagents and biotin-LC NHS ester. Surprisingly, there was also no difference in stability to biotinidase cleavage between the biotinylation reagents which contained stabilizing moieties and those that did not.

1. Introduction

Methods for specifically delivering radionuclides to cancer cells in patients, termed "Targeted Radiotherapy" are being investigated by a number of research groups. In one method under investigation, biotinylated monoclonal antibodies are first administered, followed later by administration of one of the tetrameric biotin binding proteins, streptavidin or avidin, then the radionuclide is administered attached to a biotin moiety [1,2]. This multi-step method, while promising, is complex as it involves optimization of quantities administered and time between administration of the listed reagents as well as one or two other reagents used to clear non-radiolabeled reagents from blood.

In our optimization studies of biotinylated antibodies, the length and hydrophobicity of the currently available biotinylation reagents was questioned. It was hypothesized that biotinylation reagents containing longer (e.g. 15-30Å) and more hydrophilic linker moieties might provide a more accessible biotin moiety in the aqueous environment of the blood/tumor than that of the most commonly used biotin-aminocaproate reagent, **1** (8.5 Å linker). It was also hypothesized that increased hydrophilicity might improve the antibody conjugation yields of the biotinylation reagents. Most importantly, we were concerned that the enzyme biotinidase [3] might cleave the biotin moiety from monoclonal antibodies in serum as had been observed for biotin containing small molecules [4,5]. Therefore, we synthesized a series of new biotinylation reagents as shown in Scheme 1. The new biotinylation reagents were reactive with sulfhydryl moieties (**2** & **3**) or lysine amines (**4** - **6**) and incorporated hydrophilic ether containing linkers moieties. In addition to improved water solubility, the linkers were 2 to 3 times the length of the aminocaproate moiety in **1**. In three of the new biotinylation reagents (**3**, **5**, **6**) an N-methyl biotinamide moiety was incorporated as this had been shown to block biotinidase cleavage [5].

Scheme 1: Biotinylation Reagents Used in Studies

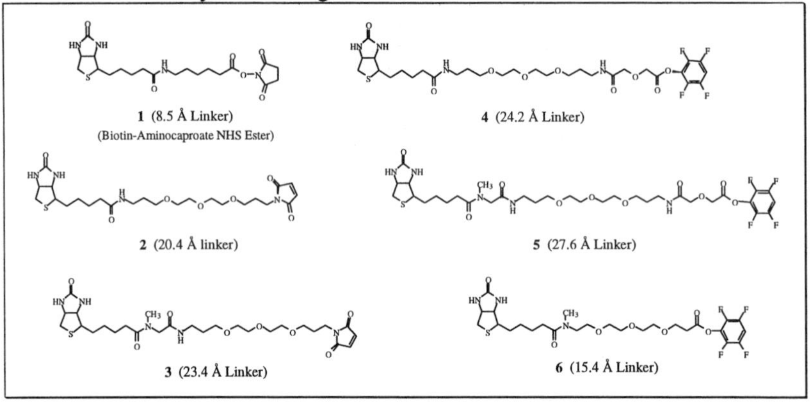

2. Methods

NHS-LC-Biotin was obtained from Pierce (Rockford, IL). Biotinylation reagents **2**, **3** and **6** were prepared as previously reported [6-8]. Biotinylation reagents 4 and **5** were prepared by reaction of the terminal amine intermediates described for the synthesis of **2** and **3** with diglycolic anhydride, followed by preparation of the tetrafluorophenyl esters with tetrafluorophenyl trifluoroacetate as previously described [8] (all NMR and MS are consistent with structures shown).

Antibody biotinylation reactions were conducted as previously described [6]. Streptavidin binding was evaluated by measuring the amount of radioiodinated and biotinylated antibody that bound with an excess of immobilized streptavidin. Biotinidase activity was measured in human serum at high dilution by passing the radiolabeled and biotinylated antibody over avidin columns (Pierce) after incubation at 37°C for two hours.

3. Results

Conjugation yields of amine reactive biotinylation reagents **4**, **5**, and **6** under a variety of conditions were essentially equivalent to those obtained with commercially available **1**. Binding of the radioiodine labeled biotinylated antibody with immobilized streptavidin was also found to be essentially equivalent. Comparison of the binding of **1** with **4** and biotinidase stabilized **5** on an avidin column after incubation with human serum indicated that biotinidase did not cleave biotin from the biotinylated antibody. Evidence for a lack of cleavage of biotin from antibody fragments biotinylated with **2** and **3** was obtained in mice biodistribution studies where equivalent amounts of radiolabeled streptavidin were co-localized in human tumor xenografts [9].

4. Conclusions

The results obtained in our investigation suggest that there are no advantages (or disadvantages) to using the new biotinylation reagents (**2** - **6**) which contain longer and more hydrophilic linker moieties for biotinylation of monoclonal antibodies than the commercially available Biotin-LC (biotin-aminocaproic acid), **1**.

5. Acknowledgement

We gratefully acknowledge funding for the research provided by the Department of Energy (DE-FG03-98ER62572) and the National Institutes of Health (CA76287).

6. Literature Cited

(1) D.J. Hnatowich, F. Virzi, F. and M. Rusckowski, *J. Nucl. Med.* **28**, 1294-1302 (1987).
(2) G. Paganelli, M. Malcovati and F. Fazio, *Nucl. Med. Commun.* **12**, 211-234 (1991).
(3) J. Hymes, K. Fleischhauer and B. Wolf, *Meth. Enzymol.* **279**, 422-434 (1997).
(4) E. Livaniou, et al., *Meth. Enzymol.* **279**, 442-451 (1997).
(5) D.S. Wilbur et al., *Bioconjugate Chem.* **8**, 572-84 (1997).
(6) D.S. Wilbur, et al., *Bioconjugate Chem.* **7**, 689-702 (1996).
(7) D.S. Wilbur et al., *Bioconjugate Chem.* **9**, 322-30 (1998).
(8) D.S. Wilbur et al., *Bioconjugate Chem.* **9**, 813-25 (1998).
(9) D.S. Wilbur et al., unreported results.

SYNTHESIS OF TWO IMIDAZOLE CYTOKINE INHIBITORS LABELLED WITH CARBON-14

Keith T. Garnes[1], *Scott W. Landvatter*[1], *Joseph Sisko*[1],
Larry A. Davis[2], *J. Richard Heys*[1] *and Jerry L. Adams*[2]

[1]*Synthetic Chemistry Department*
[2]*Medicinal Chemistry Department*
SmithKline Beecham Pharmaceuticals
709 Swedeland Road
King of Prussia, PA 19406-0939 USA

Summary: Cytokines are protein hormones that regulate immune and inflammatory responses. Compounds **1** and **2** represented novel synthetic targets of potential interest as cytokine-supressing anti-inflammatory drugs. High and moderate specific activity ^{14}C-labelled analogs were needed for drug metabolism, pharmacokinetic and formulation studies.

The key isonitrile synthetic precursor **3** was a common intermediate in the preparation of both radiolabelled analogs. This intermediate was efficiently produced in five stages in 33% overall radiochemical yield from $K^{14}CN$. Compound **1** was produced in one step from the isonitrile, while **2** required two steps. The synthetic methodology leading to **3** in ^{14}C-labelled form and the sequences leading to structures **1** and **2** are presented. Targets **1** and **2** were obtained in high radiochemical purity and specific activity.

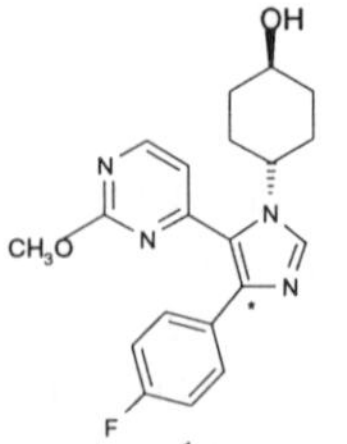

1

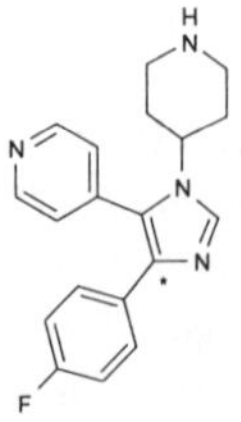

2

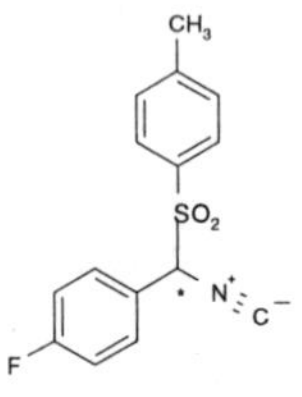

3

Introduction: 1,4,5- trisubstituted imidazoles **1** and **2** are representative of a general class of compounds which exhibit potent binding affinities for p38 MAP kinase, which appears to be involved in an inflammation regulatory pathway[1]. Disruption of the signaling pathway initiated by a variety of inflammatory stimuli is a novel approach to the treatment of inflammatory diseases such as arthritis and inflammatory skin disease[2]. ^{14}C-labelled analogs of **1** and **2** were required for

ADME studies in support of project development. The ring location of the label maximized metabolic stability.

Discussion: The synthetic strategy for the preparation of compounds **1** and **2** centered on the preparation of the key labelled isonitrile intermediate **3**. Intermediate **3** was obtained by reaction of labelled formamide **4** with phosphorus oxychloride and triethylamine in THF at 0 °C[3]. The required formamide was synthesized from labelled 4-fluorobenzaldehyde which was available in two steps from $K^{14}CN$.

4-Fluorobenzo[^{14}C]nitrile was prepared in 96.5% radiochemical yield from carrier-free $K^{14}CN$ and 4-iodofluorobenzene using the method of Sutherland[4]. The nitrile was reduced to the aldehyde in 86% RCY using DIBAL in methylene chloride, followed by methanol quench and extractive workup. The labelled aldehyde was treated with 1.1 equivalents of chloro-trimethylsilane and 2.5 equivalents of formamide in 1:1 (v/v) toluene/acetonitrile at 50 °C for 5 hours to form the bis-formamide shown below, followed by addition of p-toluene sulfinic acid (1.7 equivalents) and reaction for an additional 5 hr. Compound **4** was obtained in 66% RCY after workup. **4** underwent dehydration using $POCl_3$ as described above in 58% RCY. Thus, **3** was obtained in 33% RCY at >50 mCi/mmol from $K^{14}CN$.

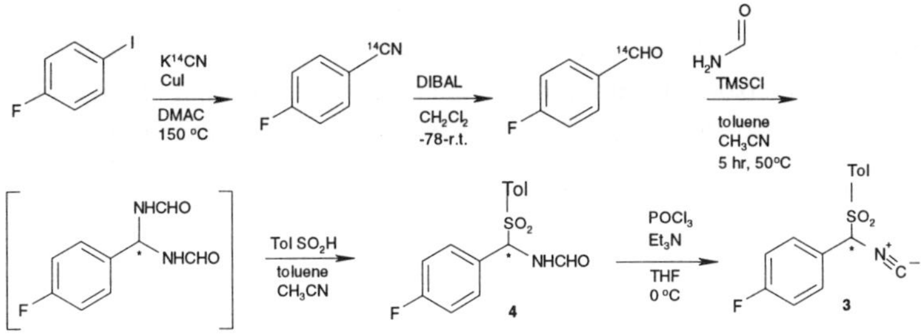

Synthetic targets **1** and **2** were prepared by adapting the method of van Leusen[5]. This method involved a base-promoted cycloadditon reaction of substituted imines with isonitriles to provide 1,4,5-trisubstituted imidazoles. Thus the condensation of isonitrile intermediate **3** (0.4 mmol, 20 mCi) with the imine **5** in DMF/acetonitrile for 20 hours at 50 °C provided **1** in 65% RCY. Recrystallization from isopropanol provided material of >99% radiochemical purity by HPLC at ca. 50 mCi/mmol.

In a similar fashion, condensation of **3** (4 mmol, 25 mCi, 6 mCi/mmol) with the imine **6** in DMF for 17 hours at ambient temperature provided **2a** in 80% RCY. Deprotection using concentrated HCl at reflux, followed by base workup provided **2** in 90% RCY. Purification by reverse phase HPLC provided material at >99% radiochemical purity by HPLC.

Imine **5** was formed *in situ* from 2-methoxypyrimidine-4-carboxaldehyde and *trans*-4-aminocyclohexanol in acetonitrile. Imine **6** was formed *in situ* from pyridine-4-carboxaldehyde and ethyl-4-aminopiperidine carboxylate in DMF.

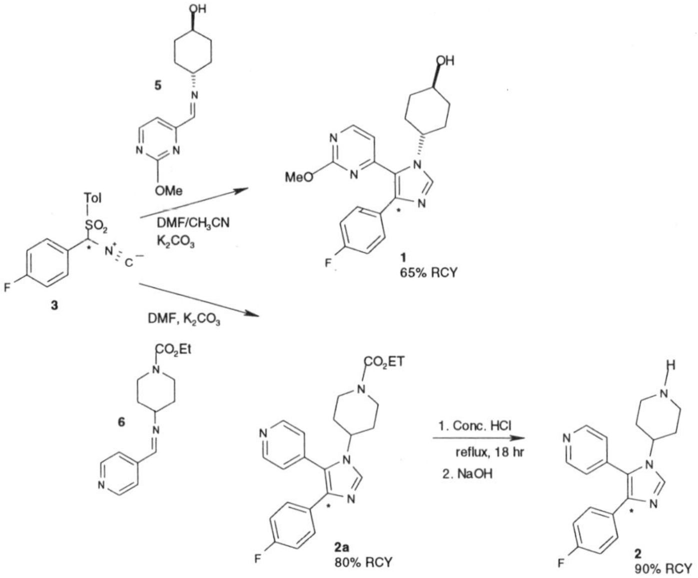

Conclusion: Described above is a versatile method of preparing radiolabelled 1,4,5-trisubstituted imidazoles from a common synthetic intermediate. Synthetic targets **1** and **2** can be obtained in good radiochemical yield at high purity and specific activity.

References:

1. Lee, J.C.; Laydon, J.L.,; McDonnel, P.C.; Gallagher, T.F.; Kumar, S.; Green, d.; McNulty, D.; Blumenthal, M.J.; Heys, J.R.; Landvatter, S.W.; Strickler, J.F.; McLaughlin, M.M.; Siemens, I.R.; Fisher, S.M.; Livi, G.P.; White, J.R.; Adams, J.L.; Young, P.R. *Nature* (London) **1994**, *372*, 739.
2. Boehm, J.C.; Smietana, J.M.; Sorenson, M.E.; Garigipati, R.S.; Gallagher, T.F.; Sheldrake, P.W.; Bradbeer, J.; Badger, A.M.; Laydon, J.T.; Lee, J.C.; Hillegass, L.M.; Griswold, D.E.; Breton, J.J.; Chabot-Fletcher, M.C.; Adams, J.L. *J. Med. Chem.* **1996**, *39*, 3929-3937.
3. Sisko, J.; Mellinger, M.; Sheldrake, P.W., Baine, N.H. *Tetrahedron Letters*, **1996**, *37*, 8113-8116.
4. Carr, R.M.; Cable, K.M.; Wells, G.N., Sutherland, D.R. *J. Lab.Comp. Radiopharm.* **1994**, *34*, 887-897.
5. van Leusen, A.M.; Wildeman, J.; Oldenziel, O.H. *J. Org. Chem.* **1977**, *42*, 1153-1159.

A SIMPLIFIED METHOD TO MEASURE GFR IN PEDIATRIC PATIENTS

SIDNEY YU

Department of Nuclear Medicine, Singapore General Hospital
Outram Road
Singapore 169608

ELIAS W ELIAS, DAVID NG, FELIX SUNDRAM

Department of Nuclear Medicine, Singapore General Hospital
Outram Road
Singapore 169608

A simplified method to measure glomerular filtration rate is described. The method requires only one blood sample from the patient and the calculation can be done using a common calculator. Results of the one sample method correlates well with the results of a published two sample method.

1 Introduction

Glomerular filtration rate (GFR) is a an important parameter in evaluating kidney function, especially in pediatric oncology. Conventional methods require multiple blood samples and take more than 3 hours to complete. This is very traumatic to young patients and time consuming to the staff performing the test. There have been many published alternative equations requiring only 1 blood sample[1-4]. However, many of them are relatively complicated. Our objective is to develop a relatively simple one sample equation to calculate GFR

2 Methods

Forty-five patients (2 to 14 yr old) were injected i.v. with Cr-51 EDTA. Their GFR were determined using a published method [1]. Blood samples were drawn 2 and 3 hr later and the plasma radioactivity was counted. The results were plotted in a semi-log scale to determine volume of distribution at time 0 (Vd_0) and effective half-life ($T_{1/2}$).

$$GFR = 0.693 \times 0.8 \times Vd_0 / T_{1/2} \quad \text{(eq 1)}$$

We also determined the volume of distribution at 2 hour (Vd_2) and calculated the ratio Vd_2/Vd_0. We then plot the graph between $T_{1/2}$ and Vd_2/Vd_0. A regression analysis resulted in the following equation :

$$T_{1/2} = 230.8\,(Vd_2/Vd_0)^{-0.9237} \quad (r = 0.9583) \qquad \text{(eq 2)}$$

This equation was merged into the eq 1 and resulted in eq 3.

$$GFR = 0.002402\,(Vd_0)(Vd_0^{-0.9237}) \,/\, Vd_2^{-0.9237} \qquad \text{(eq 3)}$$

Vd_0 of the forty patients ranged from 1396 ml to 22824 ml, with an average of 6079.53 ml. The average of $(Vd_0)(Vd_0^{-0.9237})$ with a Vd_0 ranging from 125 to 64000 ml is 1.8550. Equation 3 is simplified to eq 4

$$GFR = (0.004517)/\,Vd_2^{-0.9237} \qquad \text{(eq 4)}$$

GFR of the 45 patients were recalculated using equation 4 .
The GFR of another 25 patients were measured and compared with both equations

3 Results and Discussion

The GFR of the first group of 45 patients determined by the one-sample equation (eq 4) correlates well with the GFR determined using the 2-sample method (eq 1, r=0.9567). The average percentage difference was –1.97 %.

The GFR of second group of 25 patients determined by eq 4 also correlates well with the GFR determined using eq 1 (r=0.9848). The average percentage difference was 2.07 %.

Data from our simplified method concurred well with the usual method (r = 0.9718). Our method is comfortable for patients , less time consuming to staff and the GFR can be calculated with a common calculator.

References

1. Ham HR, Piepsz A, Estimation of Glomerular Filtration rate in Infants and in Children Using a Single-Plasma Sample Methods *J Nucl Med* **32**(1991)PP 1294-1297
2. Picciotto g, Cacace G, Cesana P, et al, Estimation of Chromium-51 Ethylene Diamine Tetraacetic Acid Plasma Clearance : a comparative assessment of simplified technique *Eur J Nucl Med* **19**(1992) pp 30-35
3. Watson WS, A simple Method of Estimating Glomerular Filtration Rate *Eur J Nucl Med* **19**(1992) p 827
4. Chantler C, Garnett ES, Parsons V, et al , Glomerular Filtration Rate Measurement in man by the Single Injection Method using Cr-51 EDTA *Clin Sci* **37**(1969) pp169-180

SN-117 AND SN-119 HIGH ENRICHMENT ASSAY (MORE THAN 95%) ATTAINMENT USING CENTRIFUGAL TECHNOLOGY

V.A. ANTSOUPOV AND V.G. AFANASIEV
SGCE 636070, Seversk Tomsk Region,

O.S. ANDRIENKO
Institute of Atmospheric Optics, Siberian Branch of Russian Sciences Academy 634055, Tomsk, Academichesky Prospect, 1,

and

I.I. ZHERIN
Tomsk Polytechnical University 634034. Lenin Prospect, 30.

The Siberian Group of Chemical Enterprises,(SGCE), has established a new facility to produce stable isotopes using centrifugal technology. Presently, SGCE is able to offer isotopes of the following elements: Tin, Selenium, Tungsten, Sulfur, Chromium, Tellurium, Xenon and Krypton.
In order to attain high enrichments (>95%) of Sn-117 and Sn-119 using Tetramethyl Tin , [$Sn(CH_3)_4$], as feed material it is necessary to account for the poly-isotopy of both carbon and hydrogen that parallels that of Tin. Since the centrifuge method separation effect depends only on differences in the mass of the feed material it follows that molecules consisting of different isotopes of elements forming this molecule and having equal mass, behave as a unified component. Consequently, a mixture can be enriched in a particular isotope only to a certain level of enrichment. This can be calculated on the assumption of an equi-probable distribution of isotopes in $Sn(CH_3)_4$ and the use of multi-nomial expressions The maximum available enrichment for Sn-117 is 91.94% and for Sn-119 is 88.4%.
Three possible variants to increase these enrichment levels have been investigated.i) replacing the methyl groups after the first enrichment cycle with methyl groups having natural isotopic abundance, ii) synthesis of Tetramethyl Tin with mono-isotopic Carbon-12 and Hydrogen-1 and, iii) redistribution of the enriched methyl groups in the enriched Tetramethyl Tin by a chemical exchange reaction.
A very simple and effective method involving methyl group exchange between molecules of gaseous $Sn(CH_3)_4$ over a catalyst has been developed and used to enrich both Sn-117 and Sn-119 to greater than 95%. Loses of less than 2% at the stage of methyl group redistribution make this method effective for enrichment of not only tin but also the isotopes of Pb and Zn.

1 Introduction

The Siberian Group of Chemical Enterprises has established a new facility to produce stable isotopes using centrifugal technology. Presently, SGCE is able to offer isotopes of the following elements: Tin, Selenium, Tungsten, Sulfur, Chromium, Tellurium, Xenon and Krypton. SGCE attaches great importance to the quality of the product and especially to attaining the maximum enrichment assay of the desired isotope.

Further details of the technology, developed by SGCE in cooperation with the Institute of Atmosphere Optics (Siberian Branch of the Russian Sciences Academy) and the Tomsk Polytechnic University are described below. This technology is implemented to get high (more than 95%) enrichment assays for Sn-117 and Sn-119, using Tin Tetramethyl ($Sn(CH_3)_4$) as feed material. The peculiarity of Sn(CH3)4 is that in parallel with Tin poly-isotopy it is characterized by the poly-isotopy of the Carbon and Hydrogen that form the methyl radicals. In the centrifuge method the separation effect depends only on the difference in the molecular mass of the feed material. Molecules having equal mass, that are comprised of the different isotopes of elements forming this molecule, behave as a unified component. Consequently, a mixture can be enriched in a particular isotope only to a certain level of enrichment assay. This level of enrichment assay can be calculated.

2 Method

Calculation of the Sn-117 and Sn-119 poly-isotopy equilibrium composition was made on the assumption of an equi-probable distribution of the isotopes in $Sn(CH_3)_4$ using the multi-nomial principle:

$$Xi = \frac{n!}{r1!.r2!\ldots rm!}.(Y1)^{r1}.(Y2)^{r2}\ldots(Ym)^{rm}$$

Where Xi - is the group enrichment assay of one type of atom, having different mass figures in the molecule

Yi - is the isotope content in the natural mixture

n - is the number of a given element in the molecule

m - is the number of given element isotopes

ri - is the number of given element with i-isotopic atoms in the molecule.

The number of molecular isotope varieties for every element is determined by the following formula:

$$Sk = \frac{(n+m-1)!}{n!.(m-1)!}$$

For the molecule Sn $(CH_3)_4$, Ssn = 10, Sc = 5 and SH = 13. Thus, the total quantity of isotope combinations for a given molecule is equal to 650, (for Sn-117 - 15 and for Sn-119 -25 correspondingly). The concentration of $Sn(CH_3)_4$ molecules of a specific isotope composition is determined by multiplication of the concentrations of the isotope components forming this molecule. The results of these calculations demonstrate that under ideal enrichment of Tin Tetramethyl, the maximum available enrichment assay of Tin-117 is 91.94%. Similarly, the maximum available enrichment assay of Tin-119 is 88.4%. Calculations were made using a special program developed on a Pentium PC in C++.

3 Discussion

Possible variants to increase the maximum available enrichment assay of a specified isotope are the following:

First Variant: Total replacement of the methyl groups in $Sn(CH_3)_4$ (after the first cycle of enrichment) by groups with natural content of C and H isotopes followed by a second cycle of enrichment. In this case, according to the calculations, the following maximum enrichment assays are attainable:

Sn-117 - 99.6% and Sn-119 - 99.4%.

Second Variant: Tin Tetramethyl synthesis using Carbon-12 and Hydrogen-1 monoisotopes. The main drawback of this method is the extremely high price of obtaining Tin Tetramethyl.

Third Variant: Methyl group redistribution in Tin Tetramethyl enriched in the desired isotope. This redistribution is performed by an exchange reaction and is followed by a further enrichment step. In this way a very simple and effective method was found: methyl group exchange between molecules of gaseous $Sn(CH_3)_4$ over a catalyst. This method was successfully implemented to produce Sn-117 and Sn-119 isotopes with enrichment assays more than 95%. The technological effectiveness and negligible losses (less than 2%) at the stage of methyl group redistribution makes this method effective for enrichment of not only Tin but also Pb and Zn isotopes.

DANDER2 FOR SIMULATIONS OF NEUTRON ACTIVATION AND SHIELDING OF NEUTRONS AND ACTIVATION γS IN IRRADIATION GEOMETRIES

WILLIAM B. WILSON AND TALMADGE R. ENGLAND
Los Alamos National Laboratory, Los Alamos, New Mexico 87545, USA
E-mail: wbw@lanl.gov, tre@lanl.gov

ROBERT E. SCHENTER
Pacific Northwest National Laboratory, Richland, Washington 99352, USA
E-mail: re_schenter@pnl.gov,

DANDER2 is a script linking of the popular DANT discrete-ordinates (Sn) multigroup neutron and γ transport code with the CINDER'90 transmutation inventory code. DANDER2 is used to describe the spatial transport of neutrons from a neutron source, the activation of the target and other materials of the geometry in the spatial neutron flux, formation of aggregate spatial decay gamma sources at each discrete time during and following an irradiation, the transport of gammas from the temporal source throughout the geometry, and resolution of the spatial, temporal dose-equivalent rate. DANDER2 has been useful for surveys of irradiation histories, materials, and dimensions defining parameters for use in more detailed and time-consuming Monte-Carlo studies.

1 Introduction

The simulation of the neutron activation of a geometry, such as a target shield, and the shielding of the neutrons and activation γs is conceptually straight forward. A radiation transport code is used to transport the neutrons and γs and the γ source is calculated with a transmutation inventory code. We have routinely used Monte-Carlo codes with the CINDER'90 transmutation inventory code[1] to investigate the radiation environment near a variety of target and shield designs. The cpu times required for Monte-Carlo calculations are excessive for survey calculations of various sources, materials and/or geometries.

The use of a fast-running discrete-ordinates (S_n) code and multigroup neutron and γ transport cross-section data greatly reduce the cpu time required for such simulations, permitting comprehensive surveys in reasonable times. Although the process is conceptually straight forward, the devil is in the details of such calculations due to the number and volume of data and file operations. We have developed a script linking of the discrete-ordinates (S_n)

code system DANT[2] with CINDER'90. The linking of the codes and data in DANDER2 is illustrated in Figure 1. The approach illustrated has been applied to spherical 1-D (ONEDANT) geometries but could be extended to 2-D and 3-D calculations with DANT.

Although this linking is general and nature, the process depends upon the availability of quality multigroup transport data with sufficient breadth of materials and energy for the accurate description of the system being simulated; the HILO-86 library,[3] describing neutrons below 400 MeV and γs below 20 MeV, has uniquely satisfied these requirements in many applications.

2 DANDER2 Script Linking

In the sequence illustrated in Figure 1 a geometry, neutron source, and irradiation-cooling history are initially defined. The result R might be the dose-equivalent rate (DER) or other flux-integrated response in a cell of the geometry external to the shield. A survey of the magnitude of this result might involve the definition of a number of problems — each with a unique choice of materials and dimensions defining the geometry, source spectrum, and history.

In preparation for a problem, the materials, dimensions, material zones, and cells of the geometry are defined in the DANT input file *odin* . The parts of *odin* preceding and following the source definition are used to form *odintop* and *odinbot* , respectively, for subsequent use in forming DANT input for γ transport calculations. The cell k, within which result R is defined, is identified in the file *selectr* , and the cells of the geometry to be activated and from which γ source contributions are to be used are listed in the file *selectq* . The materials specified in *odin* for DANT transport calculations must be completely redefined by nuclide in the *material* file for CINDER'90 use. A CINDER'90 input file is constructed for a typical cell, specifying the parameters of the inventory solution and the irradiation-cooling history; these are divided into *intop* and *inbot* , respectively, for latter use in the construction of CINDER'90 input files. The script *job.level1*, described below, must also be formed.

2.1 Level 1

In level 1, the script *job.level1* calls for clearing local storage with CLEAR-DANT, linking data libraries for DANT and CINDER'90, executing DANT to transport the neutron source, and extracting the result R within the cell k from the DANT output file *odout* with GETDERB. The standard interface file

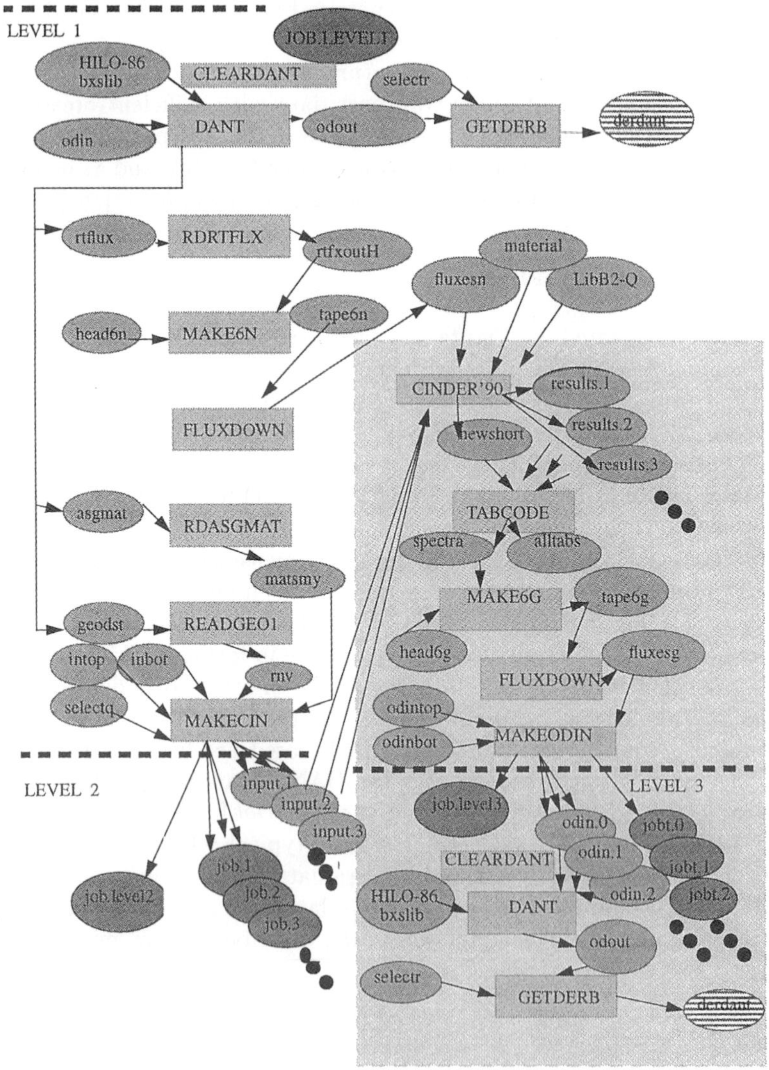

Figure 1. DANDER2 — A Script Linking of ONEDANT, CINDER'90 and Associated Codes and Data)

(SIF) *rtflux* of flux data is read by RDRTFLX — a routine by D. C. George — and reformatted in file *rtfxoutH,* which is combined with preamble data in *head6n* by MAKE6N to form the *tape6n* input file for FLUXDOWN. Here, the multigroup fluxes of all DANT cells in the multigroup structure of the transport cross-section library used in DANT are collapsed to the structure of the transmutation cross sections used in the *LibB2-Q* library of CINDER'90 and recorded in *fluxesn* . RDASGMAT reads the SIF *asgmat* and makes the simple material-by-zone summary *matsmy* . READGEO1 reads the SIF *geodst* to form the *rnv* listing of the radii, volume, and zone of each of the cells. The final task in level 1 is the automatic formation of CINDER'90 input files *input.ir*, associated job scripts *job.ir*, and the *job.level2* superscript calling for the sequential application of the *job.ir* scripts in level 2.

2.2 Level 2

In level 2 the *job.level2* superscript is read to identify the next *job.ir* script to be executed. Following *job.ir,* CINDER'90 is executed to determine the nuclide inventory in cell *ir* at each time *it* in the history; these results are listed in *results.ir* , which is read by TABCODE – along with processed library data in *newshort* – to form a variety of tabular output in *alltabs* and condensed γ spectra in *spectra* using the multigroup γ structure used for spectral data in *LibB2-Q.* The temporal γ source spectra in *spectra* are used with the preamble data of *head6g* to form the *tape6g* input to FLUXDOWN, which collapses these volumetric source spectra to the γ energy structure of *HILO-86,* forming *fluxesg.* Not shown in Figure 1 is the use of *MKODQ1* to form a cell *ir* volume-integrated source at each time *it* , which is added to the source file *qstack* . Finally, MKODIN is used to form the input files *input.it* , associated job scripts *jobt.it* , and the level 3 superscript *job.level3* .

2.3 Level 3

In level 3 the *job.level3* superscript directs the sequence of *jobt.it* scripts, each using CLEARDANT to remove files from earlier DANT calculations, executing DANT with the correct *input.it*, interrogate the DANT output *odout* for the response in cell k identified in *selectr*, and adding this response to those accumulated in *derdant* — the results of the problem.

3 Conclusions

DANDER2 clearly offers the user a fast analysis of an activated shield with a minimum of interface effort. Surveys are simplified if a common cell structure is use and a superset of materials is initially defined in the DANT input, even if not all initially used, consistent with a *material* file applicable throughout the survey. Most of the cpu time required is used in CINDER'90.

DANT is available from the RSICC at Oak Ridge National Laboratory. A manual for CINDER'90 and a document describing a survey made with DANDER2 are in preparation. CINDER'90 and DANDER2 will also be released through RSICC. The CINDER'90 library *LibB2-Q* is limited to neutron reactions for Z $<$ 103 and $E_n < 20$ MeV. A new library for neutron and proton reactions to 150 MeV for Z $<$ 85 is in preparation.

4 Acknowledgments

This research is supported by the Department of Energy under contract W-7405-Eng-36.

References

1. W. B. Wilson, T. R. England and K. A. Van Riper "*Status of CINDER'90 Codes and Data,*" Proc. Fourth Workshop on Simulating Accelerator Radiation Environments, Sept. 14–16, 1998, Knoxville, Tenn, pp. 69–79.
2. R. E. Alcouffe, R. S. Baker, F. W. Brinkley, D. R. Marr, R. D. O'Dell and W. F. Walters, "*DANTSYS: A Diffusion Accelerated Neutral Particle Transport Code System,*" Los Alamos National Laboratory manual LA–12969–M (June 1995).
3. R. G. Alsmiller, Jr, J. M. Barnes, and J. D. Drischler, "*Neutron-Photon Multigroup Cross Sections for Neutron Energies* $\leq$ 400 *MeV (Revision 1),*" Oak Ridge National Laboratory report ORNL/TM-9801 (February 1986).

UTILIZATION OF THE SULFUR ($Na_2{}^{34}SO_4$) APPLIED TO THE SOIL BY RICE AND CROTALARIA JUNCEA

P. C. O. TRIVELIN [1]

[1] *Stable Isotopes Laboratory, Center for Nuclear Energy in Agriculture - University of São Paulo, Piracicaba, Brazil*

E-mail: pcotrive@cena.usp.br

J. A. BENDASSOLLI [2], F. CARNEIRO JUNIOR [3], T. MURAOKA [4]

[2] *Stable Isotopes Laboratory, Center for Nuclear Energy in Agriculture - University of São Paulo, Piracicaba, Brazil*

[3] *Methodist University of Piracicaba, Piracicaba, Brazil*

[4] *Soil Fertility Laboratory, Center for Nuclear Energy in Agriculture - University of São Paulo, Piracicaba, Brazil*

The utilization of the fertilizer-^{34}S by rice (*Oriza sativa* L.) and crotalaria juncea (*Crotalaria juncea* L.) was evaluated under greenhouse condition, in pots with 3 kg of a Red Yellow Podzolic soil. A solution of sodium sulfate labeled with ^{34}S (14.30 ± 0.05 atom % of ^{34}S) was applied to the soil (70 mg kg^{-1} of SO_4^{2-}-S) on the 18th day after sowing both species. The shoots of the crotalaria and rice were harvested, respectively on the 72nd and 122nd day after fertilization. The concentration and the amount of the sulfur in the crotalaria were higher than in the rice, due to the higher legume requirement for this nutrient. The sulfur requirement and the short time interval between fertilization and harvest of the crotalaria resulted in a small amount of SO_4^{2-}-S mineralized in the soil and a small quantity of $^{34}SO_4^{2-}$ immobilized by soil microorganisms. Thus, the percentage of sulfur in the crotalaria derived from the fertilizer (Sdff) was higher than in the rice (%$Sdff_{crotalaria}$ = 91.3 ± 3.5%; %$Sdff_{rice}$ = 66.3 ± 0.8%). The expressive values of %Sdff in both species indicate a low rate of mineralization of SO_4^{2-}-S as a consequence of the lack of the sulfur in the soil.

Tropical soils present naturally low sulfur availability for plant [1]. In Brazilian soils with intensive agriculture there is a progressive reduction in the nutrient availability and an increasing sulfur deficiency has been verified in several crops. The lack of S in those soils is caused by crop removal, sulfate leaching and continuous use of concentrated NPK fertilizers.

Rice (*Oriza sativa* L.) and crotalaria juncea (*Crotalaria juncea* L.) were grown in triplicate pots with 3 kg of a Red Yellow Podzolic soil, three plants per pot, under green house condition. A solution of sodium sulfate (70 mg kg^{-1} of SO_4^{2-}-S) labeled with ^{34}S (14.30 ± 0.05 atom % of ^{34}S) [2] was applied on the 18th day after sowing both species. N and P as urea and monoammonium phosphate were also applied. The shoots of the crotalaria and rice were harvested, respectively on the 72nd and 122nd day after fertilization, dried, weighed and analyzed for total S and ^{34}S

abundance. Total S and ^{34}S abundance in plant material were determined by turbidimetry of barium sulfate and by mass spectrometry, respectively. The sulfur in the plant derived from the fertilizer (Sdff; % and mg per pot) or from the soil (%Sdfs =100 - %Sdff) were estimated (Table 1).

The dry matter yield of the crotalaria was lower than that of rice, once the legume was harvested 50 days before the rice. On the other hand, the concentration and the total sulfur taken up in the crotalaria were higher than in the rice, due to the higher legume requirement for this nutrient. The sulfur requirement and the short interval of time between the S fertilization and the harvest of the legume, compared to the rice, resulted in a smaller amount of SO_4-S mineralized in the soil and a smaller amount of $^{34}SO_4$ was immobilized in the microbial biomass. Thus, the value of abundance of ^{34}S and the %Sdff in the legume were higher than in the rice ($\%Sdff_{crotalria}$ = 91.3 %; $\%Sdff_{rice}$ = 66.3 %). In contrast, the sulfur in the crotalaria derived from soil was lower than that in the rice ($\%Sdfs_{crotalria}$ = 8.7 %; $\%Sdfs_{rice}$ = 33.7 %), and the quantities of S in both species derived from soil were approximately 5 and 16 mg per pot, respectively.

The expressive values of %Sdff in both species indicate a low rate of mineralization of SO_4^{2-}-S as a consequence of the lack of the sulfur in the soil.

Table 1. Dry matter yield, total sulfur, ^{34}S abundance and, sulfur in the plant derived from the fertilizer (Sdff).

Specie	Dry matter	Total sulfur		^{34}S abundance	Sdff	
	-------g per pot ------		g kg^{-1}	atom %	%	mg per pot
Crotalaria	8.7 ±0.2*	60.6 ±2,9	6.93 ±0,19	13.43 ±0.36	91.3 ±3.5	55.6 ±4.6
Rice	20.4 ±1.1	47.5 ±3,8	2.33 ±0.15	6.68 ±0.08	66.3 ±0,8	31.6 ±2.9

* mean and standard error (m ± se) with n = 3

References

1. Neptune, A. M. L., Tabatabai, M. A. and Hanway, J. J., Sulfur fractions and carbon-nitrogen-phosphorus-sulfur relationships in some Brazilian and Iowa soils. *Soil Science Society of America Proceedings* **39** (1975) pp. 51-55.
2. Bendassolli, J. A.; Trivelin, P. C. O.; Carneiro Junior, F., Sulphur stable isotope fractionation by anion exchange chromatography. Production of compounds enriched in ^{34}S. *Journal of the Brazilian Chemical Society* **8** (1997) pp.13-17.

PRODUCTION OF ^{89}SR USING (n,γ)- AND (n,p) REACTIONS

YU.G.TOPOROV, V.T.FILIMONOV, R.A.KUZNETSOV, O.I.ANDREEV, F.Z.VAKHETOV, V.N.EFIMOV, YU.L.REVYAKIN, A.S.KOROL'KOV, A.N.PAKHOMOV, A.YU.BARANOV

State Scientific Center «Research Institute of Atomic Reactors»Dimitrovgrad-10, 433510, RUSSIA

Two technologies of ^{89}Sr production: "traditional", based on ^{88}Sr(n,γ)-reaction, and alternative, based on ^{89}Y(n,p) reaction, has been discussed in the paper. The quality and suitability for radiopharmaceuticals production of two types of product (with carrier and no-carrier-added) are discussed.

For more then 20 years strontium-89 is used for palliative treatment of bone metastases. Several years ago – Metastron - strontium chloride solution produced by Amersham was approved for clinical use.

Two ways of strontium-89 production are known. First is «traditional» one, based on ^{88}Sr(n,γ)^{89}Sr reaction, taking place in thermal neutron fluxes. The second way is an alternative, based on ^{89}Y(n,p)^{89}Sr reaction, occurring in fast neutron fluxes. In both cases production of strontium-89 is accompanied by radioactive impurities production, particularly – strontium-85 and strontium-90, effecting final product quality.

Regardless of production technology, strontium-89 preparation should confirm a number of requirements, conditioned by its application as radiopharmatheutical. Its specific activity should be not less then 100 mCi/g, gamma-emitting impurities content should be not greater then 0,4%, strontium-90 impurity content should be not higher then $2 \cdot 10^{-4}$%.

Basic elements of technology flow-sheets, using both production ways, designed and applied by RIAR for strontium-89 production, are discussed in the report.

1 Experimental

To produce srontium-89 via (n,γ)-reaction the isotope enriched strontium carbonate (^{88}Sr - 99,816%, ^{84}Sr – 0,01%, ^{86}Sr – 0,05%, ^{87}Sr – 0,124%), encapsulated into quartz capsule, was irradiated in SM-3 reactor in a thermal neutron flux (flux density $1 \cdot 10^{15}$ cm^{-2}s^{-1}) for one reactor campaign (20 days). Irradiated strontium carbonate was dissolved in diluted hydrochloric acid, its activity was measured by beta-counting and gamma-spectrometry.

Pelletized yttrium oxide of natural isotope composition was used as a target for production of strontium-89 via alternative way. It was irradiated in BOR-60 fast-flux reactor by neutron flux of (1-2)$\cdot 10^{15}$ cm^{-2}s^{-1} for 60-100 days. Irradiation was followed by reprocessing, including dissolution of targets in nitric acid, precipitation of yttrium hydroxide (oxalate), sorption of strontium on ampholite resin, extraction-chromatography purification using D2EHPA column.

Strontium-90 impurity content was determined by the procedure, based on beta-counting of activity of yttrium-90 – daughter of strontium-90 decay. Aliquot of sample solution containing 5-10 mCi of ^{89}Sr was adjusted by nitric acid concentration (to 0,5 mol/l) and was passed through the chromatography column, containing teflon powder impregnated with D2EHPA, to purify from yttrium-88,90 impurities. After aging for 5-15 days the solution was passed through chromatography column with D2EHPA to isolate yttrium-90, produced by sterontium-90 decay. The column was washed out with 0,5 mol/l nitric acid to remove strontium, and yttrium-90 was washed out with 4 mol/l HNO3. The solution was evaporated to dryness, the residue was dissolved in 0,5 mol/l nitric acid and purification was repeated. The purification factor of yttrium-90 from strontium-89 was at least 10^8, providing activity of long-lived (relatively ^{90}Y) ^{89}Sr and ^{90}Sr 10-20% of isolated ^{90}Y activity. The activity was measured using $4\pi\beta$-counter during 2-3 days. It allowed qualitative and quantitative identification of yttrium-90 on the background of strontium radionuclides and other potential impurities. The reliable determination was possible at (2-3).10^{-5}% level (by activity) with the error less then 15%.

2 «Traditional» production technology

Reaction $^{88}Sr(n,\gamma)^{89}Sr$ – the general way of strontium-89 production. Its principal advantage is simplicity of irradiated target reprocessing. At high purity of target material radioactive impurity content in the irradiated material typically is not higher then allowed level. In this case, reprocessing procedure is just dissolution of strontium carbonate in diluted hydrochloric acid. The activity of impurities (relatively ^{89}Sr) in our preparation was ^{85}Sr – 0,17%, ^{86}Rb – 0,06%, ^{124}Sb – 0,003%, ^{65}Zn – 0,002%. The activity of ^{156}Eu was nearly equal to that one of ^{89}Sr (80%)/ To purify strontium-89 from this impurity we have applied extraction chromatography column with D2EHPA (similarly to the procedure of yttrium-90 isolation described above), it allowed us to decrease europium-156 content to the acceptable level.

Strontium-89 specific activity was found to be 400 mCi/g at EOB. This is one of the principal disadvantage of the «traditional» way of strontium-89 production, since production of commercially significant quantities of this radionuclide requires long-term irradiation of expensive isotope enriched target material in high-flux reactor with neutron flux density 1.10^{15} $cm^{-2}s^{-1}$ or higher. In this case expire date for strontium chloride will be less then 100 days after irradiation.

Another disadvantage is high content of ^{90}Sr impurity, it was found to be equal $1{,}5 \cdot 10^{-4}$%, that is close to acceptable limit and shortens expire period to 50-60 days.

Another disadvantages are as follows: high cost of start-up material, necessity to use specially designed target, providing safe irradiation of strontium carbonate, destroyed by thermal and radiation effects during irradiation. Besides, reactor resources available for this particular radionuclide production are limited. In total, it leads to high price for final preparation, which, in its turn, limits its clinical application.

3 «Alternative» production technology

As compared to «traditional» production technology, the alternative one, based on $^{89}Y(n,p)$ – reaction, has a number of advantages:

- possibility to produce «no carrier added» strontium-89
- low output of competitive reactions of ^{85}Sr and ^{90}Sr production
- using cheap and readily available yttrium oxide start-up material
- safety of yttrium oxide irradiation.

Theoretically, «carrier-free» strontium-89 can be produced by this way, having specific activity 28,5 kCi/g. However, strontium impurity containing in start-up material and reagents, used for reprocessing, leads to considerable decreasing of strontium-89 specific activity in the final product. According to our data, using of yttrium oxide of high purity (luminophore grade) and special grade reagents gives actual 89Sr specific activity 1500-3000 Ci/g. This is only 5-10% of theoretical value. But this is 5000-10000 times higher than specific activity of «traditional» strontium-89. So, alternative technology lifts restrictions on expire period of strontium-89 preparation considering its specific activity.

Competitive reactions producing ^{85}Sr and ^{90}Sr in an alternative technology are considerably less effective on the quality of final preparation. Strontium-85 relative activity in «carrier-free» strontium-89 was found to be $1.10^{-3}\%$, that is 170 times lower then in traditional product.

Strontium-90 impurity can be produced by two channels. First one is $^{89}Sr(n,\gamma)^{90}Sr$ reaction. Its rate is 200 times higher than that one of $^{89}Y(n,p)^{89}Sr$ reaction. Increasing of irradiation time leads to increasing of ^{90}Sr activity, according to calculations it can reach $5 \cdot 10^{-5}\%$ at irradiation period 90 days. This is twice lower than in traditional technology, and expire period can be extended up to 120 days.

The second channel of ^{90}Sr production is fission of uranium and thorium impurities in start-up yttrium. Strontium-90 content produced via fission reaction can reach $5 \cdot 10^{-5}\%$ at uranium content in yttrium oxide $2.10^{-5}\%$. So, to decrease this impurity production special purification procedure should be carried out for start-up yttrium oxide.

Accordingly to our experimental results, actual ^{90}Sr content at EOB is (3-4)$.10^{-5}\%$ for 60-70 days irradiation period.

The principal disadvantage of the alternative technology is low output of ^{89}Sr. Máximal yield was calculated to be 55-58 mCi/g of yttrium oxide for 400 days irradiation period for neutron flux density 2.10^{15} $cm^{-2}s^{-1}$. Experimental data gave 10-12 mCi/g for 60-90 irradiation period. So, production of commercially valuable quantities of strontium-89 leads to necessity of large scale irradiation of yttrium oxide. Nevertheless, this is one of features of BOR-60 reactor – availability of large volume for irradiation. For example, one target assembly contains up to 1 kilogram of yttrium oxide, allowing production of up to 10 Ci of ^{89}Sr per batch. This provides production of stock batches, allowing quick satisfaction of Customer's requirements even for typical BOR-60 operation schedule (three 90-100 days campaign per year), which is not oriented for isotope production.

Obviously, another disadvantage of alternative technology is reprocessing of large quantity of irradiated targets. But technology developed and used by RIAR is stable and full cycle is not longer then 12-15 days.

The preparation produced by RIAR has the following parameters

Chemical form	$SrCl_2$ solution in diluted HCl
Bulk activity	50-100 mCi/ml
^{89}Sr specific activity	> 1500 Ci/g
^{90}Sr content	$(0,5\text{-}2).10^{-4}\%$
Content of gamma-emitting radioactive impurities	$< 2.10^{-3}\%$
Content of non-radioactive Impurities	< 55 mg/l

4 Conclusion

Both described technologies can be used for commercial production of strontium-89. But preparation quality, produced by alternative technology, is higher then that one of traditional product. Parameters of the «no carrier-added» preparation allows its long-time storage providing its laying in a stock, thus allowing quick response for Customer requirements.

^{113}Sn AND ^{117m}Sn PRODUCTION TECHNOLOGY

O.I.ANDREYEV, F.Z.VAKHETOV, R.A.KUZNETSOV, I.V.TSELISHEV
Research Institute of Atomic Reactors, Dimitrovgrad-10, 433510, RUSSIA

Peculiarities of ^{113}Sn and ^{117m}Sn production in SM high-flux reactor, as well as short description of reprocessing technology, developed and applied by RIAR, are described in the report.

Radionuclides ^{113}Sn ($T_{1/2}$ - 115,1 d) and ^{117m}Sn ($T_{1/2}$ = 13,6 d) are widely used in medicine and research. RIAR produces these radionuclides via irradiation in various positions of high-flux SM reactor: ^{113}Sn - in a central neutron trap, ^{117m}Sn - in a hole of active core.

Tin- 113 is produced via ^{112}Sn(n,γ) reaction. Cross-section of this reaction is small (0,71 barn), so, irradiation holes with the maximum neutron flux density should be used or production of ^{113}Sn of high-specific activity. They are holes of the central neutron trap with neutron flux density up to $1{,}7 \cdot 10^{15}$ $cm^{-2}s^{-1}$.

Tin-117m can be produced via neutron capture ^{116}Sn(n,γ) or via non-elastic neutron scattering ^{117g}Sn(n,n'). Because of low neutron capture cross-section the first reaction doesn't allow production of high-specific activity ^{117m}Sn. The second reaction, being the threshold one, takes place for neutrons with energy higher then 0,5 MeV, so fast neutron fluxes should be used for production. The highest fast neutron fluxes - up to $2 \cdot 10^{15}$ $sm^{-2}s^{-1}$ are available in SM reactor in a specially designed holes in a fuel assemblies.

Calculation results on ^{113}Sn and ^{117m}Sn production while irradiation in SM reactor of tin samples, enriched with the corresponding start-up isotope to 100%, are presented on figure.

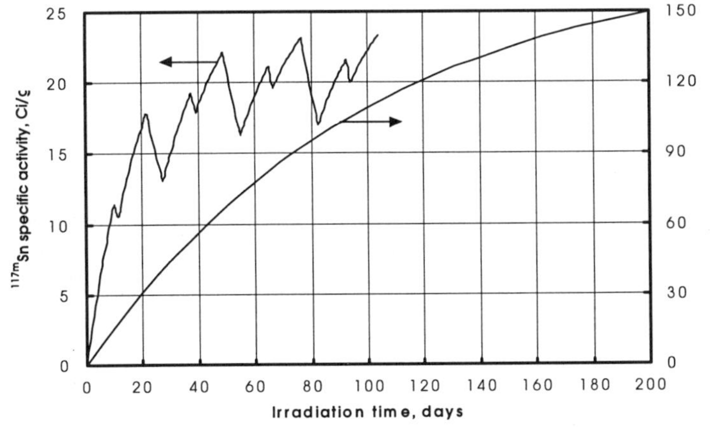

Figure. Specific activity of ^{113}Sn and ^{117m}Sn produced in SM reactor

Radionuclide output was calculated using computer code, realizing algorithm of numerical integration of the set of differential equation, describing changing of nuclide concentration at irradiation. Rates of nuclear reactions were calculated using two-group approximation. The self-shielding coefficient of ^{112}Sn nuclei was used for more correct calculation of ^{113}Sn production rate, considering contribution of non-thermal neutrons with energy, close to resonance energy of target nuclei. In case of relatively short-lived ^{117m}Sn the calculations were made for routine reactor operation schedule using effective cross-section, determined by activation experiments. As can be seen from data presented at figure, irradiation in SM reactor gives specific activity of ^{113}Sn up to 100 Ci/g and ^{117m}Sn up to 20-22 Ci/g.

Quantity and composition of radioactive impurities in irradiated tin targets depend on start-up material purity, as well as on irradiation mode. Typically, irradiated tin metal or tin oxide targets, enriched with start-up nuclides up to 90%, contain a number of antimony radionuclides ($^{122\text{-}126}Sb$), as well as radioactive impurities ^{59}Fe, ^{60}Co, ^{54}Mn, ^{65}Zn, formed from impurities containing in target. So, irradiated material should be purified from impurities.

Reprocessing (purification) technology includes the following procedures:

- dissolution of metal tin in hydrochloric acid
- ion-exchange purification from impurities
- tin reduction to metal state from chloride-citrate solutions using Al metal
- tin conversion to $SnCl_2$ or $SnCl_4$ solution.

Ion-exchange chromatography is used for purification of tin from radioactive impurities $^{122,124,125}Sb$ and ^{59}Fe. Strongly acidic cation-exchange resin KU-2 (analog to Dowex-50 resin) is used as a sorbent. Purification process is carrying out at the presence of oxidizing agent (HNO_3, Cl_2) to stabilize antimony impurity in a five-valence state. The strongly base anion-exchange resin AW-17 (analog to Dowex-1 resin) is used for concentration of tin from diluted solutions (including waste solutions). Sorption is highly effective even at the presence of citric acid or aluminum. Tin is eluted from resin by NaOH solution.

Tin reduction to metal state by cementation provides effective purification of radionuclide from impurities of alkali, alkali-earth, silicon, chromium, zinc, cobalt, manganese, other impurities. Tin metal deposits on the surface of aluminum rod immersed into solution of Sn(IV) containing complexing agent (citric acid), and after process is completed, tin is separated off the aliminum rod by etching in diluted solutions of NaOH and HCl and washed out by distilled water and ethanol. Small quantities of tin, turned from metal to solution while etching and washing can be isolated by ion-exchange chromatography.

The final form of preparation ($SnCl_4$) is produced by dissolution of tin metal in hydrochloric acid at the presence of oxidizing agent (H_2O_2, HNO_3, Cl_2).

Tin radionuclides purity produced by the described technology is not less than 99,99%.

INTRINSIC RADIOLABELING OF NUTRIENTS FOR HUMAN NUTRITION STUDIES USING ACCELERATOR MASS SPECTROMETRY

Andrew J. Clifford[1], Stephen R. Dueker[1], Elisabetta E. Fabbro[1], Yumei Lin[1], Nichole Hong[1], Michael W. Lamé[2], Henry J. Segall[2], Bruce A. Buchholz[3], John S. Vogel[1,3].

Department of Nutrition[1] and Molecular Biosciences[2] Univ of Cal, Davis, CA; and Center for Accelerator Mass Spectrometry[3], Lawrence Livermore National Laboratory, Livermore, CA

^{14}C labeling of plant nutrients coupled with detection of label by Accelerator Mass Spectrometry (AMS) represents an attractive means for investigating the metabolism of nutrients that are an intrinsic part of a food or diet [1,2]. AMS quantitates attomolar concentrations of ^{14}C-labeled substrates from biologic tissues and fluids at ethically acceptable levels of radiation exposure, eliminating the need for compounds of high isotopic concentration. Accordingly, the generic photosynthetic labeling of plants with $^{14}CO_2$ can be used to provide an inventory of difficult to synthesize phytochemicals that can be isolated and tested in their natural form. To assess the intrinsic bioavailability of nutrients as they are consumed within the diet requires labeling strategies that deliver specific, committed precursors in a biochemical pathway. These precursors can be delivered by hydroponic culture or through their leaves and stems by surface application or injection into the xylem flow. We have used some of these labeling strategies with spinach and kale, two commonly consumed leafy vegetables, to produce ^{14}C-labeled caroteneoids and folates, respectively.

PHOTOSYNTHETIC RADIOLABELING (β-CAROTENE AND LUTEIN): The carotenoid (plant pigments that serve as vitamin A precursors and biological antioxidants) content of spinach (*Spinacia oleracea*) is distributed between β-carotene and lutein. ^{14}C- β-carotene and ^{14}C-lutein was prepared by growing spinach in an atmospherically sealed labeling chamber (LC) pulsed with $^{14}CO_2$. (Fig 1). Photosynthetic labeling using $^{14}C0_2$ results in randomly labeled biomolecules and other applications have been described [3,4,5]. Several plants were grown on a 12-h light cycle under metal halide lights supplemented with tungsten illumination in a plant growth box. At 30 days, they were placed in the LC. Plants were allowed to equilibrate for one day prior to the administration of $^{14}CO_2$. A total of 50 mCi of $^{14}CO_2$ that was provided at the rate of 10 mCi/d for 5 days. Exposure was initiated on the first day by adding 10 mCi as a solution of ^{14}C-$NaHCO_3$ to 18N H_2SO_4 over a 4 hr period releasing $^{14}CO_2$. This level of exposure was repeated for the following 4 days. Following the final exposure, the plants were maintained for 72 h in LC prior to harvesting. The aerial parts of the plant were harvested and were extracted for carotenes. Final radiometric and chemical purity was determined to be greater than 98% by analytical RP-HPLC. The specific activities were 1.45 and 0.35 mCi/mmol for lutein and β-carotene, corresponding to 0.023 and 0.0056 ^{14}C atoms per molecule, repectively. One subject has been exposed to a 200 nCi (306 ug, 444,000 dpm) dose of purifed β-carotene (see Dueker et al. In this compendium). Other compounds of nutritional interest are also being purified for human testing.

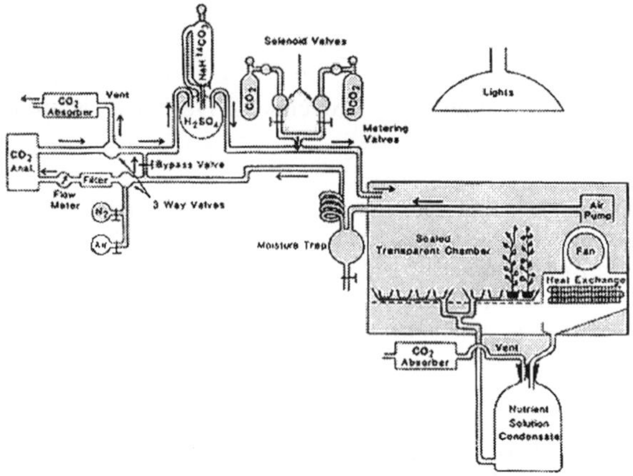

-Photosynthetic labeling produces a 'Radiopharmacy' of bioactive natural products that includes:

Vitamins
Lipids
Terpenes
Polyphenolics
Phytoestrogens
Proteins

Fig 1. Illustration of Plant Labeling Box (3)

COMMITTED PRECURSORS BY COTTON-WICK METHOD (FOLIC ACID): Folic acid (folate) is vitamin need for one-carbon transfer reactions [6,7]. Kale is a rich source of folic acid but its bioavailability from plant sources is an unsettled issue and this has led to its use as a supplement. To assess the intrinsic bioavailability of folates a kale plant (*Brassica oleracea acephala)* was grown in an open plant box as described above. At 52-d post germination the plant was administered para-aminobenzoic acid (pABA)(see fig. 2) by the cotton-wick method which entailed the insertion of a cotton thread throught the stock of the plant, below

Fig. 2: Stylized representation of folic acid. Grey area represents moeity derived from pABA.

the first set of leaves, with both ends being inserted into a vial containing ^{14}C-pABA (40 uCi; 58mCi/mmol); labeling began at 11:30 AM and 500 μL of material was taken up completely after 6 hours, at which time the vial was washed with 800 uL of water to remove residual material in the vessel and the wick. The plant was then allowed to grow for one week. Aerial parts of the plant were harvested and the bulk of the radioactivity was found to be distributed within the fastest growing shoots and the tips of the larger leaves, verifying that the dose translocation from the site of adminstrations through the plant vascular system. A 2-gr sample was ground by motar and pestle in ice-cold HPLC mobile phase, filtered and and a small aliquot (contain ~ 10 DPM of activity) was injected onto an RP-HPLC system characterized for folate separation. Fractions were analyzed for ^{14}C by AMS. The metabolite profile showed that less than ~5 of the label was associated with known folates. 95% of the activity eluted in the solvent front portion of the chromatogram, which corresponded to either free pABA or polar metabolites. Given the low incorporation of label into folate, the material was not considered appropriate for human experimentation. Further studies are underway to optimize this approach.

CONCLUSION: Photosynthetic labeling of plants with $^{14}CO2$ yields a wide inventory of low specific activity biomolecules that, given appropriate extraction methods can be purifed for use human studies using AMS. Because only a fraction of the radioactivity is associated with the target molecule(s), the plant cannot be consumed whole to assess the intrinsic bioavailability of the target compounds. Assessment of nutrient bioavailbility from intrinsically labeled plants requires the use of committed biosynthetic precursors. This method demands considerable knowledge of plant biochemistry, and experimentation with a variety of labeling techniques and growth conditions to achieve useful results. It is expected that with further experimentation, intrinsic labeling of nutrients by this method with subsequent AMS detection will facilitate investigations into the role of plant matrix on nutrient bioavailability.

References

1. Vogel, JS; Turteltaub, KW; Finkel, R; Nelson, DE. (1995) Accelerator mass spectrometry: Isotope quantification at attomole sensitivity. *Anal Chem,* 67(11):353A - 359A
2. Vogel, JS; Turteltaub, KW. (1998) Accelerator mass spectrometry as a bioanalytical tool in nutrition research. In, Advances in Experimental Medicine and Biology, edited by A.J. Clifford & H-G. Müller. (New York: Plenum) 397-410.
3. Segall, HJ; Brown, CH; Paige, DFJ. (1983) *J Lab Compd Rad*, 20: 671-89.
4. Lamé, MW; Wilson, DW; Segall, HJ. (1996) *J Lab Compd Rad,* 38, No. 12: 1053-60.
5. Dueker SR, Clifford AJ, Lame MW, Segall HJ, Buchholz BA, Vogel JS. (1998) Production of carbon-14 labeled phytonutrients in spinach for human nutrition studies employing accelerator mass spectrometry detection. *Faseb J*, 12:a210
6. Clifford, JA; Arjomand, A; Dueker, SR; Schneider, PD; Buchholz, BA; Vogel, JS. (1998) The dynaimics of folic acid metabolism in an adult fiven a small tracer dose of ^{14}C-folic acid. In Advances in Experimental Medicine and Biology, edited by A. Clifford and H-G Müller. (New York: Plenum) 239-251.
7. Cossins, EA; Chen, L. (1997) Folates and one-carbon metabolism in plants and fungi. *Phytochemistry,* 45:437-52.

BASIC TRENDS IN RADIONUCLIDE PRODUCTION OF THE V.G. KHLOPIN RADIUM INSTITUTE.

V.V. FEDOROV, B.YA. GALKIN, V.A. BYKOV, V.N. ROMANOVSKY, L.V. SOLIN, M.P. ZYKOV

V.G. Khlopin Radium Institute, 28, 2-nd Murinsky Ave., St-Petersburg 194021, Russia.

E-mail: fedvv@atom.nw.ru

A.Z. ILIASSOV, B.I. ROGOZEV

RITVERC GmbH, 28, 2-nd Murinsky Ave., St-Petersburg 194021, Russia.

E-mail: rogozev@atom.nw.ru

Traditional products and recent results in radionuclides and related products syntheses of V.G. Khlopin Radium Institute are presented. The main directions of Radium Institute's activities are the following:
- biologically active compounds labeled with tritium and carbon-14;
today Radium Institute produces about 130 items of labeled preparations and increases this list annually for 10÷15 items.
- radiopharmaceuticals for medicine: gallium-67, indium-111, iodine-123 and radionuclide generators with molibdenum-99 - technetium-99m and radium-226 – radon-224;
- alpha-, beta- and gamma-emitting radionuclide sources;
- broad variety of sealed sources for X-ray fluorescence analysis with iron-55, cadmium-109, americium-241 and curium-244;
- sources with gaseous krypton-85 for thickness gauge;
- radionuclide standards for spectrometry measurements and calibration;
- radionuclides including iron-55, cobalt-57, nickel-63, krypton-85, stroncium-85, stroncium-90, ruthenium-106, cadmium-109, tin-119m, iodine-125, iodine-129, barium-133, caesium-137, americium-241and radi-um-226 and radionuclide preparations on their base;
- Mössbauer sources on cobalt-57 and tin-119m including double-side and point sources and unique equipment for Mössbauer spectroscopy such as resonance scintillation detectors and precise data acquisition systems wit.

1 Introduction

V.G.Khlopin Radium Institute is one of the greatest and oldest scientific institutes for research in the world dealing with investigation of the radioactivity phenomena. It was established in 1922 in St. Petersburg, former Leningrad. Just then specialists of the institute began the systematic investigations of different aspects of radioactivity. Main directions of scientific activity in the Radium institute are connected with radiochemical investigations, nuclear physics and ecology including nuclear waste treatment. During past years the development and manufacturing of different types of radionuclide products became an independent and important direction of work in the institute. About 200 specialists in various fields are engaged now in this area. Now the institute has high-qualified personnel and good industrial basis in the field of radionuclide products manufacturing. All that gives us the possibility to solve rather difficult problems.

Isotope Department responsible for export delivery of significant part of radionuclide products was set up in the Radium Institute in 1974. It is acting in close collaboration with another departments of the Radium institute and first of all with Technological Department situated near St. Petersburg. Isotope Department represents quite complex organism which consists of six laboratories working in different directions.

Now the institute produces and delivers isotope products both to Russia and to tens of foreign countries.

Spectrum of radionuclide products manufacturing by our personnel is quite wide. This time we produce the following types of products, represented in the Fig. 1:

- radionuclides and preparations on its basis;
- radionuclide generators;
- biologically active compounds and preparations labeled by radio-nuclides;
- radionuclide sources of ionizing emission (including those for radionuclide devices, for X-ray fluorescent analysis and Moessbauer sources) and light sources;
- radionuclide standards (both solutions and sources);
- resonance scintillation detectors.

Besides the institute in close cooperation with Science-Engineering Centre "CENTROTECH" (Centrifugal Technologies) produces a number of high-quality high-enrichment (with isotopic concentration more than 99.5 %) stable isotopes, such as ^{30}Si, ^{58}Ni, ^{62}Ni, ^{74}Se, ^{124}Te, ^{187}Os, and other.

Detailed tests for reliability and metrological examination are followed by the manufacturing of the products mentioned above.

2 Radionuclides and preparations on their base

At the present time Radium institute produces and delivers different radionuclides (fission, cyclotron and reactor) such as gallium-67, iodine-123, cobalt-57, krypton-85, strontium-90, ruthenium-106, iron-55, nickel-63, barium-133, etc., and also compounds on their basis.

Using radon generators the institute produces pellets saturated with radon for radontherapy. Being readily soluble in water, such pellets allow to dose radon concentration in bath rather easily. The Radium Institute can deliver to medical organizations up to 10.000 pellets per year.

Experience shows that on a certain stage of investigations, the development of radiochemical processes requires checking ideas on real fuel. The Radium Institute can deliver to customers the samples of spent fuel from reactor WWR in solid state or in solution.

3 Radionuclide generators

The Radium Institute has developed and delivers to consumers solid-phase generator of radon for determination of porosity, material structure, continuity of coating, penetrability of grounds, radiotherapy and others.
The performance of generator is based on extraction of radon from solid-phase matrix, containing radium-226 and its filtration through a filter.
Content of radium-226 in generator depends at consumer's will:
1-50 mg;
Emanation coefficient - not less than 80 %;
Minimal quantity of air, necessary for complete displacement of radon - 0,55 l;
Mass - 45 kg.
Generator keeps serviceability in wide range of temperatures, is noted for safety, reliability and simplicity in operation, large resource and high purity of product for special purpose.

The institute can deliver up to 50 generators per year.

Since 1993 integrated generator of ^{99m}Tc acts in the institute using ^{98}Mo irradiated in nuclear reactor and following extraction of ^{99m}Tc in centrifugal extractors. Every week the institute delivers 3 Ci of ^{99m}Tc to 20 hospitals of St. Petersburg – it is enough for complete satisfaction of requirements of such big city (population is 5,000,000 persons).

4 Production of compounds labeled by tritium and carbon-14

This direction has been developed in the Radium institute from 1963, when a laboratory for synthesis of labeled biologically-active compounds was organized. It is one of the leaders in production of compounds labeled by tritium in Russia now. These compounds are used in molecular biology, genetics, biochemistry, medicine and agriculture. The total number of preparations exceeds 150 titles, the total activity of annual delivery is about 100 Ci, and the number of packings is nearly 3.000.
Delivery includes:
- components of nucleic acids;
- polynucleotides;
- hydrocarbons;
- modifiers of proteins;
- antibiotics;
- chemico-therapeutical preparations, etc.

List of compounds producing by Radium Institute is every year updated and 10-15 compounds are added.
Use of labeled compounds in biological experiments imposes stringent requirements upon their quality, that is of such characteristics as molar activity, radiochemical and chemical purity. Preparations produced by the Radium Institute are mostly in compliance with the world standards by their quality.
High molar activity of produced compounds is reached by means of optimizing the synthesis conditions and using multiple label. Position of label is determined by nuclear magnetic resonance spectroscopy.
Determination of chemical purity is realized by the method of gas or high-effective liquid chromatography.
Radiochemical purity is determined by paper or thin-layer chromato-graphy with subsequent measurement of radioactivity distribution on chromatogram by scintillation counting.
Radium Institute is ready to conduct works on introduction of label in new chemico-therapeutical preparations, developed by pharmaceutical companies (including works on confidential conditions).
Now the laboratory began development and deliveries of compounds labeled by carbon-14, wide used at the foreign markets. This time the number of those compounds is about 20, included benzene, ethanol, methanol, phenol, etc.

5 Radionuclide sources of ionizing radiation and light

Sealed radionuclide sources of ionizing radiation are one of the most important forms of applications of radionuclides for different fields of human activity: radiation chemistry and polymerization, radiation sterilization of medical instruments and materials, food products treatment, fault detection, neutron and gamma-ray logging, X-ray fluorescence, phase and activation analysis, neutralization of static electricity, control and automation of technological processes, radiology and metrology, and so on.
Even from the first steps of formation of the Radium Institute, the production of platinum needles filled with radium salts for oncology application was set up and continued during many years. The institute is one of the pioneers in development of industrial technology for safe sealed alpha-radiation sources, which were widely used for neutralization of static electricity charges, and also for fume detectors.
The institute remains one of the leaders in production of radionuclide sources of all types in Russia: sources for radionuclide devices (based on tritium, cobalt-57, nickel-63, krypton-85, barium-133, etc.), for X-ray fluorescent analysis (based on iron-55, cadmium-109, americium-241), Moessbauer sources (based on cobalt-57 and tin-119m) and light sources (based on tritium and krypton-85).

Minimal annual delivery is about 1000 of items, in particular:

- sources for X-ray fluorescence analysis - 700 items;
- sources based on gaseous krypton-85 for thickness gauge - not less than 100 items;
- Moessbauer sources on cobalt-57 and tin-119m - up to 200 items.

Development and certification of radionuclide sources is based on detailed investigation and tests of models, simulators and standard specimens. We realize complete cycle of tests accepted in international practice: tests on the material of "special type", classification strength tests in accordance with ISO, and also special tests - such as tests on fire in industrial conditions with temperature 1100 °C, impact of shock wave, high-speed impact against granite, concrete and other tests at customer's request. Reliability of product during service life is confirmed by corresponding resource tests and reliability tests.

Hermetic sealing of products is performed by means of modern welding methods: argon-arc welding with nonconsumable tungsten electrode and laser welding.

For control of hermetic sealing methods, method of krypton-85 inleakage, different kinds of bubbling methods and methods of oxygen extract are used.

For radionuclide production the institute has a system of light and heavy protection boxes and also a chain of 13 hot chambers, which allows to take and store radioactive materials equivalent to 100 kCi of cobalt-60, and to perform chemical operations with 10 kCi.

Highly skilled scientific, engineering-technical and operating personnel and continuous radiometric control enable to conduct all radiochemical works at St. Petersburg and Gatchina sites of the institute with ecological safety.

6 Radionuclide standards

Special attention of institute's works is paid to the problem of metrological accompaniment of products under development, metrology of ionizing radiation. In the institute an installation of 4¶ alpha-gamma(X)-coincidence - working standard of activity unit of radionuclides is designed, working standard of neutron-flux unit is designed, too, installation for transition of the photon radiation flux density is elaborated, standard solutions of radionuclides and radionuclide standard-sources (Table 1) are developed and put in production. Now the institute produces reference solutions on the basis of 86 radionuclides (from carbon-14 to curium-246); reference spectrometric gamma-standards based on 23 radionuclides, both separate sources and standard sets. Besides the Radium institute produces spectrometric alpha-standards on the basis of radium, thorium, uranium, neptunium, americium, plutonium, curium and californium, and combined spectrometric sources based on above-mentioned radionuclides. Moreover the institute produces reference radiometric sources of beta-irradiation (BERRS-sets) on the base of nickel-63, carbon-14, promethium-147, cobalt-60, thallium-204 and strontium-90. Overall uncertainty of activity is between 0.7 and 1.8 % at P=0.99 for gamma-standards and 3 % for BERRS sources. Total annual productivity of spectrometric sources of alpha-, beta-, gamma-standards is not less than 5000 units. Works in this field are supported and accompanied by publishing of catalogue - reference book: "Characteristics of emission of radioactive nuclides, used in national economics. Estimated data".

7 Resonance scintillation detectors

Traditional direction of works in the Radium Institute is also concerned with the development of effective means for measurements for gamma-resonance spectroscopy (Moessbauer effect). The unique applied possibilities of this method are not yet exhausted. The future of nondestructive express phase-analysis of materials, containing such "Moessbauer" elements as iron, tin, antimony, rare earths and others is connected with this method in different industries, including such directions as amorphous metals and alloys, microcrystalline powders, corrosion and others.

The problems of more efficiently using of nuclear gamma-resonance are solved in the Radium Institute by developing of new improved sources of gamma-quanta and high-effective radiation detectors.

At present broad variety of Mossbauer sources developed in the Radium Institute and RITVERC GmbH is available. We can offer ^{57}Co Mossbauer sources on Rhodium and Chromium matrices (10 ÷ 150 mCi) and ^{119m}Sn Mossbauer sources on SnO_2, $CaSnO_3$ and $BaSnO_3$ matrices (2 ÷ 20 mCi) with various dimensions and activities satisfying possible requests of scientists and investigators. Among last new products are recent developments of point sources with an active part diameter of 1.0 mm under activity of 5 and 10 mCi (^{57}Co) and 5 mCi (^{119m} Sn) and double sided ^{57}Co Mossbauer source. This last may be used when design of vibrator enable one to fix a source inside a hollow rod. A design like that makes it possible to have two resonance channels operating simultaneously with one source.

Developing new types of radiation detectors, we are working out two approaches.

The first one is design and production of fine scintillation detectors on the base of NaI(Tl) with optimum thickness for registration of "Moessbauer" radiation with maximum reference signal/noise. Diameters of detectors lie within the range 5-30 mm, thickness - 0,1-1,0 mm, and resolution of Mn for KX-radiation is 50 %.

The second approach involves development of high-efficient detectors on recoil-free "Moessbauer" gamma-quanta, with principle of operation based on using the proper resonance absorption. Presently, a rather wide range of resonance scintillation detectors is developed and put in production for resonances on iron-57, tin-119m and europium-151. These detectors offer the new production in gamma-resonance instrument engineering, which permit to raise sharply selectivity of resonance detection in 2-10 times.

For investigators convenience Radium Institute and RITVERC GmbH developed a device ready for Mossbauer measurements immediately. Today we are glad to present a new design of our RSDUM-01 unit. This unit permits to measure both RSD's types of detectors. RSDUM-01 consists of a Hamamatsu type photomultiplier with low dark currents, a Hamamatsu type high voltage supply, a preamplifier/amplifier with a shaper and a stabilization circuit. This last one permits to take long-term measurements with ^{57}Co-sources easily. All these components are mounted in the housing. A photomultiplier tube is suspended on spiral springs and connected to a Mossbauer spectrometer's driver.

The manufacturer performs all necessary adjustments so explorer has to plug in our RSDUM-01 and begin his measurements.In conclusion, it should be noted that accumulated during many years experience and engineering capacity of the V.G.Khlopin Radium Institute allow to propose to consumer not only radionuclide production of high quality, but also design, delivery, mounting, actuating and servicing of research equipment for hot chambers, made of stainless steel or titanium (extractors, dosators etc.), with remote control and operation. It should be pointed out, besides, that the institute takes back its products from the customer after finishing their working life for regeneration.

RADIOISOTOPE PRODUCTION WITH HORIZONTALLY-ORIENTED SOLID TARGETS

ALFRED WONG, WILLIAM Z. GELBART, NIGEL R. STEVENSON AND RAY WONG

TRIUMF, 4004 Wesbrook Mall, Vancouver, B.C., V6T 2A3, Canada

A new target station is being designed at TRIUMF for a small beam impact angle (5 degrees or less) with horizontally oriented targets. Small beam impact angles allow the beam to spread over a larger target area, while reducing the required thickness of plated materials and maintaining adequate heat transfer. The horizontally oriented targets (as opposed to the presently used vertical targets) take full advantage of the inherent beam stability in the vertical plane. This paper discusses the overall design of the target station, along with developments implemented to improve reliability, enhance accessibility, and increase radiation hardness. Modular assemblies of the system including the manipulator, irradiation chamber, and a novel approach to transfer systems are also discussed.

Introduction

The Applied Technology Group at TRIUMF operates two cyclotrons (30 MeV TR30 and 42 MeV CP42) on behalf of MDS Nordion for medical radioisotope production. With an increasing demand for radioisotopes and irradiation at smaller beam impact angles, the present targetry system could no longer provide an adequate environment to meet the requirements. Departure from the old target station design concept was necessary; and a new target station was designed, combining the improvements made on the older system with almost two decades of operating experience.

The new target station is composed of modular components, which allow for easy interchangeability of parts. Use of only metals and ceramics improves the reliability of operation and keeps maintenance at minimum levels. To further minimize radiation damage and allow full accessibility, the mechanical pump, as well as the valves and controls, are all placed outside the target vault. High purity aluminum is employed whenever possible in the fabrication of components to keep activation low, allowing for shorter cool-down periods before access.

The three main modules; vacuum chamber, manipulator, and the landing terminal are discussed in more detail below.

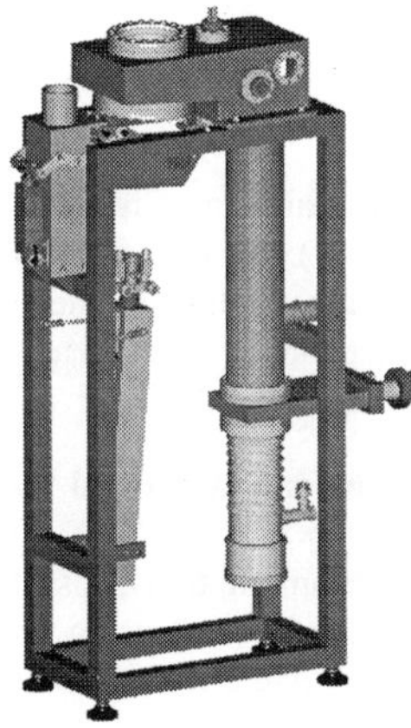

Figure 1: Target Station

Vacuum / Irradiation Chamber

Figure 2: Vacuum Chamber (with side walls removed)

The vacuum / irradiation chamber has its own vacuum system and is isolated from the beam line vacuum by a gate valve. The target is placed horizontally to take advantage of the inherent beam stability in the vertical plane. The less stable horizontal position (determined by the extraction foil position rather than the cyclotron magnetic field) is projected onto the full width of the target.

Beam collimators, located close to the target, allow precise shadowing of the target plated area. They confine the oval cross section of the beam to the rectangular active area of the target. To ensure even spread of the beam, ten to twenty percent of the beam energy is evenly distributed on the four collimators. Each of the collimators can take the full thermal load of the beam in case of a sudden beam position shift.

A viewing port on top of the chamber permits full visibility of the target, even during irradiation. This feature is especially useful for target alignment during which a scintillating target is monitored to determine the beam coverage.

Manipulator

The manipulator is used to retrieve the target (from the pneumatic transfer carrier 'rabbit') and to place it in the irradiation chamber. Water connections on the head provide cooling to the target during irradiation. At the end of the process, the procedure is reversed to allow the return of the irradiated target to the hot cell.

Transfer System and Landing Terminal

A pneumatic transfer system is used to transfer the target between the target station and the hot cell using pressurized air. A permanent magnet, placed under the end cover of the rabbit, triggers detectors that indicate the rabbit's progress along the transfer line.

The landing terminal of the pneumatic transfer system uses a dynamic braking system, utilizing the pressure build-up in front of the rabbit, combined with an injected air from below, to bring the rabbit to a soft landing. Once landed, the rabbit is oriented and held in place inside the receive housing by permanent magnets inside both the rabbit and the landing terminal. The target placing sequence is then initiated, beginning with the opening of the landing compartment, followed by the engagement of the target and the manipulator, and the insertion of the target into the irradiation chamber.

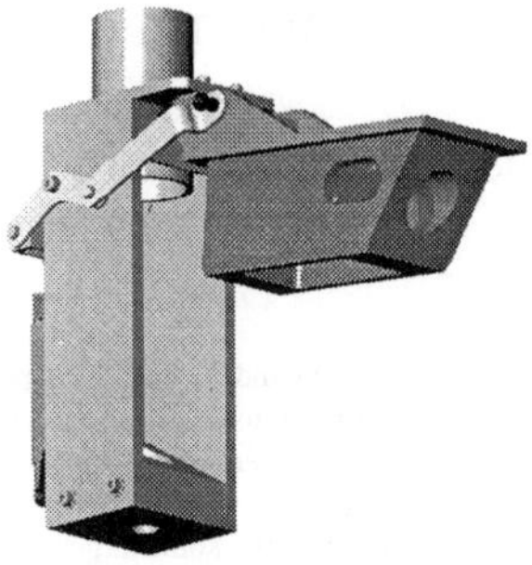

Figure 3: Landing Terminal

Conclusion

The new target station shall be able to handle the increasing production demand of MDS Nordion for years to come. Benefits of the design, such as small station footprint, permits multiple stations to be placed inside an existing target cave, allowing for production expansion as required. Features such as the high target and component visibility, and micrometric adjustments to the beam collimators, put this design well ahead of the current system.

References

1. Gelbart, W. Z., Stevenson, N. R., Ho, W., Bakhtiari, S., "High Current Encapsulated Target and Target System for Radioisotope Production," Proceedings of 7th Workshop on Targetry and Target Chemistry, 1997, Heidelberg, Germany
2. Gelbart, W. Z., Stevenson, N. R., Johnson, R. R., Nortier, F. M., Orzechowski, J., Cifarelli, F., "High Current Radioisotope Production with Solid Target System," Proceedings of the 1993 Particle Accelerator Conference
3. Gelbart, W. Z., Stevenson, N. R., "High Current Targetry for Radioisotope Production," Targetry '91, Paul Scherrer Institute
4. J. J. Burgerjon et al., "A High-Power Target System for Radioisotope Production," Proc. 11th Int. conf. on Cyclotrons and their Applications, 1986, Tokyo
5. G. Pinto et al., "Target Systems for Radioisotope Production," Proceedings of 7th Conf. on the Applications of Accelerators in Research and Industry, 1982, Denton, Texas

INVESTIGATION OF THERMAL-MECHANICAL PROPERTIES OF SOLID TARGETS FOR RADIOISOTOPES PRODUCTION

RAY WONG, WILLIAM Z. GELBART, NIGEL R. STEVENSON AND ALFRED WONG

TRIUMF, 4004 Wesbrook Mall, Vancouver BC V6T 2A3, Canada

As the demand for radioisotope production increases, the design of the solid targets used for irradiation must be improved in order to take advantage of higher beam currents. To do so, the target's thermal and mechanical properties, and the effects of heat deposition and thermal-stress must be optimized. Finite element methods are used to analyze the steady state response of the solid target under the heating from beam bombardment, and the mechanical stress resulting from coolant pressure. Results of the analysis are presented and discussed.

1 Introduction

The beam, delivered from the cyclotron, is intercepted by the target placed at seven-degree angle. Water-cooled silver target face supports the electroplated target material. The deflection (resulting from the water pressure and the irradiation chamber vacuum) and surface temperature limit the maximum current of the 30 MeV proton beam to 400 μA (12 kW). With ANSYS, a finite element package, the present target configuration is used as a reference to design two new targets. The objectives of the designs are to:

1. Increase overall cooling efficiency by matching the water flow to the thermal load.
2. Improve target mechanical strength to allow further increase of water flow.

Double Gaussian heat flux is used to model the heat dissipated by the proton beam. All the calculations are for a bare silver face in view of the large variety of target materials.

2 Present Target (Figure 1)

The presently used target is composed of a copper body soldered to a silver face. Water (at 20°C, 16 L/min, 8 bar) is fed through 16 channels (0.8 mm x 2.5 mm cross-sectional area and 52 mm long) to provide cooling.

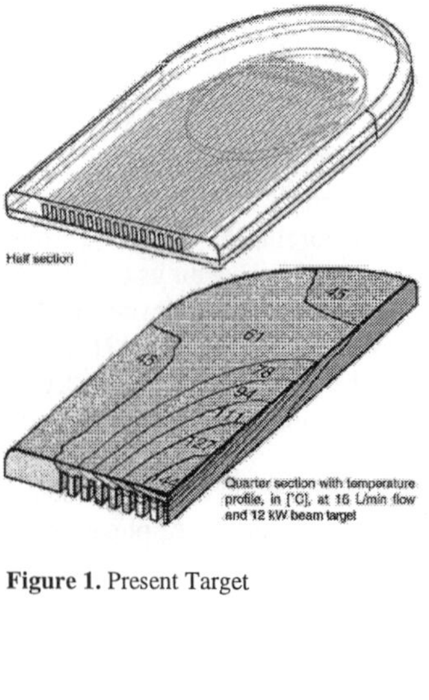

Figure 1. Present Target

3 No-fin Target (Figure 2)

In the no-fin target, all water channels are removed to strengthen the target face. A bevel (24.8 mm x 0.65 mm cross sectional area and 7.25 mm tapering to 5 mm long) is formed in the copper body to maximize the water velocity in the middle region, where cooling is needed most.

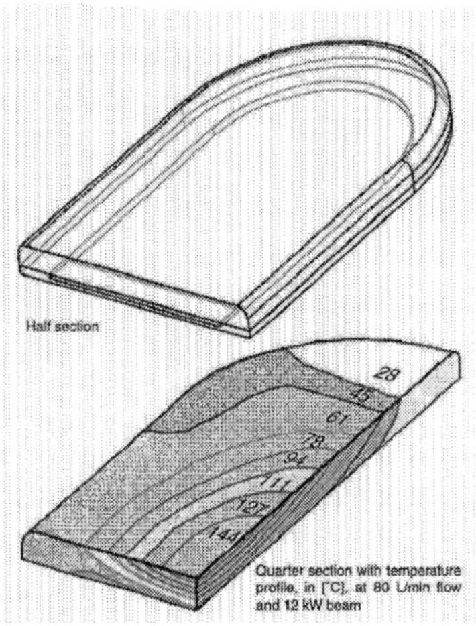

Figure 2. No-fin Target

4 Super Target (Figure 3)

The "super target" has higher cooling efficiency and strength than the present target. Cooling channels are limited to the middle 11.5 mm x 30.7 mm of the target. "Fingers" in the copper body sit in the middle of the channels forming a 0.2 mm U-shaped gaps, to boost the water velocity.

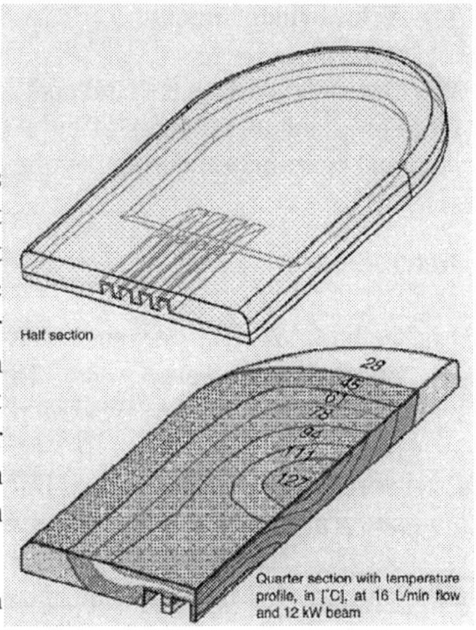

Figure 3. Super Target

5 Comparison

Comparing the deflection of the three targets at 8 bar, the no-fin target and the super target are 20 and 15 times stronger than the present target, respectively. (Table 1) This dramatic increase in strength permits higher water pressure to be used to increase water flow in the new design models. The super target running at 48L/min, can achieve similar surface temperature as the present target with double the beam power (24kW compared with 12kW). (Table 2)

Table 1. Deflection and temperature at 8 bar, 16 L/min and 12 kW

Target	Deflection	Maximum Surface Temperature
Present	102 μm	145°C
No-fin	5 μm	258°C
Super	7 μm	134°C

Table 2. Maximum surface temperature at various beam powers and increased flow

Target	Flow	Deflection	Maximum Surface Temperature			
			At 6 kW	At 12 kW	At 18 kW	At 24 kW
Present	16 L/min	102 μm	83°C	145°C	208°C	271°C
No-fin	80 L/min	125 μm	86°C	152°C	218°C	283°C
Super	48 L/min	63 μm	54°C	89°C	123°C	158°C

6 Conclusion

Provided installations of higher pressure pumping systems, the super target will be capable of handling much higher beam power while keeping the surface temperature at allowable levels. The results obtained are still preliminary, as experimental testing remains to be done to verify the simulated results. Dimensions and the shape and cross section of water channels as well as the targets, could be further optimized for more evenly distributed temperature profiles with lower "hot spot." Transient analysis remains to be performed to investigate the effect from instantaneous heating.

7 Acknowledgements

We thank Steven Rogak (University of British Columbia), Doug Latornell (M.D.S. Nordion), and all members of the Applied Technology Group in TRIUMF for their help and contribution during the design process.

References

1. Gelbart, W. Z., Stevenson, N. R., Ho, W., Bakhtiari, S., "High Current Encapsulated Target and Target System for Radioisotope Production," *Proceedings of 7th Workshop on Targetery and Target Chemistry*, 1997, Heidelgerg, Germany
2. Nortier, F. M., Stevenson, N. R., Gelbart, W. Z., "Investigation of the thermal performance of solid targets for radioisotope production," *Nuclear Instruments & Methods in Physics Research Section A*, 1994, pp. 236-241
3. Nortier, F. M., Stevenson, N. R., "Thermal Analysis of the TRIUMF Solid Targets", *Symposium on Isotope Production and Applications in Medicine, Science and the Environment*, Vancouver, Canada, 1993
4. J. J. Burgerjon et al., "A High-Power Target System for Radioisotope Production," *Proc. 11th Int. conf. on Cyclotrons and their Applications*, 1986, Tokyo

THE TRANSPORT OF CARBOHYDRATES AND CELL CONSTITUENTS IN PLANTS USING AMS DETECTION OF CARBON-14 TRACERS

P. B. ROBERTS AND K.KIM

Institute of Geological & Nuclear Sciences, Lower Hutt, New Zealand
E-mail: p.roberts@gns.cri.nz

Two examples are provided of the application to plant research of accelerator mass spectrometry (AMS) used to detect carbon-14 labelled biomolecules at ultra-trace levels. The activity of labelled carbohydrates in single plant cells of different types has been measured following $^{14}CO_2$ feeding of barley leaves. Salicylic acid transport within a full-sized kiwifruit vine has also been demonstrated. The great sensitivity of AMS opens up new possibilities for the use of isotopic labels in plant research.

1 Introduction

Labelling of whole plant tissues and plant sections following $^{14}CO_2$ feeding has increased understanding of carbohydrate assimilation and transport in laboratory plants. This paper demonstrates that AMS detection of the label makes possible the study of labelled carbohydrates in individual plant cells.

AMS also provides new opportunities for studies of material transport in large, economically important plants such as kiwifruit vines. Local infection of plants by certain pathogens induces systemic acquired resistance (SAR) that is accompanied by an increase in salicylic acid (SA) levels. Labelling experiments plus AMS detection can investigate whether SA production is not only a necessary part of SAR but is also the cause of the primary systemic signal.

2 Methods

A section of barley leaf was fed with 3.7MBq (100μCi) of $^{14}CO_2$ over 45 minutes. Samples of cell sap (10-200pl) were withdrawn from individual cells using an oil-filled microcapillary. To obtain an ion beam during AMS measurement, enough tributyrin was added to achieve 2mg total carbon. In separate SAR experiments 100μl of SA at 0.2mM and 0.7Bq (18.9pCi) activity were spotted onto 9 leaves on separate canes of two kiwifruit vines. Three canes were removed for analysis at 3 days after inoculation, and again after 7 and 14 days. Leaves and cane segments adjacent to the inoculated leaf (see Figure) were sub-sampled and submitted for AMS analysis. The percent organic carbon of dry tissue was 42%. Procedures for AMS analysis of ^{14}C activity have been described elsewhere [1].

3 Results and Discussion

The activity measured in the barley cell sap samples was corrected for the amount contributed by tributyrin (1.5fCi/mg total carbon) and normalised to the volume of sap removed for each cell type sampled. The values obtained were 29.5, 2.5 and 2.2 fCi/pl for mesophyll, bundle sheath and epidermal cells respectively, and compared to 0.3fCi/pl for cell sap from non-labelled mesophyll cells. If cell extracts were pooled into a single sample, a good linear relation (r^2=0.985) was found between the number of cell extracts pooled and the activity measured. Separate experiments indicated that the activity was associated with glucose. The results show that carbohydrate flux can be studied at the level of individual plant cells and cell types.

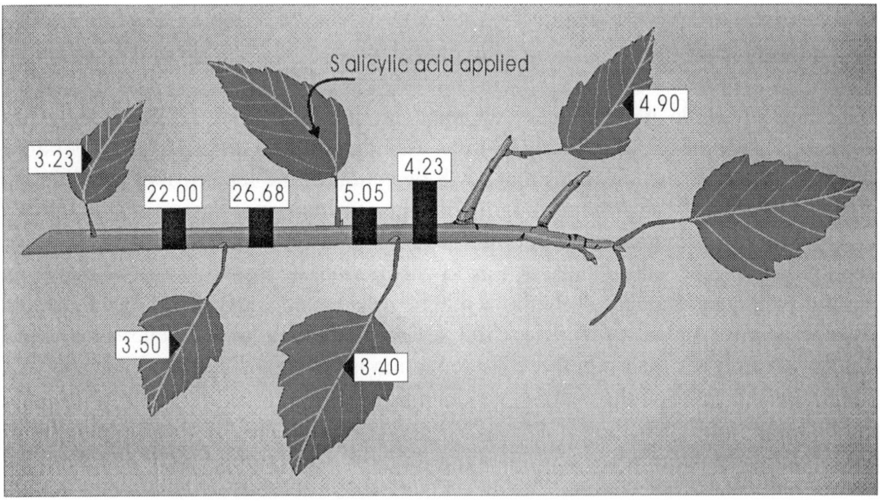

Figure 1. The activity of labelled salicylic acid in fCi/mg dry weight. The natural abundance value (unlabelled vine) was 2.75fCi/mg dry weight.

Fig. 1 shows the activity in the leaf and cane segments three days after a nearby leaf of a kiwifruit vine was treated with ^{14}C-labelled SA. There was an approximate doubling in the activity over natural abundance levels throughout the tissues studied except in the two cane segments in the direction of the main trunk. These contained about 10 times more activity. Results were similar 7 and 14 days after treatment. The data indicate that SA and its metabolites were being transported up the cane towards the main trunk and were not accumulating significantly in any leaf tissue. Separate measurements (not shown) confirmed the induction of SAR in the leaf tissue. Thus, it appears that SAR is induced in the leaf tissue of kiwifruit vines without SA accumulation, and SA is not likely to be the main elicitor of SAR.

4 Acknowledgements

We thank staff of the School of Biological Sciences, University of North Wales, Bangor and of HortResearch, New Zealand, for permission to use unpublished data, and staff of our Institute for assistance with AMS measurements.

References

1. Roberts P.B. and Kim K., Carbon-14 labelling studies at ultra-high sensitivity in agricultural and horticultural research. *this volume* (1999) pp. 305–307.

MAPPING OF PHARMACEUTICAL DEGRADATION PRODUCTS IN TABLETS USING RADIOTRACERS - A STUDY WITH ^{14}C-LABELLED LEVORMELOXIFENE

CHRISTIAN FOGED AND JACOB S. VALSBORG

Novo Nordisk A/S, Isotope Chemistry, Novo Nordisk Park, DK-2760 Maaloev, Denmark

HENRIK OLSEN AND PETER G. NIELSEN

Novo Nordisk A/S, Structural Chemistry, Novo Nordisk Park, DK-2760 Maaloev, Denmark

STELLA R.RASMUSSEN

Novo Nordisk A/S, Product Development, Novo Nordisk Park, DK-2760 Maaloev, Denmark

1 Introduction

Radiolabelling provides a very sensitive and specific method of locating and measuring particular compounds. It has thus become a preferred method for mapping the metabolic pathways of compounds in biological systems. Previously we have shown that radiolabelled material can be used to map the degradation products of the new drug Gabitril® in solid dosage form[1]. In the present study we apply this methodology on levormeloxifene, a new partial oestrogen receptor agonist. We show that radio-mapping can used as a supplement to conventional analytical methods to investigate the degradation pathways and mass balance of levormeloxifene in tablets. We applied stressed conditions (oxygen atmosphere, high pressure and temperature) to mimic the degradation processes that take place under standard storage conditions of the tablets.

2 Result and Discussion

The ^{14}C-labelled version of levormeloxifene ((-)-1-(2-{4-[(3R,4R)-7-methoxy-2,2-dimethyl-3-phenyl-4-2-[^{14}C]chromanyl]phenoxy}ehyl)-pyrrolidine, hydrogen fumarate) (figure 1) was synthesised essentially following the procedure described in the literature for the synthesis of the racemic mixture [2] followed by a chiral purification step to achieve the right stereo isomer.

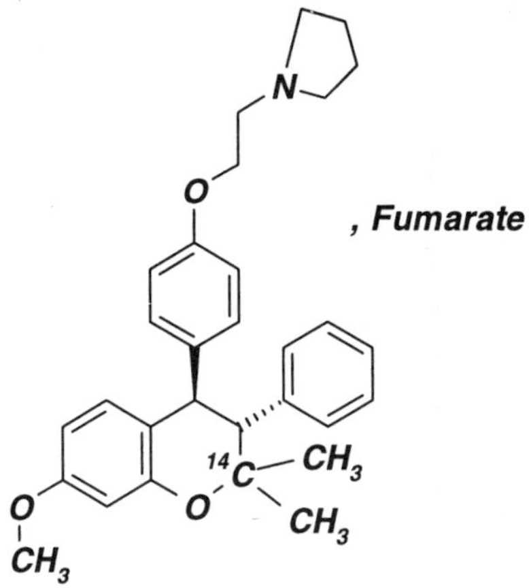

Figure 1. Structure of the ^{14}C-labelled levormeloxifene ((-)-1-(2-{4-[(3R,4R)-7-methoxy-2,2-dimethyl-3-phenyl-4-2-[^{14}C]chromanyl]phenoxy}ehyl)pyrrolidine, hydrogen fumarate)

The radioligand was diluted with non-radioactive levormeloxifene to achieve bulk material with a suitable specific radioactivity. This was done by dissolving the compounds in ethanol, adding fumaric acid followed by precipitation of the fumarate salt of ^{14}C-labelled levormeloxifene. The dried bulk material was then pulverised in a mortar. The radiochemical purity was >99% determined by radio-HPLC and the specific radioactivity was 25 kBq/mg, determined by scintillation counting. Radioactive tablets were prepared and exposed to stressed conditions in a Mini Bench Top Reactor.

A pilot study using non-radiolabelled tablets and HPLC-UV detection showed that super enhanced condition (40 hours at 80 °C and 12 bar of O_2) could mimic the degradation processes in the tablet that take place under standard storage conditions. Radioactivity counting was applied to the stressed radiolabelled tablets to examine whether all degradation products were extracted from the tablets using the extraction procedure developed for analysis of tablets. Sulfuric acid (0.1 M) was added followed by extraction with methanol. By filtration of the insoluble residue it was shown that less than 0.4% of the radioactivity was left in the solid residue after extraction of the degraded tablets. Comparisons between the radioactivity in the supernatant and the start tablets showed that no volatile degradation products escaping detection were produced.

Since it is very important to determine the real content of the degradation products, knowledge about the UV response factors must be achieved when using UV-HPLC as the purity method. Normally this requires available reference compounds. In our study the UV response factors could be examined using RP-HPLC with subsequent fractionation by comparing the UV response with the molar content obtained by scintillation counting. This is illustrated in figure 2. From these data the response factors of all major degradation products could be calculated.

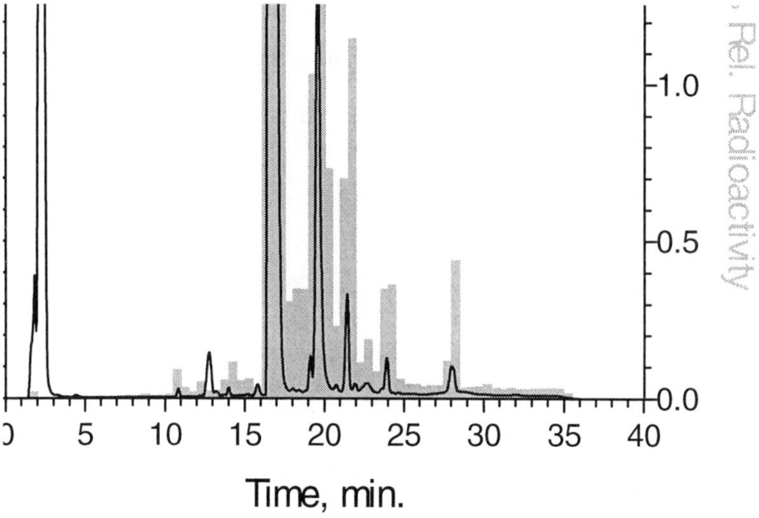

Figure 2. Superimposed UV/radioactivity profiles of degraded (40 hours at 80 °C and 12 bar of O_2) [^{14}C]levormeloxifene in tablets of 1.25 mg strength. Levormeloxifene has a retention time of 17 min.

We also used the radioactive tablets to check the UV-HPLC assay method developed to quantify the content of Levormeloxifene in the start tablets. In order to avoid the commen problem of matrix interference on the scintillation counting (quenching) we applied standard addition for the assay determinations. Thus extracts of placebo and tablets containing ^{14}C-labelled levormeloxifene were spiked with different amounts ^{14}C-labelled levormeloxifene and subjected to liquid scintillation counting. The results are shown in Figure 3. The negative intercept to the x-axis corresponds to the amount of ^{14}C-labelled levormeloxifene in the test sample. This potency was in excellent agreement with the values determined by a conventional UV based HPLC method.

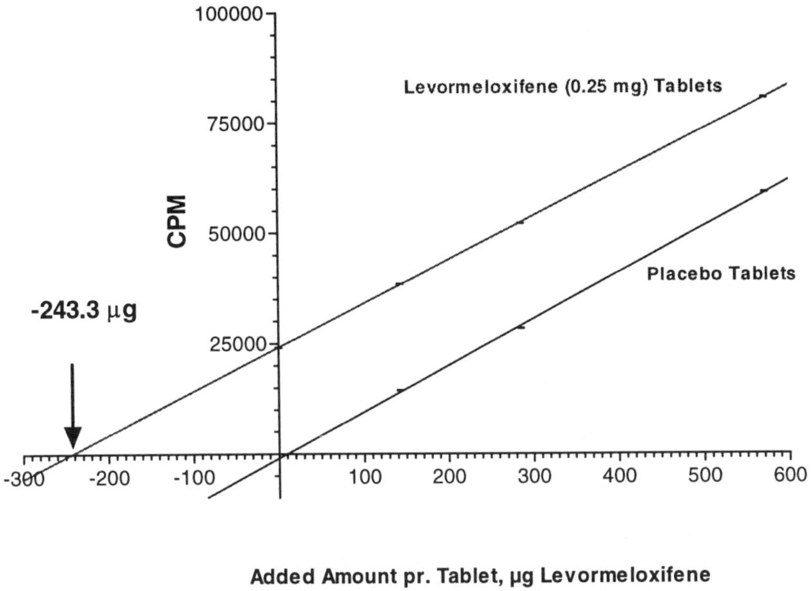

Figure 3. Plots of added amount of levormeloxifene to placebo and 0.25 mg tablets versus CPM (corrected for background radiation).

3 Conclusions

The present study demonstrate that radiolabelled tracers provides a valuable tool to determine the assay values and the fate of the degradation products in solid pharmaceutical products.

References

1. Olsen H., Foged C., Hansen L. and Nielsen P. G. *Mapping of Pharmaceutical Degradants of Gabitril Using Radiotracers*, Anal.Chem. 1996, **68**, 4076-4079.
2. Kole P.L., Ray S. *Synthesis of trans-2,2-Dimethyl-3-phenyl-4-(p-β-pyrrolidinoethoxy)phenyl-7-methoxy-[2-^{14}Cchroman,* J. Labelled. Comp. Radiopharm.. 1978, **16**, 373-375.

SYNTHESIS AND STABILITY OF SUBSTANCE P ANTAGONISTS [^{14}C]CJ-11,974-01 AND [^{14}C]CJ-11,972-01

K. S. ZANDI[*], S. A. MILLER[*], K. E. MCCARTHY[*], W. W. MASSEFSKI[#], A. KAMEL[&]
[*]Radiochemical Synthesis, [#]NMR Spectroscopy, [&]Drug Metabolism Development
Pfizer Central Research, Eastern Point Road, Groton, CT 06340

E-mail: Kathleen_S_Zandi@groton.pfizer.com

CJ-11,974-01 and CJ-11,972-01 are structurally related substance P antagonists currently in development (-01 indicates the HCl salt). The synthesis of radiolabelled analogs was completed to aid in full ADME characterization. A straightforward route to both compounds was developed via directed lithiation/metal halogen exchange and carbonation. Conversion to the benzylic amine was accomplished by one of two methods. In the case of [^{14}C]CJ-11,974-01, the carboxylic acid chloride was treated with a chiral amine followed by amide reduction, and for [^{14}C]CJ-11,972-01, conversion to the aldehyde was followed by reductive amination. While both [^{14}C]CJ-11,974-01 and [^{14}C]CJ-11,972-01 are quite stable in solution, when stored as a solid, [^{14}C]CJ-11,974-01 degrades to one major degradation product over a relatively short time period. The carbon-14 labelled degradation product was isolated from low specific activity material and identified by HPLC/MS/MS and NMR to be an isopropyl peroxide. Studies were performed to identify the factors responsible for the oxidative degradation of [^{14}C]CJ-11,974-01, which included salt form, storage conditions and salt formation solvent. Of all the variables studied over a three week period, only a change in the salt form prevented this oxidative degradation.

1 Introduction

Structurally related CJ-11,974 and CJ-11,972 are substance P antagonists currently in development. The synthesis of radiolabelled analogs was completed to aid in full ADME characterization. Both compounds could be prepared from CP-102,706 and the corresponding carbon-14 labelled benzoic acid.

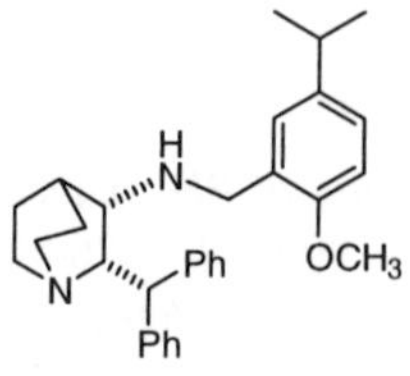

CJ-11,974

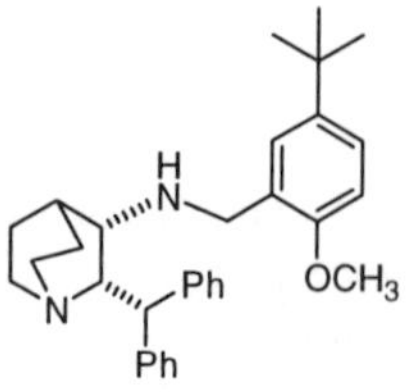

CJ-11,972

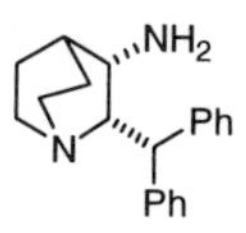

CP-102,706

2 Experimental

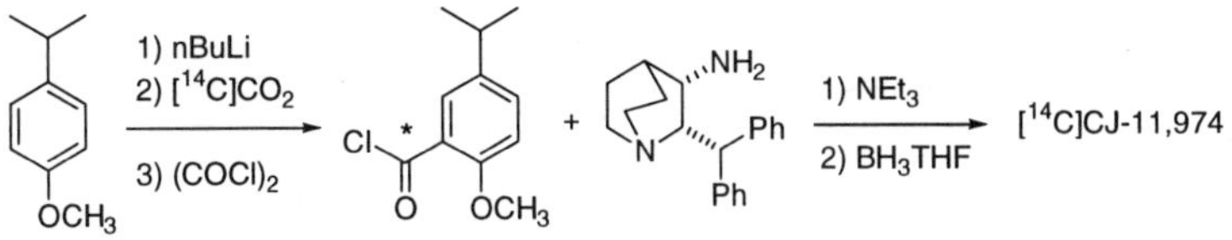

Carbonation of the readily available anisole, treatment of the acid choride with CP-102,706 and borane reduction gave [^{14}C]CJ-11,974 in 53% yield based on [^{14}C]CO_2.

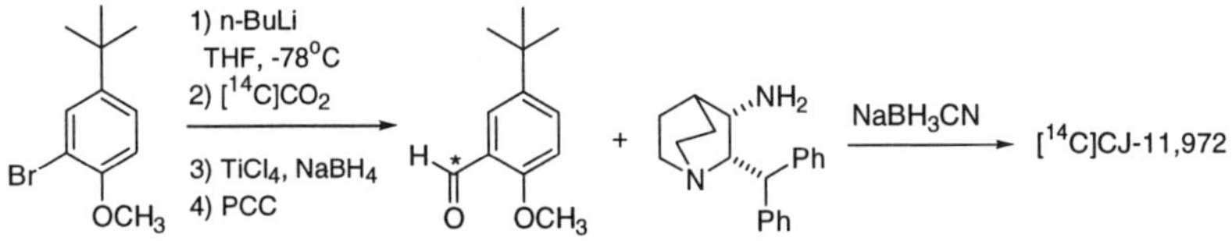

Similarly, carbonation of the readily available bromoanisole followed by two-step conversion to the aldehyde and reductive amination with CP-102,706 gave [^{14}C]CJ-11,972 in 38% yield based on [^{14}C]CO_2.

3 Results and Discussion

[^{14}C]CJ-11,974-01 is stable in solution but degrades to one major degradant when stored as a solid. Solid state degradation of [^{14}C]CJ-11,972-01 is not observed.

Table 1. Percent [^{14}C]CJ-11,974-01 degradant as a function of storage conditions

Storage	Temp	1 week	3 weeks
ethanol	5°C	<0.020	<0.020
water	5°C	<0.020	<0.020
solid	5°C	0.157	0.455

HPLC/MS/MS analysis suggested the degradant product is the isopropyl hydroperoxide. The degradant was isolated from low specific activity [^{14}C]CJ-11,974-01 (2.2 g, 0.8 mCi, 0.2 mCi/mmol) containing 3% of the degradant by successive reverse phase (ACN/0.05M NH_4OAc) and normal phase (2% methanol/chloroform/0.1% NH_4OH, then 25% acetone/hexanes) chromatographies. The structure of the purified degradant was confirmed to be the isopropyl hydroperoxide by NMR spectroscopy.

Several stability studies were carried out in an effort to identify controlling factors in this oxidative degradation. [^{14}C]CJ-11,974-01 (~25 mg, 0.4 mCi/mg) was stored under different conditions. The purity of each sample was evaluated weekly by

HPLC analysis. Storage of solid [^{14}C]CJ-11,974-01 under inert atmosphere or at lower temperatures does not significantly slow the degradation. The presence of BHT in the tetrahydrofuran used in this salt formation step slightly slows degradation. The citrate salt was prepared (citric acid, isopropanol) and was shown to be resistant to oxidative degradation in the solid state.

4 Conclusion

The syntheses of structurally related Substance P antagonists [^{14}C]CJ-11,974-01 and [^{14}C]CJ-11,972-01 have been developed. The route involves either directed lithiation or metal halogen exchange of an appropriately substituted anisole derivative, followed by carbonation with [^{14}C]CO_2. [^{14}C]CJ-11,974-01 was prepared by treatment of the benzoic acid chloride with chiral amine CP-102,706, followed by amide reduction. [^{14}C]CJ-11,972-01 was prepared by conversion of the benzoic acid to the benzaldehye, followed by reductive amination with CP-102,076.

[^{14}C]CJ-11,974-01 is quite stable in solution, but degrades to one major degradation product when stored as a solid. HPLC/MS/MS analysis indicated that the degradant product was the isopropyl peroxide derivative. This degradant was isolated from a sample of low specific activity [^{14}C]CJ-11,974-01 by successive reverse phase and normal phase chromatographies. The purified degradant was confirmed to be the isopropyl hydroperoxide by NMR spectroscopy. Attempts were made to identify the factors responsible for the oxidative degradation of [^{14}C]CJ-11,974-01, which included salt form, storage conditions and salt formation solvent. Of all variables studied, only a change in the salt form prevented this oxidative degradation.

COMPARISON BETWEEN BLOOD PLASMA MODELS (ECCLES) FOR ^{153}SM AND ^{166}HO COMPLEXED TO DIPHOSPHONATE LIGANDS (POSSIBLE RADIOTHERAPEUTIC BONE AGENTS) AND THEIR *IN VIVO* BEHAVIOUR AS STUDIED IN NON HUMAN PRIMATES.

JR ZEEVAART, NV JARVIS, WKA LOUW, JM WAGENER.

Nuclear Technology, Atomic Energy Corporation of South Africa Ltd, PO Box 582, Pretoria, 0001, South Africa.

IC DORMEHL, RJ MILNER.

University of Pretoria, South Africa.

^{153}Sm complexed with the bone seeking ligand ethylene-diaminetetramethylene phosphonate (EDTMP) is effective in pain palliation therapy for patients suffering from metastatic bone cancer. Using the ECCLES code, blood plasma models for this ligand with Sm^{III} and Ho^{III} explained the considerably lesser bone uptake of ^{166}HoEDTMP as compared with ^{153}SmEDTMP. Consequently our search for a more effective bone therapeutic radiopharmaceutical involving ^{166}Ho continues. This isotope may have advantages because of its more energetic β particle. In this quest a further three more phosphonate ligands have been synthesised viz. APD (1-hydroxy-3-amino-propylidene-diphosphonic acid), APDDMP (APD-dimethylenephosphonic acid) and PEI-MP (polyethyleniminomethyl phosphonic acid). Formation constants for the complexation of Ca^{II}, Mg^{II}, Zn^{II}, Sm^{III} and Ho^{III} with all of these ligands were calculated using potentiometry, polarography and linear free energy correlations. For PEI-MP (molar mass fraction >10 kDa) the functionalised monomer units were considered to represent a ligand. The program ECCLES used the calculated constants to predict the speciation of Ho^{III} and Sm^{III} with these four ligands in blood plasma.

From these models predictions concerning bone-uptake and side-effects were made. The resulting *in vivo* behaviour found in the animal models confirmed the predictions made by ECCLES and showed that ^{166}HoEDTMP, ^{166}HoAPD and ^{166}HoAPDDMP are unsuitable as therapeutic bone agents. The ECCLES predictions for both ^{153}Sm- and ^{166}Ho-PEI-MP *in vivo* are presented. Valuable information on how a future radiopharmaceutical should be designed have been gained.

1 Introduction

Gratifying results have been obtained locally using ^{153}Sm-EDTMP in pain palliation for metastatic bone cancer [1]. Baboon tests with ^{166}Ho-EDTMP confirmed the blood plasma models constructed for Ho^{III} and EDTMP [2] using the ECCLES code. The latter gives insight into the *in vivo* behaviour of radiopharmaceuticals. The ECCLES model showed that the bone-seeking ligand, EDTMP, competes poorly with citrate in blood plasma for Ho^{III} resulting in lower bone uptake of ^{166}Ho. In the search for a cure for metastatic bone cancer, ^{166}Ho may be useful as it is a more energetic ß-emitter than ^{153}Sm. Therefore, a study of other bone-seeking ligands, HEDP, MDP [3] and APD [4] as well as ligands designed for Ho^{III}, APDDMP and the polymer PEI-MP have been undertaken. The last three ligands are discussed herein. For PEI-MP (molar mass fraction between 20 and 30 kDa) the functionalised monomer units were considered to represent a ligand. [5]

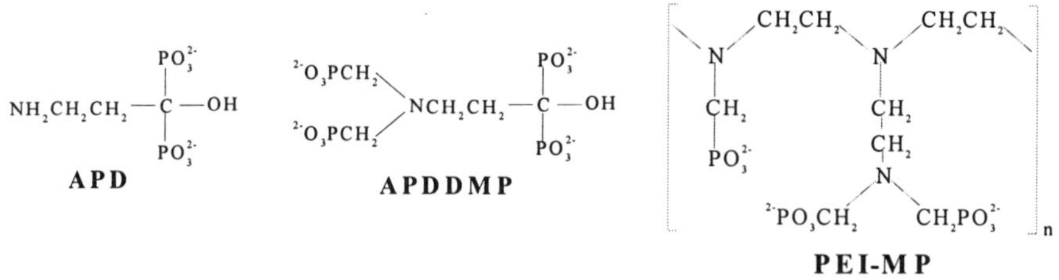

Formation constants for the complexation of Sm^{III}, Ho^{III} and other important blood plasma metal ions (Ca^{II}, Mg^{II} and Zn^{II}) with all of these ligands were calculated using potentiometry, polarography and linear free energy correlations. The program ECCLES used these constants to predict the speciation of Ho^{III} and Sm^{III} with these three ligands in blood plasma. From these results found bone-uptake as well as side-effects in animal studies could be predicted and explained in retrospect.

2 Results and discussion

The formation constants calculated for the three ligands is published separately. [4, 5] An interesting aspect which arises is the unusual species M_2L. Evidence for this species is well documented [4]. The APD constants are one order of magnitude higher than MDP and HEDP [3], indicating amino-group participation in complexation. The same result was achieved for EDTMP.[2]

The complexation of APDDMP involves more conventional species while PEI-MP favours binuclear species with all the metal ions.

Results of the ECCLES modelling for Ho^{III} including APD show that 58% of the Ho^{III} remains bound to APD, almost all of it in the species MLH. This is a better retention of Ho^{III} than EDTMP although it is weaker than EDTMP with Sm^{III}.[4] This is confirmed by the baboon test; The γ-ray image of the baboon three hours after injection with ^{166}Ho-APD, indicates good bone uptake but high liver uptake. This is due to the MLH species which is neutral (not charged) causing colloid formation (or precipitation) and consequently high liver uptake. The *in vivo* selectivity of APD for Ca^{II} explaining its reported [7] usefulness in bone reconstruction. As it can be shown that the speciation of Sm^{III}-MDP and Sm^{III}-HEDP

[3] is similar to Ho^{III}-APD the predominant neutral species MLH explains the high liver uptake of these combinations found elsewhere [8].

To overcome this liver uptake, charged groups were added to APD resulting in a ligand like APDDMP. ECCLES calculated that only 11.4% of Ho^{III} will remain bound to APDDMP while the corresponding figure for Sm^{III} is 58.3%. The predominant species in both cases are MLH and ML - both charged. The γ-ray image of the baboon four hours after injection with ^{166}Ho-APDDMP shows no liver uptake but little bone localisation. For ^{153}Sm-APDDMP (Photo 3) however good bone uptake is visible although not comparable to ^{153}Sm-EDTMP [2]. The reason for the lesser stability of these metal ions with APDDMP is the latter's affinity for Ca^{II}. Thus in future the methylene phosphonate arms on APD could possible be replaced by methylene amides to increase the ligand's ability to complex Ho^{III}.

This Ca^{II} affinity is again visible in PEI-MP. According to ECCLES, no Ho^{III} or Sm^{III} will remain bound to the ligand in blood plasma. The idea behind PEI-MP was to utilise the increased permeability of cancer tissue to introduce a polymer of a particular size, which would preferably enter the damaged tissue of a osteo-sarcoma and subsequently increase the tumour to healthy bone uptake ratio. (20 to 30 kDA proved to be the optimum for ^{153}Sm-PEI-MP). The comparative uptake of ^{99m}Tc-MDP (a radiopharmaceutical used to image secondary bone metastasis) with ^{153}Sm-PEI-MP in the same osteosarcoma containing dog. Although ^{153}Sm-PEI-MP shows poor tumour uptake, no uptake is visible in healthy bone as is the case with ^{99m}Tc-MDP.

3 References

1. A. S. Alberts *et al.* Sixth Biennial Congress of the SA Society for Nuclear Medicine, Midrand, South Africa, September, 1994.
2. N. V. Jarvis, *et. al., J. Chem. Soc. Dalton Trans.*, 1995, p. 1411
3. J. R. Zeevaart, *et. al., S.Afr.J.Chem*, 1977, **50**(4), p. 189
4. J. R. Zeevaart *et al., J. Inorg. Biochem.*, 1999, **73**, p. 265
5. J. R. Zeevaart, N. V. Jarvis and J. M. Wagener, *to be published.*
6. N. V. Jarvis and J. M. Wagener, *Talanta*, 1995, **42**, p. 219 and refs therein.
7. J. H. Lin, *Bone*, 1996, **18**, no. 2, p. 75
8. J. Simon et. al., *Patent 899734*, Dow Chemical Company.

4 Acknowledgements

The authors thank the Atomic Energy Corporation for permission to publish this work.

Session K

Applications of Isotopes in Environmental Studies

ISOTOPE STUDIES ON COASTAL ENVIRONMENT

S.M. RAO

Isotope Group, Bhabha Atomic Research Centre, Mumbai 400 085, India
E-mail: smrao@magnum.barc.ernet.in

Isotopes have been playing an ever increasing role in studies on coastal environment in India. These primarily deal with sea water intrusion, either active or in paleotimes and sediment transport on seabed. Environmental isotopes have shown that whereas there is active seawater intrusion into the coastal aquifer north of the city of Chennai, the coastal aquifer salinity in Orissa and West Bengal is largely attributable to Flandrian transgression of the sea during the Holocene (~8000 BP) and also to an earlier transgression in the late Pleistocene (~ 24,000 BP). The isotopes also helped in identifying a potential freshwater aquifer sandwiched between the two saline water bodies. Dynamics of coastal sediments on the Indian coastline are very complex and controlled by the monsoons. Isotope tracers proved to be unique tools for studying the bed-load dynamics to plan dredging operations in ports. The tracer techniques have helped in the choice of sites for disposal of dredged sediments in a number of ports. They have also shown the reversal of direction of littoral drift on the east-coast during the south-west and north-east monsoons. The Hoogly river which is the inland approach channel for the Calcutta Port presents many challenges for navigation. Isotope tracers have been extensively used to investigate many such problems.

1 Introduction

Coastal regions are more thickly populated than any other geographical locations for a variety of reasons; land fertility, fishing and maritime trade. The population in the coastal regions is expected to continue to rise. With the advent of sea level rise due to global warming, increasingly greater attention is being paid to coastal environment.

Sea water intrusion into coastal aquifers is an area of large interest and this has been extensively studied using natural isotope tracers at several places along the east coast of India[1] .

Navigational channels in the ports need to be dredged often to keep them open for shipping around the year. Dredging is an expensive activity and choice of a dumping site for dredged silt dictates the economy of dredging. Such choice depends on a thorough understanding of bed-load dynamics, particularly during the monsoons. Tracer methods have been well developed to determine the direction and quantum of bed-load transport in any given location.

2 Coastal Aquifer Salinity

Analyses of environmental isotopes present in groundwater and their interpretation along with geochemical and hydrogeological data lead to valuable information such as the existence and location of natural recharge to the aquifer(s), source of salinity, origin of groundwater and residence time or "age" of the water.

Coastal aquifer salinity is influenced by the marine environment and the degree of influence is dependent on local geohydrology, level of exploitation, climate changes in the geological past and deposition history of coastal sediments.

3 Case Studies in Coastal areas

3.1 Minjur Aquifer, Chennai

There has been increasing salinisation of groundwater in the Minjur aquifer, north of city of Chennai on the east coast of India. The aquifer is exploited for city water supply. The obvious mechanism of salinisation is intrusion of sea water due to pumping activity. This simple explanation was suspect because of the presence of pockets of hypersaline groundwater close to the coast. The possibility of entrapped marine connate water could not be ruled out.

A large number of water samples were collected from piezometers, tubewells and surface sources and analysed for environmental isotopic (D, ^{18}O, ^{34}S and ^{3}H) as well as major and minor ionic species. The isotopic composition and hydrochemistry of the hypersaline groundwaters indicate that they are not marine connate waters, but probably owe their salinity to highly saline salt-pan water infiltrating to the deeper zones [2].

3.2 Southern Mahanadi Delta, Orissa

The deltaic land is represented by vast plains of fluvial, lacustrine and marine origin intersected by less prominent rivers. The subsurface geological studies (Figure 1) show that the basement comprising rocks of Archaen age deepens southward. The depth varies from about 20 m near Jagadalpur to more than 600 m near Puri, a distance of 30 km. The dark gray sediment horizons were deposited in a near-shore shallow, marine basin. An unfossiliferous yellow-brown sediment horizon is sandwiched between two dark gray sediment horizons indicating a prominent change in the climate during the evolution of the delta.

A large number of piezometers, drilled in the area, reveal the presence of multiaquifer system down to the depth of about 300 m. Fresh, brackish and saline

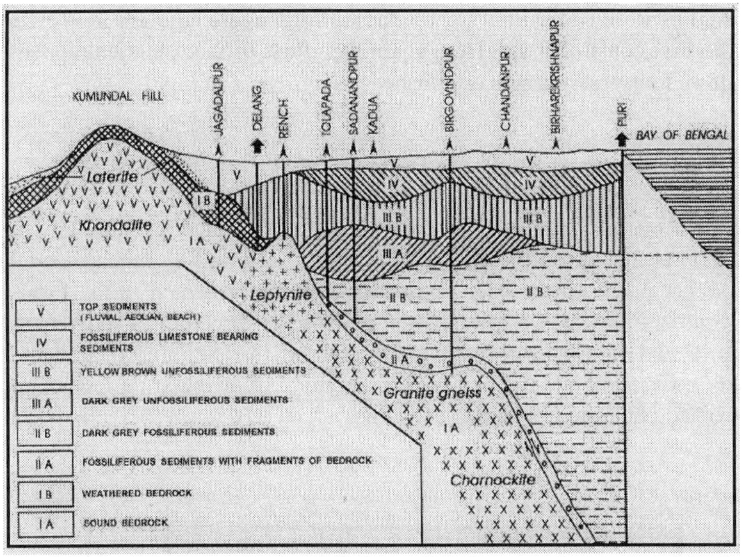

Figure 1 Schematic representation of the subsurface geology along Delang-Puri sector

groundwaters occur in a rather complicated fashion. Both lateral and depthwise variation in salinity is noticed. The shallow aquifers (depth <50m) occur under phreatic conditions and deeper aquifers change from leaky confined to artesian flowing type with increasing depth. The flow of groundwater is towards southeast in Delang area and towards south in Puri area, all towards the sea.

A number of surface and groundwater samples were analysed for environmental isotopes (D, ^{18}O, ^{34}S, ^{3}H and ^{14}C) as well as minor and trace ion species. It was concluded from the isotope geochemical study [3] that

a) Shallow groundwater is relatively young and is recharged through flood Plains and outcrops of basement rocks.
b) Saline groundwaters in the intermediate aquifer (depth – 50 to 100 m) are largely due to Flandrian transgression during the Holocene (based on ^{14}C ages of groundwater and selected fossils).
c) Deep, fresh and young groundwaters form a well developed aquifer which receives recharge through basement rocks. This could form the main source of groundwater for exploitation.
d) The deep, saline groundwaters are marine waters entrapped during Late Pleistocene interglacial stage about 24,000 BP.
e) Very deep aquifer occurring in weathered basement and basement rocks may also form a potential aquifer for exploitation. The water here is very old.

Similar studies were carried out [4] on coastal Midnapore aquifers in West Bengal, also on the east coast of India. Here again the effect of seawaters entrapped during the Flandrian transgression was established.

4 Bed-load Transport Studies in Ports

The field method involves deposition of the tracer material (normally Sc-46 in glass matrix) at a selected location on the seabed followed by repeated tracking of the radiotracer patch on the bed. The data is usually presented in the form of iso-activity contours and such a presentation reveals the direction of movement of the tracer. A 'Count rate balance method [5] is used to determine quantitatively the velocity of movement and the bed-load discharge. Following are a few examples of studies carried out along the Indian coastline.

4.1 Karwar

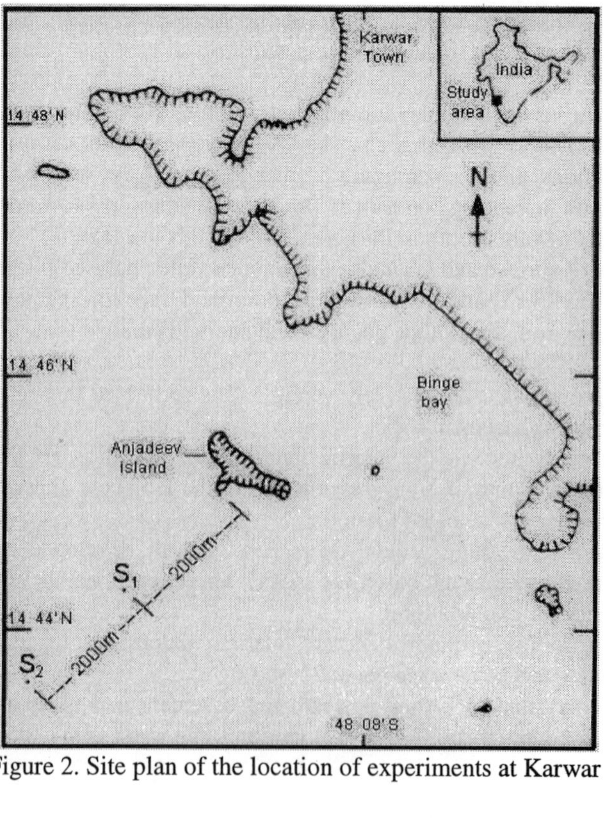

Figure 2. Site plan of the location of experiments at Karwar

As part of the development of Karwar port, two sets of radiotracer experiments were carried out [6]. The Engineers India Ltd. had selected two possible locations (Figure 2) for disposal of silt dredged from the proposed turning bay (Binge Bay) based on their model studies; one at a distance of 2 km south west of Anjadeev island and another at further distance of 2 km in the same direction.

The results (iso-activity contours) indicated that if the dredged material is dumped at the first point (i.e. 2 km from Anjadeev), it will find its way back to the proposed navigational channel and the Binge Bay. The second point (i.e. 4 km from Anjadeev) is more suitable as the sediment transport at this point is away from the coast.

4.2 Chennai Port

The port has an enclosed harbour and extensive dredging of 9 km long navigational channel is needed to maintain a depth of 22m. The littoral transport is affected by both the north-east and south-west monsoons and an optimum choice of the dumping site for the dredged material is vital for operational economy. There is no better way of making such a choice under the complex climatic conditions than carrying out a radiotracer study of bed-load dynamics.

A radiotracer study [7] was carried out some years back by making a Sc-46 glass tracer injection at a point selected 6 km north of navigational channel. The selection was made on a model based investigation in a hydraulic research station. Monitoring the tracer patch during the north-east monsoon and quantitative interpretation of tracer data indicated, that the channel is not resilted by dredged material and the direction of transport reverses from southward to northward in March.

Similar studies have been carried out in many ports on the Indian coastline. Special mention should be made of the studies carried out in the Hoogly river [8] which is the navigational channel from the Bay of Bengal to the Calcutta port. Bed-load dynamics of this 100 km long channel are very complex and isotope tracer studies were of special importance in planning and implementing dredging programs.

5 Conclusions

The foregoing shows the importance of environmental isotopes in studies on interaction between sea water and coastal aquifers. The information is of great value in planning for exploitation of groundwater.

The role of radioactive tracers in planning dredging operations in ports is well demonstrated in many ports on the Indian coastline.

References

1. Rao S.M., "Isotope Hydrogeology Studies on Coastal Aquifers", Applications of Radioisotopes and Radiation in Industrial Development (ed: D.D. Sood et al), NAARRI, Mumbai (1998) p 301-306
2. Rao S.M. et al, "Isotope Studies on Seawater Intrusion and Interrelations Between Water Bodies - Some Field Examples" , Proc. IAEA Symp. On Isotope Techniques in Water Resources Development, Vienna (1987) pp 403-425
3. Kulkarni K.M. et al, "Drinking Water Problem in Coastal Orissa, India Identification of past transgressions of seawater by isotope investigation" Isotope Techniques in the Study of Environmental Change, IAEA, Vienna, (1998) p 293-306
4. Shivanna K. et al, "Isotope Geochemical Evidence of Past Seawater Salinity in Midnapore Groundwater", Proc. Symp. On Applications of Isotope Techniques in Studying Past and Current Environmental Changes in the Hydrosphere and the Atmosphere, IAEA, Vienna (1993) p 199 – 212
5. Krishnamurthy K., Rao S.M., "A Theory for the Quantitative Estimation of Bedload Transport using Radioisotopes", Isotopenpraxis, B. Jahrgang, Heft H-12/1972 pp 455-597
6. Yelgaonkar V.N. et al, "Investigation of Bedload Transport off Karwar Coast, Southwest of India", Indian Journal of Marine Sciences, Vol. 26, March 1997, pp 91-94
7. Pant H.J. et al, "Sediment Transport Study at Madras Port Using Radiotracer Technique", Journal of Applied Hydrology, Vol. II, No. 4, 1989, pp 47-53
8. Pendharkar A.S. et al, "Radiotracer Applications in Bedload Transport – Case Studies at Calcutta Port", Isotopes and Radiation Technology in Industry [ICARID-94] (Ed. S.M. Rao, K.M. Kulkarni), NAARRI, Bombay (1994), pp 287-299

ASSESSMENT OF EROSION AND ACCRETION IN CATCHMENT AREAS BASED ON ^{210}Pb AND ^{137}Cs CONTENTS IN SOIL AND SEDIMENT

P. S. HAI, P. N. SON, N.N. DIEN

Dalat Nuclear Research Institute, Nguyen Tu Luc Str., Dalat, Vietnam
E-mail: npdept@hcm.vnn.vn

V. H. TAN, P. D. HIEN

Vietnam Atomic Energy Agency, 59 Ly Thuong Kiet Str., Hanoi, Vietnam

Measurements of ^{210}Pb and ^{137}Cs in soil and sediment using alpha and gamma spectrometry methods were done to assess the extent of soil erosion and sediment accretion for three water reservoirs in Lamdong province, South Vietnam. ^{137}Cs depth profiles in soil were measured for 11 virgin sites across the three watersheds providing an inventory (± standard deviation) of 330 (± 67) Bq m^{-2}. The radionuclide was found mainly in the 10 cm top layer. ^{137}Cs and unsupported ^{210}Pb were measured in 15 sediment cores. For the reservoir constructed in 1961 to supply water for the 200 MW hydropower plant Danhim, the two indicators gave the same sedimentation rates ranging from 0.7 to 1.6 cm.y^{-1}. For the two other reservoirs, Daham and Tuyenlam, which were constructed 12-14 years ago for irrigation purpose, the sedimentation rates at different locations calculated using the ^{210}Pb dating method vary from 0.5 cm y^{-1} to 1.4 cm y^{-1}. The radionuclide ^{137}Cs could not be used for sediment dating in the case of recently constructed reservoirs. However, the comparison of ^{137}Cs in soil and sediments allowed to establish the dominant source of recent sediment deposited in the reservoirs.

1. INTRODUCTION

About 30% of the 330,000 km^2 Vietnam's land area are covered by forests and woodland. The most prosperous forest and woodland area in Vietnam is the Central Highland (11° - 16°N, 107° - 108°E, 150 - 1500 m asl), which accounts for large amounts of the country outputs in natural rubber latex, precious timbers, coffee and the various crops. The Central Highland is home to ethnic aborigines who make living using mainly a slash-and-burn farming practice. During the last two decades, land reclamation has been extensively going on by the people immigrating from other areas of the country under government-supported programmes to establish "new economic zones". As a result, reforestation is not keeping pace with deforestation, and large areas of fertile basalt red soil have become more susceptible to erosion under heavy monsoon rainfalls.

During the last two decades, the nuclear tests-derived radionuclide ^{137}Cs and the naturally occurring isotope ^{210}Pb have successfully been used in studies of soil erosion and sediment accretion processes [1,2]. These techniques have been applied to three watersheds in the southernmost province of the Central Highland, Lamdong (11°20' - 12°20'N; 107°15' - 108°45'E), and the results are reported in this paper. The combined measurements of the two isotopes allowed to obtain more useful information providing insights into the erosion and sedimentation processes.

1 THE STUDY AREAS

The relevant characteristics of the three reservoirs and their watersheds (Fig. 1) are given in Table 1. The high rainfall is associated primarily with the period of time from May to October. The soil is rather rich in humus (3.5%) and clay minerals (30% average). Natural forests cover about 40% (560,000 ha) of the Lamdong land area, but about 15,000 ha are estimated to be vanished every year. Meanwhile, only 10,000 ha new forests were established during the last years.

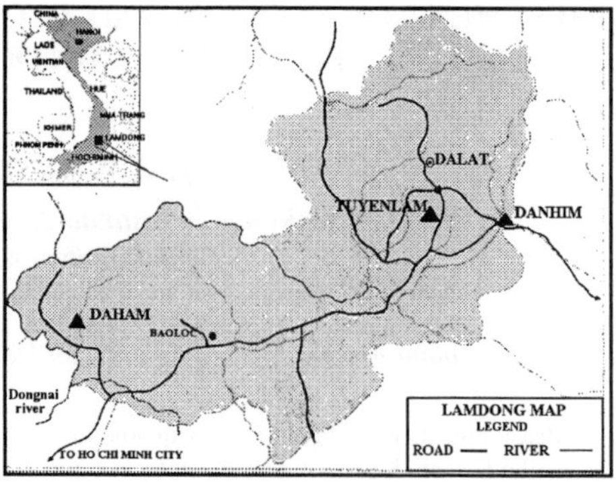

Figure 1. The map of Lamdong and the location of reservoirs investigated.

Table 1. Relevant characteristics of the three reservoirs and their catchments.

Name of reservoir	Time of construc	Original volume (m^3)	Elevation (m)	Catch. area (km^2)	Annual rainfall (mm)	Annual inflow (m^3)	Purpose
Tuyenlam	80-87	10.6×10^6	1377	32.8	1721	27.88×10^6	irrigation, tourism
Daham	86-88	2.8×10^6	140	19	1600	16.15×10^6	irrigation
Danhim	60-62	163×10^6	1042	775	1574	6.59×10^8	electricity

2 METHODS

In order to investigate the behaviour and the depth profiles of ^{137}Cs in soil, soil cores (30 cm deep, 10 cm diameter) were taken from virgin sites across the three catchment areas. The average ^{137}Cs content in soil layers and the mean inventory were determined from 60 cores taken from 11 virgin sites around the three watersheds. The dominant source of sediment deposited in reservoirs was established based on the comparison of ^{137}Cs concentrations in sediment with ^{137}Cs depth profiles in soils.

The sedimentation rates were estimated based on both ^{137}Cs and ^{210}Pb indicators present in sediments, which were collected at the selective sites using a 5 cm diameter corer after water depth surveys had been carried out by echo sounder.

Sediment and soil samples were measured for ^{226}Ra and ^{137}Cs by using a 30% p-type intrinsic germanium detectors. For this analysis, all samples prepared as a fine homogeneous powder were cast using polyester resin in the desired geometry. The content of ^{210}Pb was determined through it's daughter ^{210}Po with the assumption of secular equilibrium. Radionuclide ^{210}Po was analyzed using silicon surface-barrier detectors with 450 mm^2 active area.

3 RESULTS AND DISCUSSION

Characteristics of ^{137}Cs in undisturbed soils

The depth distribution of ^{137}Cs activity in the reference sites showed that the maximum depth of ^{137}Cs penetration is 20 cm for all sites. The majority (> 80%) of ^{137}Cs is concentrated in the top 10 cm of the soil profiles with a peak in activity concentration at about 2-5 cm depth. Below the peak, the ^{137}Cs activity exhibits an approximately exponential decline with depth.

The mean ^{137}Cs inventory (± standard deviation) in the three watersheds calculated from 60 cores, is 330 Bq.m^{-2} (± 67 Bq.m^{-2}), that is much less than the values obtained from areas of higher latitudes in Europe [3], North America [4] and Australia [5]. More detailed studies of ^{137}Cs content in soil cores gave the average ^{137}Cs concentrations in the 0-2 cm, 0-6 cm and 6-30 cm soil layers as in Table 2.

Table 2. Mean ^{137}Cs content in soil layers (in Bq.kg^{-1}).

	Virgin sites (60 cores)	Hillslopes, woodland (36 cores)
0-2 cm	3.7	
0-6 cm	3.4	2.1
6-30 cm	0.5	0.4

Sedimentation rates

The Danhim reservoir, which was constructed in the early 1960's, is very suitable for applying and comparing the two radiometric dating techniques based on ^{137}Cs and ^{210}Pb. Figure 2 shows a sediment chronology for a typical core from Danhim reservoir determined by the CRS ^{210}Pb dating model [2]. This result is supported by the ^{137}Cs data.

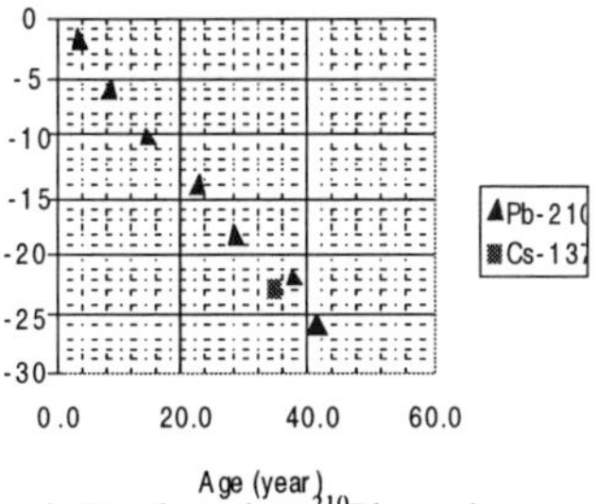

Figure 2. Depth against ^{210}Pb age in a core from Danhim reservoir with the depth dated to 1963 by the ^{137}Cs data

For the two recently constructed reservoirs, Tuyenlam and Daham, the sedimentation rates were estimated by ^{210}Pb technique. The information on sediment deposited in the reservoirs since their construction as well as sediment yields from the reservoir's catchment is given in Table 3.

Table 3. Sedimentation rates and sediment yields for the three reservoirs.

Name of reservoir	Sedimentation rate (cm.y^{-1})	Sediment volume (m^3)	Mass of sediment (tone)	Sediment yields (t.km^{-2}.y^{-1})	Method applied
Danhim	0.7 - 1.6	1,300,000	1,040,000	36	Cs-137, Pb-210
Tuyenlam	0.5 - 0.8	187,000	106,000	308	Pb-210, CRS model
Daham	0.8 - 1.4	133,000	101,000	560	Pb-210, CRS model

The mean ^{137}Cs content in the top 15 cm of the sediment which was deposited since 1975 for Danhim and since the construction for Tuyenlam and Daham is given in Table 4. Because of insignificant ^{137}Cs fallout inputs after 1975, the sediment layer of the top 15 cm would contain ^{137}Cs associated primarily with the eroded source material. Moreover, the abundance of fine particles < 20μm, which contain the majority of ^{137}Cs in soil and sediment [6], is not much different in soil (28%) and sediment. Thus, the low value of ^{137}Cs in the sediment deposited in Danhim and the right branch of Tuyenlam compared to that in the surface soil suggested that erosion of the stream channels and rills is the main process contributing to the sediment deposited in the reservoirs. Meanwhile, the sediment deposited in Daham and the left river branch of Tuyenlam has an average ^{137}Cs concentration as high as that in the surface soil. This suggests that, the dominant source of sediment in Daham reservoir and the left branch of Tuyenlam is surface materials from the slopes of the basin, rather than materials eroded from the stream channels.

Table 4. The mean ^{137}Cs content and the abundance of fine particles < 20μm in sediment

	Danhim	Tuyenlam		Daham
		Left branch	Right branch	
Inventory ($Bq.m^{-2}$)	400 - 600	400 - 500	100 - 200	600 - 1400
0-15 cm layer	1.2 $Bq.kg^{-1}$	2.6 $Bq.kg^{-1}$	1.1 $Bq.kg^{-1}$	3 - 9 $Bq.kg^{-1}$
< 20μm fraction	20 %			32 %

4 CONCLUSION

In the absence of long-term fluvial sediment monitoring programmes in Vietnam, the radiometric techniques based on ^{137}Cs and ^{210}Pb have provided an effective tool for sediment dating and for studying the erosion-sedimentation relationships in catchment areas. Particularly, the combined use of the two indicators clearly enhanced the advantage of the radiometric methods.

Within an area of about 10,000 km^2 of the Lamdong province, the extent and mechanism of erosion in the three watersheds are quite different. Even within one watershed, the differences in the magnitude and mechanism of erosion processes were revealed between the catchments of the right and the left river branches as in the case of Tuyenlam reservoir.

References

1. Walling D.E., He Q., 1992. *Hydrobiologia* 235/236, 219-230.
2. Appleby, P.G., Oldfield, F., 1978. *CATENA*, Vol. 5, 1 - 8.
3. Owens, P.N. et al. 1996. *Appl. Radiat. Isot.*, Vol. 47, No. 7, pp. 699-707
4. Sutherland, R.A. 1991. *Soil Technology*, Vol. 4, p. 33-50.
5. Loughran, R.J. et al. 1992. *Hydrol. Proc. 6*, 145-158
6. Hai P.S., Hien P.D., et al., *IAEA-SM-354/200P*, Monaco, 5-9 Oct., 1998

ORIGIN AND DYNAMICS OF SOIL ORGANIC MATTER AND VEGETATION CHANGES IN A FOREST-SAVANNA ECOTONE, SOUTHERN BRAZILIAN AMAZON REGION

L.C.R. PESSENDA[1]

[1] ^{14}C Laboratory, Centre for Nuclear Energy in Agriculture, University of São Paulo, SP, Brazil
E-mail: pessenda@cena.usp.br

R. BOULET[2], V. ROSOLEN[3], R. ARAVENA[4], S.E.M. GOUVEIA[1]
[2]IRD-France, Geosciences Institute - University of São Paulo, São Paulo SP, Brazil
[3]Geosciences Institute - University of São Paulo, São Paulo SP, Brazil
[4]Environmental Isotopes Laboratory, University of Waterloo, Waterloo, Ontario, Canada

This paper presents a study concerning the origin and dynamic of soil organic matter and palaeovegetation changes in a transition forest-savanna in the southern Brazilian Amazon region

The study area is located in the southern border of the Amazon Sedimentary Basin, Amazon State. The soils are Podzólico Vermelho Amarelo (Ultisol, USDA) on plateaus and hydromorphic lateritic (Plintic Gleysols) on depressions.

A trench of 3 meters height and length of 90 meters was made in the forest-savanna transition area. This trench includes a depression covered by savanna vegetation.

Charcoal found in the trench provided radiocarbon dates in the range 3.810 and 4.770 years BP. For a peat sample located in the depression the age was 1.650 years BP.

Carbon data show a significant increase of carbon content in the shallow soil (1.9 to 3.1%) from forest ecosystem to sites representing savanna ecosystem. In the depression, the carbon values were high as 10%, indicating an area of organic matter deposition.

^{13}C and ^{14}C data from soil samples indicate that the organic matter is of forest origin and it has been deposited since at least the last 5.000 years in the depression areas located in the savanna ecosystem. The ^{13}C pattern observed in the soil organic matter profiles at the forest and savanna sites indicate a predominance of C_3 vegetation in the early Holocene. About 7.000 to 4.000 years the data show the influence of C_4 plants, indicating forest regression associated with a drier climate than the present. The more recent portion of the ^{13}C records indicates the expansion of the forest, and probably the return to a more humid climate.

1 Introduction

In the southern border of the sedimentary Amazon basin, under humid tropical climate, the upper part of the landscape is constituted by large plateaus, which contain several small topographic depressions. The soils are generally ferralitic soils, with hydromorphic characteristics in the depressions, which are flooded by at least 6-7 months during the rainy season. The present vegetation cover is represented by a patchwork of predominantly tropical forest and discontinuous natural savanna areas [1].

This paper presents a study concerning the origin and dynamic of soil organic matter and vegetation changes occurred during the late Pleistocene and Holocene, in a transition forest-savanna on lateritic soils in the southern Brazilian Amazon region. The research approach was based on evaluation of soil carbon content, determination of $\delta^{13}C$ of soil organic matter and radiocarbon dating of buried charcoal and rich organic horizon samples, collected in distinct soil locations covered by forest and savanna vegetations.

2 Study Site

The study area is located in the southern border of the Amazon Sedimentary Basin, (Figure 1) in the km 79 of BR 319 (8°10'S; 63°51'W), Amazon state. The vegetation cover is represented by natural tropical forest-savanna contact.

A trench of 3 m depth and 90 m length was studied in the vegetation transition, with a positive elevation of 2 m from forest in the plateau to savanna in the depression (Figure 1). The soils are Ferralitic Cambissol (point A to point B), Plintic Gleysols (B to C) and Humic Plintic Gleysols (C to E) [2]. The vegetation cover is tropical forest (A), a transition forest-savanna (B) and from C to E the savanna ecosystem.

3 Material and Methods

The sampling collection from soil surface to 290 cm depth, were made in the location A, B, C, D and E (Figure 1). The soil fraction less than 0.200 mm were used for $\delta^{13}C$ analyses. Charcoal fragments were collected directly from the trench wall, from 5 cm to about 60 cm depth, in distinct vertical positions. In the depression (E), two samples of a rich organic horizon were collected at 0-10 cm and 20-30 cm depth A detailed description of the chemical treatment for soil and charcoal samples can be found in [3].

The ^{14}C analyses on charcoal fragments and rich organic horizon samples were carried out at the Radiocarbon Laboratory, Centro de Energia Nuclear na Agricultura (CENA), following the standard procedure for liquid scintillation counting [4]. The ^{14}C on small charcoal fragments were carried out at the Isotrace laboratory of the University of Toronto, employing the AMS technique.

Radiocarbon data are reported as percent modern carbon (pmc) and radiocarbon ages as years BP.

The $\delta^{13}C$ analysis and total organic carbon on soil samples were carried out at the Environmental Isotope Laboratory, University of Waterloo. $^{13}C/^{12}C$ data are expressed in δ (‰) units relative to the PDB standard.

4 Results and Discussion

4.1 *Chronology*

The radiocarbon dates of charcoal samples range between 3.810 and 4.770 years BP and in the rich organic sediment modern in the 0-10 cm and 1.650 years BP at the 20-30 cm depth [5]. A generalized age-depth profile for soils (humin fraction) in the study area can be represented by an age of 5.000-6.000 years for the 90-100 cm soil horizon and 10.000-12.000 years for the 190-200 cm soil depth [6, 7].

4.2 *Carbon profiles*

The carbon data show the typical profiles of decreasing carbon content with depth [5]. The shallow soil horizons (0-10 cm) show carbon content of 1.9 to 3.1% in the sites representative of forest ecosystem (A, B). Slightly higher carbon content was observed at similar depths in the sites (C, D) representing the savanna ecosystem. However a significant change to higher carbon values as high as 10% is observed in location E in a depression, that is also under savanna vegetation. The higher carbon content could be associated with transport of organic matter from the relative higher forested areas and deposited in the depression. The hypothesis of the origin of the organic matter will be evaluated using the $\delta^{13}C$ data.

4.3 *$\delta^{13}C$ Data*

A significant carbon isotopic difference is observed in the shallow soil horizon (0-10 cm) between the sites representing the forested and the savanna areas (Figure 1). The latest represented by site C show a $\delta^{13}C$ value of -16‰ compared to values of -28.1‰ and -25‰ for the forested areas (sites A and B). This pattern is related to the predominance of C_4 plants in the savanna vegetation and is well documented in the Amazon region [6, 7].

However, the sites D and E, that are also under savanna vegetation, show a $\delta^{13}C$ trend toward the forest soils. Site D shows a $\delta^{13}C$ value of -19.5‰, compared to site E located in the middle of the depression, that shows a $\delta^{13}C$ value of -22.5‰. This has to be related to a major influence of C_3 plants as a source of the soil organic matter, probably from the forest vegetation. Therefore, the ^{13}C data support the hypothesis, based on carbon content, that the majority of the organic matter preserved in the depression areas was transported into these areas from the forested areas, probably during periods of intense rains.

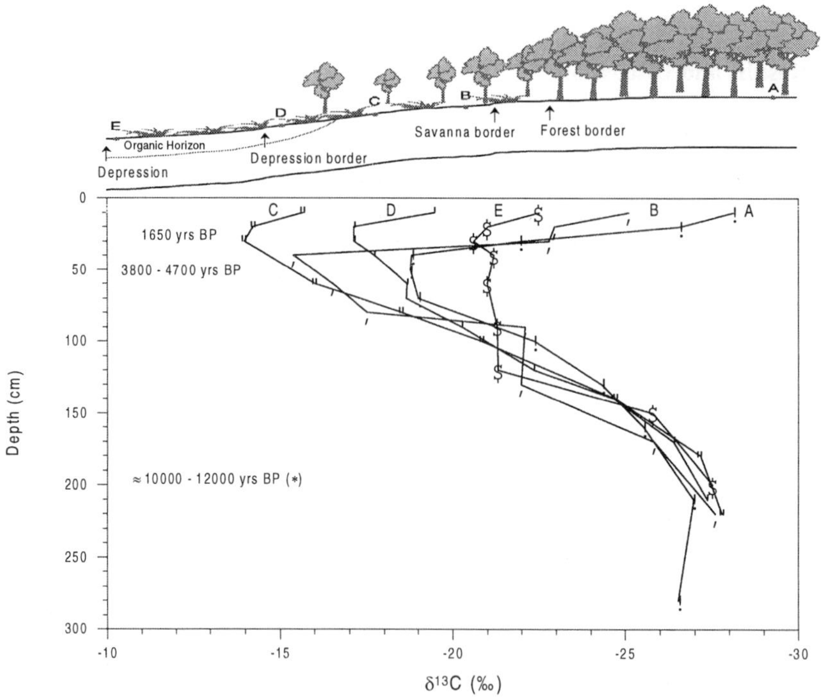

Figure 1 - $\delta^{13}C$ values of soil organic matter under distinct locations in the trench in relation to soil depth. The radiocarbon datings were obtained from soil organic horizon and buried charcoal samples. (*) Average values obtained from soil (humin) samples in the horizon 190-200 cm in the Amazon region [6, 7].

Based on the $\delta^{13}C$ depth profiles (Figure 1), it is also possible to postulate that organic matter of forested origin has been deposited since at least the last 5.000 years in the depression areas located in the savanna ecosystem. These profiles show $\delta^{13}C$ values around -26‰ in the lower part of the profiles in all sites indicating that during the period 10.000-12.000 years [6, 7], the study area was covered by forest vegetation. A significant change toward more enriched $\delta^{13}C$ values is observed at sites A, B, C and D reaching values as high as -14‰ in the 40-50 cm soil horizon interval. This trend is a clear indication of a major influence of C_4 plants during the time represented in the soil horizon between 130 and 30 cm, reaching a maximum in the 40-50 cm soil interval. However, the ^{13}C enrichment trend is not so pronounced at site E reaching only values around -21‰ in the soil interval 130-30 cm, indicating that most of the organic matter deposited during the time represented by this soil horizons was of C_3 origin (forested vegetation). Similar vegetation changes based on ^{13}C data had also been documented in the southern (Rondonia state) Amazon region [7].

5 Conclusions

Carbon data show a significant increase of carbon content, from relative higher sites representative of forest ecosystem (1.9 to 3.1%) to sites representing savanna ecosystem located in a depression, where the organic matter was deposited and shows carbon values high as 10%. The origin of the organic matter was associated to transport from the forested areas by water during rainy periods.

The hypothesis of the origin of the organic matter from forest to savanna ecosystem was confirmed by using ^{13}C data of soil samples. More depleted values (-19.5‰ to -22.5‰), related to a major influence of C_3 plants, were found in the savanna sites in the depression. In the site under savanna located in an elevation slightly higher, the value was significantly more enriched (-16‰) showing the predominance of C_4 plants. Under savanna border and forest ecosystem, the $\delta^{13}C$ values were characteristics of C_3 plants (-25‰ and -28.1‰).

^{13}C and ^{14}C data from soil samples indicate that the organic matter of forested origin, has been deposited since at least the last 5.000 years in the depression areas located in the savanna ecosystem. The ^{13}C pattern observed in the soil organic matter profiles indicate a predominance of C_3 vegetation plants for at least 12.000 years in the study area. About 7.000 to 4.000 years the data show the influence of C_4 plants, indicating forest regression associated with a drier climate than the present. The more recent portion of ^{13}C records indicates the expansion of the forest, and probably the return to climate similar the present.

6 Acknowledgements

We gratefully acknowledge financial support from São Paulo Foundation For Research (FAPESP) grant no. 97/11733-4 and PRONEX/FINEP no. 41.96.0938.00.

7 References

1. Brasil. Ministério das Minas e Energia. Projeto Radambrasil. Folha SB 20 Purus. Rio de Janeiro, 1978. 566p. (Levantamento de Recursos Naturais, 17).
2. Rosolen, V., Fritisch, E., Boulet, R., Melfi, A.J., Lamotte, M., Grimaldi, M., Eschenbrenner, V. and Magat, Ph., 1998. Soil differentiation in a toposequence characteristic of the low plateau landscapes of the southern part.

3. Pessenda, L.C.R., Valencia, E.P.E., Martinelli, L.A. and Cerri, C.C., ^{14}C measurements in tropical soil developed on basic rocks, *Radiocarbon* **38** (1996b) pp. 203-208.
4. Pessenda, L.C.R. and Camargo, P.B., Datação radiocarbônica de amostras de interesse arqueológico e geológico por espectrometria de cintilação líquida de baixa radiação de fundo, *Química Nova* **14** (1991) pp. 98-103.
5. Pessenda, L.C.R.; Boulet, R.; Aravena, R.; Rosolen, V.; Gouveia, S.E.M.; Lamotte, M., Origin and dynamics of soil organic matter vegetation changes during the Holocene in a forest-savanna transition zone, southern Amazon state, Brazilian Amazon region, *Geology*, submitted.
6. Gouveia, S.E.M., Pessenda, L.C.R., Aravena, R., Boulet, R., Roveratti, R. and Gomes, B.M., Dinâmica de vegetações durante o Quaternário recente no sul do Amazonas, indicada pelos isótopos do carbono do solo, *Geochimica Brasiliensis* **11** (1997) pp. 355-367.
7. Pessenda, L.C.R., Gomes, B.M., Aravena, R., Ribeiro, A.S., Boulet, R. and Gouveia, S.E.M., The carbon isotope record in soils along a forest-cerrado ecosystem transect: implications for vegetation changes in the Rondonia state, southwestern Brazilian Amazon region, *The Holocene* **8** (1998) pp. 631-635.

Session L1

Production and Applications of Short-Lived Radioisotopes

OPTIMISATION OF THE SINGLE PASS GAS PHASE SYNTHESIS OF ^{11}C-METHYL BROMIDE FROM ^{11}C-METHANE.

K.R. BUCKLEY, J.M. HUSER, J. LU, M.J. ADAM, T.J. RUTH
UBC/TRIUMF PET Program, 4004 Wesbrook Mall, Vancouver, BC, V6T 2A3, Canada

^{11}C-methyl bromide may provide a fast and facile route from ^{11}C-methane to ^{11}C-methyl triflate for use as a radioactive precursor in the synthesis of radiopharmaceuticals for PET. The gas phase production of ^{11}C-methyl iodide has been reported by others (1,2) as both single pass and recirculating methods, wherein the resultant ^{11}C-methyl iodide is used directly in the labelling of pharmaceuticals. The use of a recirculating method for analogous production of ^{11}C-methyl bromide has also been reported (3,4) wherein the resultant methyl bromide was subsequently converted to methyl triflate prior to use as a radioactive labelling precursor. Since the reactivity of the bromine free radical is greater than that of the iodine free radical we investigated the single pass production of ^{11}C-methyl bromide on route to ^{11}C-methyl triflate. Radio-gas chromatography is a vital tool in determining the relative production efficiencies of ^{11}C-methyl bromide and the multi-brominated species ^{11}C-dibromomethane and ^{11}C-bromoform. The conditions required to obtain the maximum yield of ^{11}C-methyl bromide result in some of the multi-brominated species being produced. We will report on the radio-gas chromatography results obtained in our work and the implications on the use of the single pass gas phase synthesis of ^{11}C methyl bromide to produce ^{11}C methyl triflate.

1 Introduction

The ^{11}C methylation of compounds is a common method for the production of radiolabelled pharmaceuticals for use in PET. In 1996 Larsen and Link independently reported the production of methyl iodide via catalytic reduction of $^{11}C\text{-}CO_2$ to methane followed by a radical chain reaction with iodine at high temperature conducted entirely in the gas phase as both recirculating (2) and single pass (1). This technique has the advantages of simplicity, speed, and high specific activity. In 1998 Steel reported the recirculating production of methyl bromide and it's subsequent conversion to methyl triflate (CH_3OTf).

The reaction of methane with a halogen is a chain reaction which can proceed beyond a single halogenation. Larsen's implementation of the gas phase method involves recirculating unreacted methane while continually trapping out methyl iodide. In this mode conditions can be used to achieve a low single pass conversion without multi-iodination yet greater overall yield is realized by allowing multiple passes. We felt the improved labelling yield with methyl triflate would allow it's production through the simpler, though perhaps less efficient conversion of CH_4 to

CH_3Br in a single pass. The greater reactivity of the bromine radical providing a more efficient single pass conversion than that of the iodine radical.

2 Method

The apparatus is shown in Figure 1. ^{11}C-CH_4 was trapped on Porapak Q (5) at LN_2 temperature and subsequently flushed with helium. Pyridinium tribromide (6) was placed in a quartz tube which went on to extend through an oven approximately 6 inches long. The pyridinium tribromide was at room temperature or heated as high as 100°C. The oven temperature was varied between 550 and 700°C. The helium gas stream was varied between 30 and 400 ml/min. At the exit of the oven an Ascarite (6) trap removed bromine. For the purposes of these experiments the various brominated species were trapped in acetone cooled in a dry ice/isopropanol bath. These samples were measured for total radioactivity and analyzed by gas chromatography to determine relative yields of the brominated species. For each test approximately 220 mCi of ^{11}C-CH_4 at end of bombardment was produced via the $^{14}N(p,\alpha)^{11}C$ reaction on the TR13 cyclotron (7).

Gas chromatography was performed on a HP 5840A chromatograph using an Alltech C-5000 column, 1/8" x 8' SS, with 1% AT-1000 on Carbograph. Separation conditions for methyl bromide (CH_3Br), dibromomethane (CH_2Br_2), and bromoform ($CHBr_3$), were achieved with standards made in house (CH_3Br) or purchased (6). Helium flow was 10 ml/min with a temperature program of 5 minutes at 50°C, ramping at 30C/min to 200°C, and 10 minutes at 200°C. CH_3Br appears at 4.8 mins, CH_2Br_2 at 10.7 mins, and CH_3Br at 15.8 mins. Carbon tetrabromide, CBr_4, was not investigated.

3 Results

Optimal conditions for the yield of total radioactivity where in conflict with the optimal conditions for the preferential production of methyl bromide. The most radioactivity, 107 mCi (EOB), was produced with the pyridinium tribromide at 50°C, oven at 660°C, and flow 390 ml/min. This yielded 60% CH_3Br, 37% CH_2Br_2, and 2%$CHBr_3$. The optimal preferential production of CH_3Br was achieved with the pyridinium tribromide at 24°C, oven at 660°C, and flow 390 ml/min. This yielded 56 mCi with the following distribution, 73% CH_3Br, 26% CH_2Br_2, and 1%$CHBr_3$. (Figures 2 and 3). Not all tests were reproducible, two runs under the apparent same conditions gave different total yields and inverse distributions between CH_3Br and CH_2Br_2, 39:2 mCi vs 9:56 mCi respectively.

4 Conclusion

The single pass conversion of methane to methyl bromide may offer a simple route to methyl triflate. However, optimal production of CH_3Br is concomitant with a production of CH_2Br_2 and a high flow rate. This high flow may complicate the subsequent production of CH_3OTf. If so, then methyl triflate may be achieved by the recirculating reaction of bromine as reported by Mock.

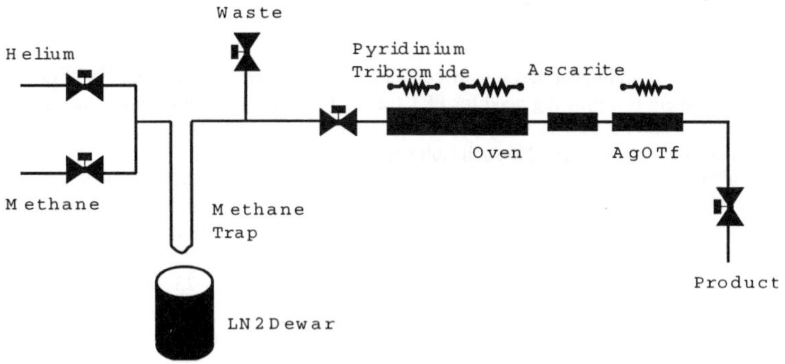

Figure 1: Schematic of apparatus.

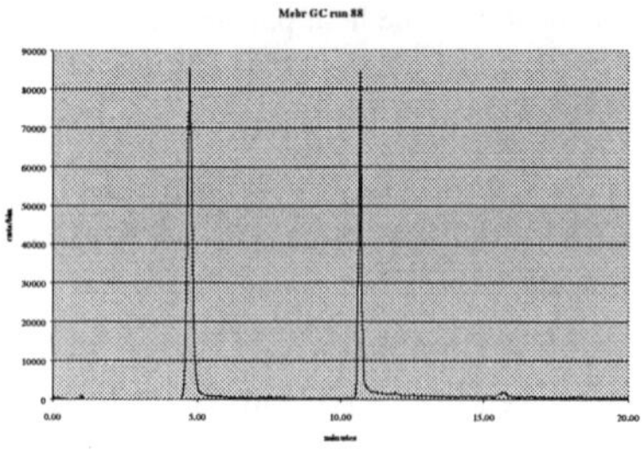

Figure 2: Best overall yield.

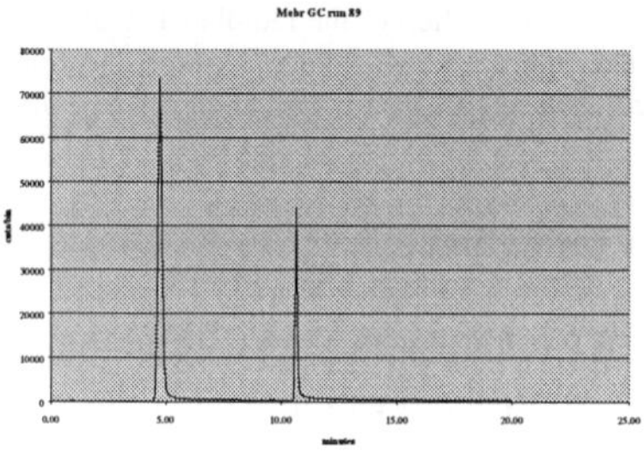

Figure 3: Best distribution of yield.

5 References

1. J.M. Link et al., J. Lab. Comp. Radiopharm. 37 (1995) 77 & Nuc. Med. Biol. 24, (1997) 93.
2. P. Larsen et al., J. Lab. Comp. Radiopharm. 37 (1995) 73.
3. C.J. Steel et al., J. Nuc. Med. 39:5 (1998) 237
4. B.H. Mock et al., Nuc. Med. Biol. 26, (1999) 467.
5. Alltech Associates, Inc. Deerfield IL 60015 USA.
6. Fisher Scientific Nepean, ON. K2E 7L6, CANADA
7. EBCO Technologies, Richmond BC. V6X 2A4, CANADA

TUMOR IMAGING WITH PET USING THE SHORT-LIVED PRASEODYMIUM-140 FROM THE ^{140}ND /^{140}PR IN VIVO GENERATOR

STEFAN K. ZEISLER AND DIRK W. BECKER

German Cancer Research Centre, Department of Radiochemistry
Im Neuenheimer Feld 280, D-69120 Heidelberg, Germany
E-mail: zeisler@dkfz-heidelberg.de

Most common PET radiopharmaceuticals contain the short-lived isotopes ^{11}C, ^{13}N, ^{15}O and ^{18}F. However, there is a growing need for PET tracers labeled with radionuclides that have a half-life of up to several days to study physiological processes requiring a longer observation time, such as tumor metabolism. No such isotopes are currently available.
The ^{140}Nd/^{140}Pr radionuclide generator provides an excellent alternative to long-lived positron emitters. ^{140}Nd decays into the short-lived daughter ^{140}Pr which has a high positron emission rate of 50 %. Radiopharmaceuticals may be labeled with ^{140}Nd, while the in situ produced ^{140}Pr serves as the actual imaging agent.
^{140}Nd was produced from praseodymium foils by irradiation with 23 MeV protons via ^{141}Pr(p,2n)^{140}Nd. The product nuclide was separated with a yield of 35 % by cation exchange using α-hydroxyisobutyric acid as an eluent. The specific radioactivity was sufficient for protein labeling.
We synthesized a stable human serum albumin–diethylenetriaminepentaacetic acid (HSA-DTPA) conjugate labeled with the ^{140}Nd/^{140}Pr generator and injected it into tumor-bearing rats. Accumulation of the radiolabeled albumin in the tumor started one hour p. i. and reached a maximum after 48 hours. No translocation or in vivo loss of ^{140}Pr occurred. The results demonstrate the great potential of this new diagnostic method.

1 Introduction

Most PET radiopharmaceuticals are biologically active compounds or drugs labeled with the short-lived isotopes ^{11}C ($T_{1/2}$ = 20.4 min), ^{13}N ($T_{1/2}$ = 10.0 min), ^{15}O ($T_{1/2}$ = 2.0 min) and ^{18}F ($T_{1/2}$ = 109.7 min). However, there is a growing need for PET tracers labeled with radionuclides that have a half-life in the order of several hours to a few days which can be used to study physiological processes requiring a longer observation time, such as tumor metabolism. Longer-lived PET radioisotopes that have been proposed are ^{64}Cu ($T_{1/2}$ = 12.7 h), ^{71}As ($T_{1/2}$ = 65.3 h), ^{73}Se ($T_{1/2}$ = 7.1 h) and ^{76}Br ($T_{1/2}$ = 16.1 h) [1], although their application may be limited due to unfavorable decay characteristics.

The ^{140}Nd/^{140}Pr system provides an ideal in vivo radionuclide generator for diagnostic applications with PET. ^{140}Nd ($T_{1/2}$ = 3.37 d, EC 100 %, no γ) decays into the short-lived daughter ^{140}Pr ($T_{1/2}$ = 3.4 min, $E_{\beta+}$ = 1.1 MeV) which has a high positron emission rate of 50 %. ^{140}Nd can be produced by the ^{141}Pr(p,2n)^{140}Nd reaction and separated by ion exchange chromatography [2].

Human serum albumin (HSA) dominates the plasma proteins of man and plays an important role in the protein metabolism of tumors. HSA can be used as a carrier for antitumor drugs as well as for diagnostic or therapeutic radioisotopes. The long biological half-life of albumin facilitates tumor accumulation [3]. HSA labeled with $^{140}Nd/^{140}Pr$ allows to quantify tumor metabolism with PET over several days.

2 Experimental

The excitation functions and thick target yields were determined as described previously [4]. The ideal energy range for the production of ^{140}Nd via $^{141}Pr(p,2n)^{140}Nd$ was found to be 23 MeV to 14 MeV. Irradiations were performed for one hour at 2 μA on a 1 mm praseodymium metal foil. 260 MBq (7 mCi) could be obtained. The purification of ^{140}Nd was performed by cation exchange at room temperature. A column of 12 mm i.d. and 600 mm length was filled to a height of 500 mm with Dowex 50W-X8 resin (100-200 mesh) in NH_4^+ form. A resin batch containing Pr^{3+}/Nd^{3+} from the dissolved target foil was poured onto the top of the separation column. The ion exchanger was then eluted with a solution of α-hydroxyisobutyric acid (0.4 M, adjusted to pH 4.0 with ammonia) at a flow rate of 0.6 mL/min. The heavier lanthanides were eluted faster than the lighter ones, i. e. neodymium left the column before praseodymium.

Fractions were collected until the amount of radioactivity in the eluate decreased significantly. The combined fractions containing ^{140}Nd were acidified and passed through a small Dowex 50W-X8 cation exchanger in H^+ form (100-200 mesh; column 5 mm i.d., 50 mm length) to remove the α-hydroxyisobutyric acid. The ^{140}Nd deposited on the resin was eluted with 6 N hydrochloric acid. The final chemical form of the product is $NdCl_3$ in aqueous solution. The entire purification procedure took about 25 hours.

HSA was labeled rapidly with the produced $^{140}Nd/^{140}Pr$ in vivo generator as follows. 7.4 MBq (200 μCi) of product solution were mixed with 0.2 M sodium citrate and then added to a previously prepared HSA-DTPA conjugate dissolved in 0.17 M sodium bicarbonate. Unbound tracer and excess sodium citrate were separated in a centricon ultrafiltration C30 unit [3].

3 Results and Discussion

Gamma spectroscopy of the product confirmed that no radioactive contaminants were present. The amount of neodymium carrier and possible praseodymium traces were determined by neutron activation analysis of a sample taken from the product. No praseodymium could be detected. As expected, the amount of neodymium carrier was equal to the concentration of Nd in the praseodymium target material (50 ppm). ^{140}Nd was isolated in 35 % yield.

After one injection of the labeled $^{140}Nd/^{140}Pr$-DTPA-HSA conjugate into ACI rats (black agouti) bearing Morris hepatoma, tumor metabolism could be visualized for a period exceeding 2 days. Radioactivity started to accumulate in the tumor tissue 1 hour p.i.; 4 hours p.i., most of the radioactivity was still found in the circulation (Figure 1). 50 hours p.i., the radioactivity was mainly located in the tumor (Figure 2).

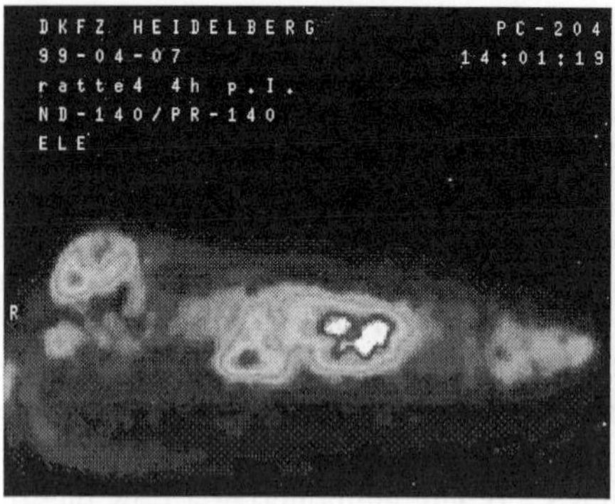

Figure 1: PET Scan of a Morris hepatoma bearing rat 4 h p.i.

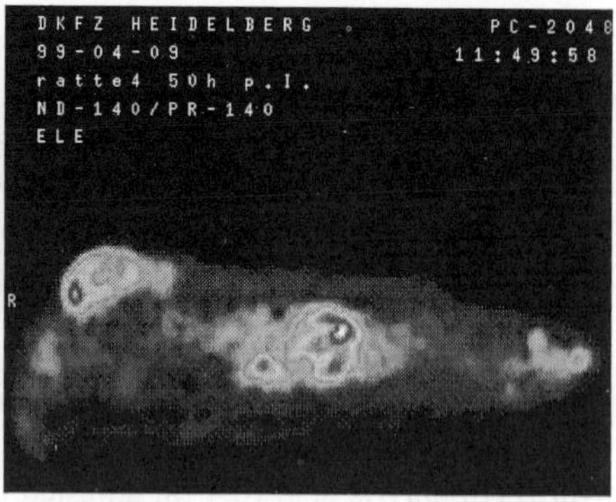

Figure 2: PET Scan of a Morris hepatoma bearing rat 50 h p.i.

The new $^{140}Nd/^{140}Pr$ in vivo generator provides an excellent method for quantifying protein metabolism in tumors. Further optimization is in progress in order to supply this radionuclide for clinical trials.

References

1. Stöcklin G. and Pike V.W. (Editors) Radiopharmaceuticals for Positron Emission Tomography; Methodological Aspects, Kluwer Academic Pub., 1993
2. Choppin G. R. and Silva R. J., Separation of the Lanthanides by Ion Exchange with Alpha-hydroxy Isobutyric Acid, *J. Inorg. Nucl. Chem.* **3** (1956) 153 – 154
3. Stehle G., Sinn H., Wunder A., Schrenk H. H., Schütt S., Maier-Borst W., Heene D. L., The loading rate determines tumor targeting properties of methotrexate-albumin conjugates in rats, Anti-Cancer Drugs **8** (1997) 677 – 685
4. Zeisler S. K. and Becker D. W., Production of the $^{140}Nd/^{140}Pr$ Radionuclide Generator for Biomedical Studies, *J. Lab. Comp. Radiopharm.* **42** (1999) 921 – 923

STUDIES OF NITROGEN ABSORPTION AND ASSIMILATION IN PLANTS BY USE OF ^{13}N

DEV T. BRITTO AND A. D. M. GLASS

Dept. of Botany, University of British Columbia, Vancouver BC V6T 1Z4, Canada
E-mail: dtbritto@interchange.ubc.ca

HERBERT J. KRONZUCKER

Dept. of Plant Sciences, University of Western Ontario, London ON N6A 5B7, Canada
E-mail: kronzuck@julian.uwo.ca

The short-lived radiotracer ^{13}N ($t_{½}$ = 9.98 min.) has been used to precisely measure inorganic nitrogen fluxes in a variety of plant systems. Analysis of the efflux component from roots and leaves, in particular, provides a method by which internal pool sizes of NO_3^- and NH_4^+ may be non-invasively estimated. Estimates of the half-life of exchange, or turnover, of inorganic N within the plant cell are also provided by efflux analysis, with the unexpected result that this parameter remains constant against a backdrop of changing flux components. This behaviour is contrasted with that of the half-life of exchange for other ions such as potassium, which show strong variability dependent upon external conditions.

1 Introduction

Compartmental analysis by efflux (CAE) is used to study the transport of molecules across membranes in a wide variety of biological systems [1]. Within the field of plant physiology alone, this method has shed light on ion fluxes and compartmentation of ionic pools in species as diverse as unicellular algae [5], grasses [3,6,9,11], *Arabidopsis* (a model plant used for studies in plant molecular biology) [8], and both coniferous and flowering trees [2]. One component of the analysis that has been somewhat neglected in higher plant studies is the cytosolic half-life of exchange ($t_{½,c}$) of the traced ion, which is distinct from the half-life of a decaying radiotracer in that its first-order kinetic character is the result of biological flux processes rather than events of nuclear chemistry. $t_{½,c}$ is important because it yields an estimate of the turnover of the traced ion within the cell cytosol, a compartment poised at the interface between the organism and its environment.

2 Methodology

In plants, CAE is most commonly performed on roots, although it may be possible to carry out similar, if more invasive, analyses on excised leaf tissue [7]. Tissue adapted to a bathing solution is labeled for a specified time period in a chemically equivalent solution containing a tracer that is typically radioactive (*e.g.* ^{13}N, ^{86}Rb or ^{42}K), but may also be stable (*e.g.* ^{15}N). The tracer is then eluted from the external and cytosolic compartments of the plant (and potentially from other compartments, depending on the duration of labeling), into a timed series of unlabeled aliquots of bathing solution. Tracer activity resulting in the aliquots, and tracer remaining in the eluted tissue(s) are measured, and these data are analyzed along the lines of Walker and Pitman [10] and Siddiqi *et al.* [9].

3 Discussion

Ammonium (NH_4^+) and nitrate (NO_3^-), the two major forms of inorganic nitrogen taken up by plants, have been shown by numerous ^{13}N and other studies to be transported by distinct molecular systems. Responses of root N influx activities (ϕ_{oc}) to variations in environmental concentrations are typically linear in the low-concentration range but become increasingly independent of external conditions in moderately higher ranges (*i.e.* they become saturated) [2,8]. It can be shown [10] that the flux of an ion into a pool is related to the size of that pool (Q) and the turnover of the ion within it ($t_{½}$) by the equation

$$Q = \frac{t_{½}\,\phi_{oc}}{\Omega} \qquad (1)$$

where Ω is a proportionality coefficient. However, experimentation shows that the cytosolic turnover time of neither NH_4^+ nor NO_3^- varies significantly across an extensive range of external conditions (both saturating and non-saturating) [9,11]. This indicates that the relationship between flux and pool size is directly linear, and therefore that the cytosolic concentrations of ammonium and nitrate are not maintained homeostatically, at least in the plant systems for which the parameters in equation 1 are known.

This condition may be exclusive to nitrogen. It does not hold, for instance, in the case of another essential macronutrient, potassium (K^+), which is thought to be maintained at a relatively constant level inside the cytosol (for osmotic and metabolic reasons), even while influx into that compartment from the external solution is variable. A lack of linearity relating influx with pool size, then, should argue for compensatory variations in $t_{½,c}$ for potassium, and indeed this is verified in a study with ^{86}Rb, a widely-used tracer analogue for K^+ [6]. Another ion whose cytosolic levels are strictly maintained (for signal transduction reasons, in this case) is calcium (Ca^{++}); however, Ca^{++} fluxes respond to non-saturating increases in externally provided Ca^{++} [4], leading to the prediction that half-lives of cytosolic calcium exchange should be variable. In practice, however, this parameter is not likely to be directly estimable by CAE given that cytosolic Ca^{++} pools are known to be in the nanomolar range; from equation 1 and influx values determined experimentally [4], it follows that $t_{½,c}$ values for calcium must be on the order of much less than one second.

To conclude, it appears that while it is important for the plant to maintain ions such as K^+ and Ca^{++} at constant levels within the cytosol of its root cells, in the case of inorganic N, there appears to be a highly variable, environment-dependent level of accumulation. Perhaps this reflects a strategy evolved by plants to maximally acquire this most frequently limiting element, even when exposed to an abundant supply.

References

1. Jacquez, J. A., Compartmental analysis in biology and medicine. 2nd Ed. (University of Michigan Press, Ann Arbor, 1985).
2. Kronzucker, H. J., Siddiqi, M. Y. and Glass, A. D. M., Conifer root discrimination against soil nitrate and the ecology of forest succession. *Nature* **385** (1997) pp. 59-61.
3. Lee, R. B. and Clarkson, D. T., Nitrogen-13 studies of nitrate fluxes in barley roots. I. Compartmental analysis from measurements of ^{13}N efflux. *J. Exp. Bot.* **37** (1986) pp. 1753-1756.
4. Loneragan, J. F. and Snowball, K., Rate of calcium absorption by plant roots and its relation to growth. *Australian J. Agric. Res.* **20** (1969) pp. 465-478.
5. Malhotra, B. and Glass A. D. M., Potassium fluxes in Chlamydomonas reinhardtii: II. Compartmental analysis. *Plant Physiol.* **108** (1995) pp. 1537-1545.
6. Memon, A. R., Saccomani, M. and Glass, A. D. M., Efficiency of potassium utilization by barley varieties: The role of subcellular compartmentation. *J. Exp. Bot.* **36** (1985) pp. 1860-1876.
7. Raven J. A. and Farquhar, G. D., Methylammonium transport in *Phaseolus vulgaris* leaf slices. *Plant Physiol.* **67** (1981) pp. 859-63.
8. Rawat, S. R., Erner, Y., Kronzucker, H. J., Silim, Salim N., Schjoerring, J. K., Siddiqi, M. Y., and Glass, A. D. M., *AtAMT1* gene expression and NH_4^+ uptake in roots of *Arabidopsis thaliana*: evidence for regulation by root glutamine levels. *Plant J.* **19** (1999) pp. 143-152.
9. Siddiqi, M. Y., Glass, A. D. M. and Ruth, T. J., Studies of the uptake of nitrate in barley. III. Compartmentation of NO_3^-. *J. Exp. Bot.* **42** (1991) pp. 1455-1463.
10. Walker, N. A. and Pitman, M. G., Measurement of fluxes across membranes. In *Encyclopedia of Plant Physiology, Vol. 2, Part A*, ed. by U. Luttge and M. G. Pirman (Springer Verlag, Berlin, 1976), pp. 93-126.
11. Wang, M. Y., Siddiqi, M. Y., Ruth, T. J. and Glass, A. D. M. Ammonium uptake by rice roots. I. Fluxes and subcellular distribution of $^{13}NH_4^+$. *Plant Physiol.* **103** (1993) pp. 1249-1258.

Session L2

Labeling Compounds and Other Applications of Tritium

TRITIUM LABELING FOR INVESTIGATION OF STRUCTURE-FUNCTION RELATIONSHIP IN TOBACCO MOSAIC VIRUS

E.V. LUKASHINA, G.A. BADUN AND V.M. FEDOSEEV

Div. of Radiochemistry, Chemistry Dept., Moscow State University, Russia
E-mail: lena@inorg.chem.msu.ru

E.N. DOBROV

Dept. of Biochemistry of Plant Viruses, Belozersky Institute of Physico-Chemical Biology, Moscow State University, Russia

L.A. BARATOVA

Dept. of Chromatography, Belozersky Institute of Physico-Chemical Biology, Moscow State University, Russia

The catalytic thermal dissociation of tritium gas on a hot tungsten wire is the convenient way for generation of atomic tritium. During the bombardment of solid targets by tritium atoms from gas phase labeling takes place in the thin surface layer. This phenomenon is used for structural investigations of biological objects.

In the present work intact virions of tobacco mosaic virus (wild strain U1 and mutant ts21-66, which has two substitutions in the amino acid sequence of the coat protein and is characterized by temperature sensibility and ability to induce necroses on some plant hosts) were exposed to atomic tritium labeling for determination of differences in the organization of coat proteins in virus particles. After labeling and purification the wild-type and mutant coat proteins were hydrolyzed by trypsin and specific radioactivity of individual amino acid residues in each peptide was determined. Several reliable differences in intramolecular tritium distribution in the tobacco mosaic virus coat proteins were found and their relationship to some changes in mutant coat protein higher order structure which can lead to changes in function is discussed.

1 Introduction

Tritium labeling is one of many ways for structural investigation of biological objects. The method of tritium planigraphy is based on the high reactivity of tritium atoms produced via thermal dissociation of tritium molecules on a heated tungsten filament. During the bombardment of solid targets by tritium atoms from gas phase labeling takes place in the thin surface layer. The distribution of radioactivity among different parts of the object shows their accessibility for tritium atoms and allows to make suggestions about the spatial structure [1].

In the present work we used the method of tritium planigraphy for comparative investigation of the spatial structure of intact virions of two strains of tobacco mosaic virus (TMV): wild strain U1 and mutant ts21-66, which has only two substitutions in the amino acid sequence of the coat protein, but it is characterized by temperature sensibility and ability to induce necroses on some plant hosts [2].

2 Materials and methods

TMV particles were exposed to atomic tritium labeling. The label was introduced by means of bombardment with tritium atoms heated to a temperature of 2000 • at the setup shown in Figure 1. TMV suspensions were sprayed on the liquid nitrogen-precooled inner surface of the reaction vessel. Liquid nitrogen (77 •) was also used for thermostating the vessel in the process of labeling.

Obtained samples were purified with lyophilization and ultracentrifugation. Isolation of the coat protein from the labeled TMV was carried out with cold acetic acid. Then coat proteins were hydrolyzed by trypsin, and mixtures of 12 peptides were separated by HPLC. Tryptic peptides were hydrolyzed into individual amino acids and analyzed on a Hitachi-835 amino acid analyzer in a standard mode of protein hydrolyzate analysis. Detection of radioactivity of amino acids was carried out by Canberra Packard 150TR flow scintillation analyzer.

As a result of the experiment we determined the specific radioactivity of individual amino acid residues in each peptide.

3 Results and discussion

TMV is the type member of the tobamo-virus groupe, rod-shaped viruses 300 nm long and 18 nm in diameter with a central hole of diameter 4 nm (Fig. 2a). Identical protein subunits (molecular weight 17,500) form a right-handed helix protecting RNA.

Figure 2b illustrates the structure of the coat protein in wild TMV particles according to X-ray fiber diffraction data [4]. Our experimental data for wild TMV coat protein are in well agreement with these data.

There are no such data for the mutant strain, and on the basis of our experimental data for mutant TMV we suggested some changes in mutant coat protein structure. In the mutant virus particle accessibility for tritium atoms of the N-terminus and of the loop between RS and RR alpha-helixes decreases, and we observe change of labeling for C-terminus (some residues become more accessible and other less accessible). At the same time accessibility for tritium atoms of residues, which are located at the surfaces of hydrophobic interactions between protein subunits, increases. So we can suggest that in mutant virus particle change

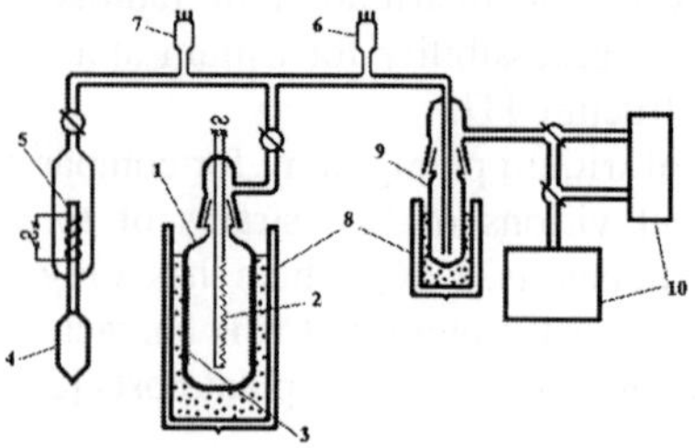

Figure 1. The setup for tritium labeling: 1 – reaction vessel, 2 – tungsten wire, 3 – target (sample for labeling), 4 – vessel with tritium gas, 5 – palladium tube, 6 and 7 – vacuum detectors, 8 – vessels with liquid nitrogen, 9 – catcher, 10 – vacuum pumps.

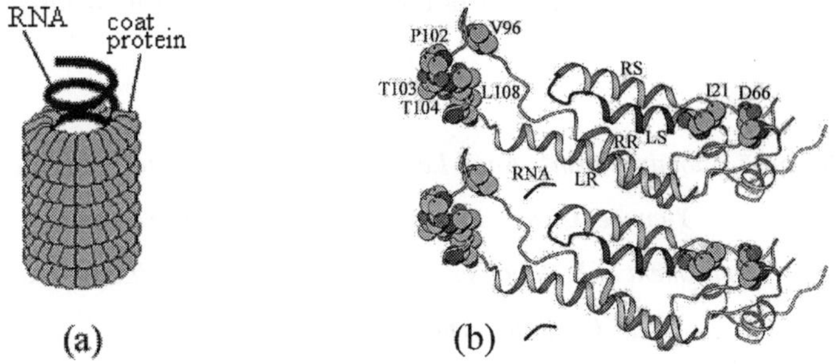

Figure 2. Tobacco mosaic virus. (a) Schematic representation. (b) Structure of the coat protein in wild TMV particles. Backbone structures of 2 protein subunits are shown. Alpha-helixes are marked LS, RS, LR, RR (left and right slewed, left and right radial), N (N-terminal) and C (C-terminal). Also localization of amino acid substitutions in mutant TMV (I21• T, D66• G) and several residues of the internal loop between RR and LR alpha-helixes (which are tritium labeled in our experiments) are shown. The viral axis is vertical, to the left of the Figure. Figure 2b was prepared using MOLSCRIPT©.

of conformation of polypeptide chain takes place. Consequently the system of intersubunit contacts changes and mutant virus particles become more friable. It is interesting to note that we observed labeling for the internal loop between RR and LR alpha-helixes. This fact was described earlier [3] and explained by the penetration of tritium atoms inside virus particle through channels between protein subunits. In mutant TMV accessibility for tritium of these residues increases, and it confirms our suggestion about more friable structure of mutant TMV particles.

Probably these changes of structure lead to an appearance of new properties for mutant coat protein and a capability of plant hosts with N'-gene to recognize mutant TMV and to fight with it. So the method of tritium planigraphy allows to get information about structure-function relationship.

References

1. Bogacheva E.N., Gol'danskii V.I., Shishkov A.V., Galkin A.V. and Baratova L.A., Tritium planigraphy: from the accessible surface to the spatial structure of a protein. *Proc. Natl. Acad. Sci. USA* **95(6)** (1998) pp. 2790-2794.
2. Dobrov E.N., Abu-Eid M.M., Solovyev A.G., Kust S.V. and Novikov V.K., Properties of the coat protein of a new tobacco mosaic virus coat protein ts-mutant. *J. Prot. Chem.* **16(1)** (1997) pp. 27-36.
3. Goldanskii V.I., Kashirin I.A., Shishkov A.V., Baratova L.A. and Grebenshchikov N.I., The use of thermally activated tritium atoms for structural-biological investigations: the topography of the TMV protein-accessible surface of the virus. *J. Mol. Biol.* **201** (1988) pp. 567-574.
4. Namba K., Pattanayek R. and Stubbs G., Visualization of protein-nucleic acid interactions in a virus. Refined structure of intact tobacco mosaic virus at 2.9 Å resolution by X-ray fiber diffraction. *J. Mol. Biol.* **208** (1989) pp. 307-325.

Uniform Tritium Labeling of Combinatorial Libraries for Affinity Selection Screening

Bruce Surber, Shomir Ghosh[x], Anne-Laure Grillot[x], Jyoti Patel[x], Charlotte Woodall[x], Yuanwei Chen[x], Lin Yi[x], Irini Zanze[x], and Ye Yao**

**Department of Metabolism, Radiochemistry and Cellular Toxicity*

[x]Combinatorial Chemistry Department

Abbott Laboratories, D46V, AP9, 100 Abbott Park Road, Abbott Park, IL, 60064-6114.

In high-throughput affinity selection screening, mixtures of compounds are passed through a column containing a stationary protein target. Compounds of high affinity remain bound to the protein target while weaker binders are washed away. This is useful for screening large mixtures of compounds such as mix and split combinatorial libraries when the small amount of individual compounds can be detected. One way to facilitate detection is to label every compound in the library.[1] Herein is described the labeling of 50x50x50 mix and split libraries (50 pools of 2500 compounds each) that were created on solid phase support beads such that the final step before cleavage from the bead was a simple reaction with a tritium-labeled reagent. The labeling chemistry was developed using single-compound models. After the labeling reaction, the products were washed with a standard battery of solvents and dried. Samples were then cleaved to determine the extent and quality of reaction by scintillation counting, radio-HPLC, and mass spectroscopy. To date, two combinatorial libraries have been uniformly labeled for affinity selection screening using solid phase synthetic procedures. One was labeled by acetylation of 2° amines (10mg pools at 0.5μmol/mg) using PyBroP-activated CT_3COOH (25mCi per pool at 7.9Ci/mmol) in DCM (0.5mL) in the presence of DIEA to yield beads that were labeled at 86μCi/mg on average (4% yield, cleaved). Reactions were terminated after 2h at room temperature to avoid reactions that resulted in an unacceptable level of polar radioactive impurities. Secondly, a library of sulfonamides

(10mg pools at 0.5μmol/mg) was N-methylated with CT_3I (500mCi/pool at 8Ci/mmol) overnight at room temperature in DCM (1mL) in the presence of an excess of DIEA to give beads at 1300μCi/mg on average (32% yield). Model reactions with $NaBT_3CN$ reductive methylation of amines and $NaBT_4$ ketone reduction also showed promise as means of creating libraries of compounds uniformly tritium-labeled. Thus, models of 3 different libraries containing 2° amine functionality (10mg of beads at ~0.5μmol/mg) were each treated with 37% formaldehyde (50μL) and acetic acid (15μL) in DMA (300μL). This was followed by a solution of $NaBT_3CN$ (300mCi at 13Ci/mmol) in DMA (0.5mL) and the reactions were shaken overnight to give beads labeled at 538, 834, and 1134μCi/mg (25, 38, and 52% yields, res.). Finally, two models of a ketone library (10mg each at 0.55μmol/mg) suspended in dioxane (0.4mL) each received 500mCi of $NaBT_4$ (71Ci/mmol) as a solution in 0.1M NaOH (120μL). These reactions resulted in beads labeled at 3.47 and 4.26mCi/mg (35 and 44% yields, res.) Scaffold structures cannot be revealed at this time.

---

[1] Bruce Beutel, Annual Reports in Medicinal Chemistry-32, 1997, Sect. VI, George L. Trainor, ed., Chapter 26: Discovery and Identification of Lead Compounds from Combinatorial Mixtures, pp.261-268.

Session L3

Novel Applications of Isotopes and Opportunities for Technology Transfer

Contraband Detection based on $^{13}C(p,\gamma)^{14}N$ *

R. RUEGG, B.F. MILTON[1], D. DALE, D. BISHOP, B. ROBERTS, P.W. SCHMOR
TRIUMF, Vancouver, BC, Canada

J. SREDNIAWSKI, S. MELNYCHUK, J. RATHKE, E. KAMYKOWSKI
Advanced Energy Systems, Inc., Medford, NY, USA

E-mail: ruegg@triumf.ca

The Contraband Detection System (CDS) developed by TRIUMF and Advanced Energy Systems (a group formerly from Northrop Grumman) is a system designed for the detection of explosives and other contraband utilizing Gamma Resonance Absorption (GRA). The CDS is based on the resonant absorption by ^{14}N of gammas produced using $^{13}C(p,\gamma)^{14}$N. This reaction uses 1.75 MeV protons to produce gammas having an energy of 9.17 MeV. By measuring both the resonant and non-resonant absorption using detectors with good spatial resolution, and applying standard tomographic techniques, we are able to produce 3D images of both the nitrogen partial density and the total density. Together these images may be utilized to determine if small amounts of nitrogen based explosives are present.

In order for the system to be practical, GRA scanning requires an intense source of protons. However this proton source must also be very stable, have low energy spread, and have good spatial definition. These demands suggested a tandem as the accelerator of choice. We have therefore constructed a 2 MeV H^- tandem optimized for high current (10 mA) operation, while minimizing the overall size of the accelerator. This has required several special innovations, which will be reviewed in this paper.

1 Introduction

An extensive study of GRA for contraband detection was conducted by the TRIUMF-Grumman team in 1994/95. This study concluded that the $^{13}C(p,\gamma)^{14}$N reaction was very promising for the detection of nitrogen based explosives. Previous TRIUMF studies established that RFQs and cyclotrons were not suitable for the demands placed on beam quality for the CDS. These include the production of a 10 mA proton beam at 1.75 MeV with an energy stability of better than 6 kV, energy spread of less than 25 keV and a divergence of 12 mrad. The most viable alternative was an electrostatic accelerator incorporating existing ion source and acceleration column technologies, and new high voltage generator and stripper technologies to reduce the overall size while increasing proton current outputs to record levels.

* Initiated by Northrop Grumman Advanced Technology and Development Center, Bethpage, NY, USA

[1] present address: Creo Products, Burnaby, BC, Canada

2 The CDS Tandem System

The layout of the CDS System is presented in figure 1. This system was designed to perform as a laboratory proof of principle experiment to allow for the testing and development of components. The CDS uses a high brightness CUSP H^- ion source coupled with a short matching section as an injector to the tandem. This is followed by a high energy beam transport section capable of analyzing the beam and producing the specified beam parameters at the target. The high energy beam line and target have been constructed such that the resonant gamma fan lies in a horizontal plane in the region of the object inspection platform. This platform is capable of rotating and elevating the object in order to acquire tomographic information. Immediately down stream of the platform is the BGO block detector system. The output of each detector segment is collected and then converted to gamma position information [1].

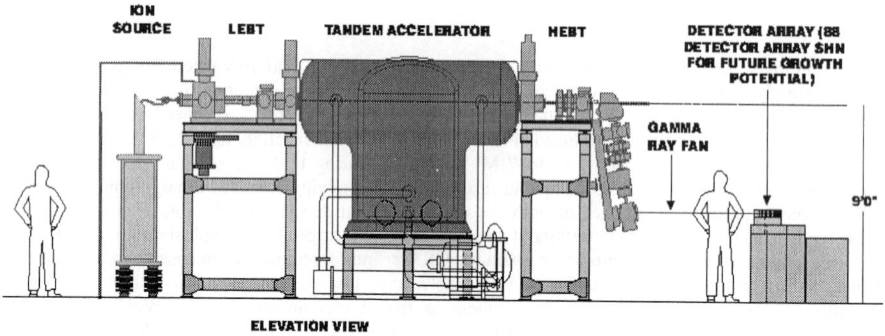

Figure 1: CDS Tandem Accelerator System

3 The CDS Tandem

Reducing the overall tandem size means increasing electric fields. High voltage breakdowns increase rapidly with increased electric fields. Therefore, a conservative 50kV/inch gradient in the acceleration region was chosen. The containment vessel is pressurized to 55 PSI with SF_6, which allows the distance between the high voltage terminal and the pressure vessel to be reduced by a factor of greater than 7 versus air.

Calculations show that foil stripper lifetimes at the design currents is in the order of a few minutes as the energy loss in the foil at 1 MeV is large. A water vapour stripper has been developed that ensures good differential pumping between the stripper chamber and the acceleration columns to minimize interaction between the beam and residual gas [2]. To further reduce the possibility of free electron acceleration the stripper is biased, as well as an electron suppression ring located at each end of the terminal. A closed loop re-circulating system using a turbopump and water reservoir provides gas to the stripper. In addition to the stripper vacuum equipment, the terminal contains two magnetic quadrupole triplets with asymmetric steering, a cooling system for the segmented collimator diagnostics, a Programmable Logic Controller (PLC), and power supplies. The terminal PLC communicates with a ground potential PLC through fibre optics. Compressed helium gas for the operation of the cryoheads is brought from the ground potential compressor to 1MV through nylon tubing wound around the insulator legs.

The preliminary CDS design study identified a power supply development at the University of Waterloo that was capable of meeting the 6.25 cubic feet space limitation [3].

Power to operate the high voltage terminal power supplies, pumps, and controls is provided by a 1 MV isolation transformer. This transformer operates as a normal one to one laminated steel core transformer in a racetrack configuration. The vertical legs of the core are segmented into 20 Mylar insulated sections. These Mylar insulators introduce a large gap in the magnetic circuit that causes poor transformer coupling. The coupling is improved by placing compensation boards on each section. These boards store energy when there is no current flow in the secondary winding and return the energy to the magnetic circuit when the rate of field change is the greatest.

The High Voltage Power Supply (HVPS) consists of a variable DC power supply, a resonant mode inverter, a transformer and rectifier doubler. The inverter converts a DC level to a 100kHz sine wave using a LC resonant circuit with the primary of the transformer forming part of the inductance. Printed circuit boards on an insulated gap ferrite core form the transformer. Each board contains 32 secondaries, each with a rectifier and doubler that are connected in series to obtain 12.5 kV per board. The HVPS uses the same technique for compensating for flux loss with three boards equally spaced in the transformer. Regulation of the power supply has been accomplished using a fast control circuit [4].

During testing it was readily apparent that while the high voltage power supply was capable of producing the required output current into a resistive load, it was not capable of accepting variable beam loads from 0 to 10 mA without reconfiguring the compensation boards. The low voltage transformer also required extensive development to achieve the desired voltage holding capability, and was identified as a weak link due to high field gradients (up to 50 kV across 0.025 inch Mylar). A parallel program was initiated to build a second-generation power supply system to meet these demands.

4 Developments and Results

The second generation power supply system incorporates both a 60 Hz AC power source and the 1MV high voltage supply into one unit. The AC power source uses the same high frequency ferrite core transformer technique as the existing HVPS to get DC power to the terminal. An inverter is then used to convert the DC to 60Hz AC.

Magnetic modelling was used to examine the effects of core leg spacing and gaps due to the Mylar insulation. These results were in turn incorporated into an electrical model of the transformer where the compensation characteristics were optimized. The redesigned system improves compensation by integrating compensation components with the secondary windings rather than on separate boards.

The prototype power supply uses a 100 kHz inverter low duty cycle device adapted for the continuous duty CDS application. Two new, continuous duty, inverters were designed and built. The high voltage inverter continues to operate at 100 kHz, the second inverter, for the AC power source, operates at 30 kHz to prevent cross talk between the two systems.

The AC power source has achieved 2.8 kVA at 30 kHz into a resistive load. The 30 kHz to 120 VAC 60 Hz inverter needs to be designed and built. The high voltage power supply has been tested to 848 kV at 3mA. Measurements show that the compensation remains within the specified tolerance for a load range of 2000 to 100 Mohms.

The CDS Tandem has accelerated various beam intensities at different energies. Initial tests were conducted with low currents, e.g. 100 μA at 1.54 MeV verified that the transmission was virtually 100 % with acceleration column pressures in the low 10^{-7} Torr. Full energy acceleration at 1.76 MeV with 1.4 mA proton beam on a high current ^{13}C target [5][6] has also been achieved with transmissions greater than 90 %. The measured 4-sigma energy spread using a thick ^{13}C target is 38 keV. In other tests, a 0.85 mA proton beam was accelerated to 1.84 MeV, 2.2 mA at 1.59 MeV, and with a nitrogen gas stripper, 0.80 mA at 1.28 MeV was achieved. The highest terminal voltage obtained to date is 935 kV.

5 Status

Advanced Energy Systems has de-commissioned the CDS at the Northrop Grumman facility in Bethpage, NY and it has been shipped to Brookhaven National Laboratory (BNL). Advanced Energy Systems and BNL intend to re-commission the system for dual purpose Boron Neutron Capture Therapy and Gamma Resonant Absorption development. TRIUMF anticipates that should this project be funded a renewed effort will evolve to complete the second-generation power supply system.

6 Acknowledgements

The CDS team consists of three groups and has included many people over the past 7 years. Advanced Energy Systems, Inc., Scientific Innovations, and TRIUMF wish to thank the many people who have contributed to this project. This work was conducted under USAF contract #F0865094C0097 and FAA contract #DFFA0397D00012.

7 References

[1] J. Sredniawski, "A New Proof-of-Principle Contraband Detection System", Office of National Drug Control Policy Conference in Nashua, N.H., 1995.
[2] B. Milton, et. al., "A High Current Tandem Accelerator for Gamma-Resonance Contraband Detection", 31st Symposium of Northeastern Accelerator Personnel Conference in Aachen, Germany, October 1997.
[3] Dr. J. Cross, ERDL, Nova Scotia, Canada, private communication.
[4] D. Dale, et. al., "The Control System for a 2 MeV Tandem Accelerator used for Contraband Detection", International Conference on Accelerator and Large Experimental Physics Control Systems in Beijing, China, October 1997.
[5] J. Rathke, et. al., "Engineering Design of a Continuous Duty Gamma Production Target for the Contraband Detection System", Particle Accelerator Conference in New York, 1999.
[6] S. Melnychuk, et. al., "Development of a Thin Film 9.17 MeV Gamma Ray Production Target for the Contraband Detection System", Particle Accelerator Conference in New York, 1999.

RADIO-CHROMATOGRAPHY UNDER GPL-CONDITIONS

G. DIETZEL, R. GRUGEL AND E. RAUSCH

raytest Isotopenmessgeräte GmbH, Benzstr.4, D 75334 Straubenhardt, Germany,

email: gdietzel@raytest.de

The radioactivity detection generates digital output signals in counts/time. All HPLC peak integration systems accept only analog data input signals o-1V. Therefore, all manufacturers of radioactivity flow detectors convert counts/ time into an analog voltage signal 0-1V using a conventional or microprocessor controlled ratemeter. GLP requires categorically storage of raw data in order to enable the recalculation of the results at any time. A recent development of a new radioactivity HPLC flow detector and HPLC system fulfills the requirements of GLP.

1 Introduction

The German Government has published the legal requirements of „Good Laboratory Praxis" (GLP) and „Data Processing" in 1996. Most industrialized countries in Europe, America and Asia have already or are going to introduce the same or very similar, legal regulations for GLP. What does this mean for radioactivity detection in chromatography?

2 Statistics on Radioactivity Signal

Radioactivity detectors for example in HPLC are different from conventional detectors for UV-absorption, fluorescence, conductivity, reflective index etc. Radioactivity is defined as number of decay events/time in Becquerel (d/s). A sensitive radioactivity detector registers single events/time. For a particular, defined radioactivity the obtained number of events/time are not constant but statistically distributed.

A continuous record of small activities my appear as „grass" showing big statistical variations of small numbers of events/time, which may not show any indication of a peak originated by a radioactive fraction. Conventional peak integration system will either find every statistical variation as a separate peak or not find wide spread fractions with many statistical data points.

Furthermore the measurement of low radioactivities with small numbers of events/time is not zero when the activity is zero. But there is background, which is produced by cosmic rays, natural activities in the environment and small contaminations in the construction material. Often background can not be separated from the activity of the sample. In principle one has to subtract the background from the total number of measured events/t in order to determine only the number of events originated by the radioactive sample. The statistical variation of the subtraction can produce „negative“ peaks, because the subtracted background may be higher than the individual statistical value of the sample activity. Nobody accepts negative radioactivity peaks. Experienced radio chromatography users select a wide region of interest at the beginning of the chromatogram, when no sample activity can be present. The same can be done at the end of a chromatogram. The mean background line will be calculated and displayed in the curve of the chromatogram.

3 Flow Correction

In radioactivity flow detection the obtained number of events/ time is:

Counts = A (d/min) * e * V (ml) / F (ml/min)

This will influence the variation of the obtained radioactivity signal strongly.

The statistical variations of the small radioactivity fraction can be improved by using greater interval times or bigger cell volumes. But that will deteriorate the chromatographic resolution, the peaks will be smeared out and two peaks of narrow fractions can not be resolved anymore.

4 Ratemeter

Conventional integrator inputs of peak-integration-systems offer for UV- or other detectors analog inputs of 0 - 1V. In order to convert a digital radioactivity signal into an analog voltage signal for continuous flow detection, a ratemeter has to be used. In case the range is selected to small, the radioactivity signal may override the range and the peak can not be integrated. In case the range may be selected to large the analog output signal may be smaller than the resolution of the integrator.

Sophisticated smoothing algorithms have been developed for the reduction of the statistical variations, which may look reasonable for one particular set of parameters. Experienced users of radio chromatography are selecting one set of parameters of flow rate, flow cell volume, scintillator, efficiency, range, interval time, smoothing etc., which enables an acceptable record of a radio-chromatogram for a particular sample. In case a ratemeter is used to convert the digital radioactivity signal into an analog output, the dimension of radioactivity is lost. The output of a ratemeter is mV. The integral of a ratemeter peak is mV * s. For one set of parameters you may be able to produce one calibration curve, in order to convert the peak integrals of mV*s into c/s *s = counts. For every changed parameter of flow rate, cell volume, efficiency, background, range, interval time, smoothing etc. a new calibration curve would have to be prepared. Nobody likes to do that and all of that will not satisfy professional chromatographic presentation and GLP requirements.

Radio-HPLC-systems with 2 PC`s and 2 programs or 1 PC with 2 programs running simultaneously produce 2 separate set of data, one for the conventional HPLC and one for radioactivity. This again is not acceptable for GLP.

5 Digital Data Transfer and Control of HPLC

Therefore a new development has been completed for radio chromatography applications under GLP. Gina-NT is a sophisticated HPLC control and evaluation interface and program, which contains for example all software features to perform a complete control and evaluation of a HPLC system of Hewlett Packard HP 1100: Automatic Sample Injection Control, Gradient Profile Control, Detector Control for UV, Recording of Diode Array Spectra, Fraction Collector Control, External Valve Control for Waste Collection, Decay Correction, Internal and External Standardization, Batch Printing of Chromatograms and Result Tables.

In addition to the summary of all features of the conventional HPLC HP 1100, Gina-NT has several specific features for Radio-Chromatography. The beta and the gamma radioactivity flow detectors, transfer the measured numbers of counts/time directly on-line in digital form to the recording computer and integration system. The time interval can be as small as 100 ms. The range is as big as 3 million counts. The histogram type presentation of radio chromatograms is replaced by a mathematical peak curve, which fits into the data points of a peak, in order to form a real peak shape. But all of the peak display algorithms do maintain the raw data and the correct value of the peak integral. Wrong selection of range, interval time and/or smoothing is impossible. The display is set automatically to the best possible form at any time during or after the run without changing the quality, dimension and calibration of the digital data.

In order to complete the GLP conformity, flow cells are equipped with their own memory chips, which contain all parameters influencing the recording like cell volume, efficiency etc. As soon as a chromatogram run is started, the chip is automatically red and all parameters are used, printed and stored with all raw data. There is no way to forget or lose any data or parameter, which might be required for reevaluation of raw data in order to reproduce reported results at any time later. All relevant data like study names, animal ID, time point, equipment ID, measurement data, sample ID, standard ID, parameters, settings, comments etc. everything is organized and stored under the firm sequence of data structure.

The validation of the system by the user is easy at any time. A suitable, very long time stable standard is inserted in the flow cell position, the calibration program is started and the results are reported and stored.

6 Conclusion

The first radioactivity HPLC system working under GLP is available now.

PRODUCTION OF TUNGSTEN-188 AT OAK RIDGE NATIONAL LABORATORY

J.A. KLEIN, F.F. (RUSS) KNAPP, M. DU, S. MIRZADEH, A.L. BEETS

Isotope and Nuclear Medicine Programs, Oak Ridge National Laboratory

Rhenium-188 (Re-188) shows great promise as a versatile, therapeutic radioisotope for a variety of applications including tumor therapy, the palliative treatment of intractable bone pain resulting from metastasis to the skeleton, the inhibition of restenosis of the arteries following high-pressure balloon angioplasty, tumor therapy with peptides and antibodies, and the treatment of rheumatoid arthritis of the synovial joints. The tungsten-188 (W-188) parent is produced by the irradiation of enriched W-186 oxide targets in the Oak Ridge National Laboratory High Flux Isotope Reactor (HFIR). The HFIR has one of the highest thermal fluxes in the world and is therefore unique in its ability to produce W-188. To optimize W-188 production, the use of solid metallic W-186 is being evaluated, since the target density can be substantially increased. Pressed and sintered metallic targets have been prepared, irradiated for a full cycle in the HFIR at a flux of 1 2 x 10^{15} neutrons/cm^2/s, successfully processed, and analyzed. Production yields of 3.2 mCi/mg have been achieved.

A wide variety of beta-emitting radioisotopes have been considered for medical applications. Currently rhenium-188 (Re-188) is undergoing clinical trials at more than two dozen facilities throughout the world (Table 1). Due to its relatively short half-life (16.9), Re-188 is most conveniently obtained from the tungsten-188 (W-188)/Re-188 generator (Figure 1 and 2). Because of the long useful generator shelf-life (3 to 6 months), the high daily elution yields (>60%), and a chemistry similar to the very widely used pertechnetate, Re-188 is proving to be a cost-effective radioisotope for broad use.

Traditionally, the W-188 parent has been produced by the irradiation of enriched granular W-186 oxide targets in the Oak Ridge National Laboratory (ORNL) High Flux Isotope Reactor (HFIR) at a thermal neutron flux of 2–2.5 x 10^{15} neutrons/cm^2/s. The specific activity of W-188 averages 4 to 5 mCi/mg W-186 for a one-cycle, 24-day irradiation. Since yields of radioisotopes produced by the double neutron capture process (as W-188 is produced from W-186) are proportional to the square of the flux, the reactor neutron flux is an extremely important factor. The HFIR has one of the highest thermal fluxes in the world and is therefore unique in its ability to produce W-188 (Table 2).

Tunsten-186 is generally irradiated in the HFIR, using a quartz ampule inserted into the hydraulic tube (HT) facility, which provides space for nine "rabbit" tubes. However, recently six new Peripheral Target Positions (PTP), holding a maximum total of 48 additional rabbits, have been installed and can be utilized for full-cycle

Table 1. Clinical and research protocols using rhenium-188.

Inhibition of Restenosis	Perth, Australia Antwerp, Belgium Bonn, Germany Dresden, Germany Munich, Germany Ulm, Germany Seoul, Korea (2) Taichung, Taiwan Taipei, Taiwan Knoxville, USA Los Angeles, USA New York, USA	Palliation of Metastatic Bone Pain	Antwerp, Belgium Canterbury, England Bonn, Germany Dresden, Germany Essen, Germany Athens, Greece Szeged, Hungary Seoul, Korea (2) Montevideo, Uruguay
Synovectomy	Seoul, Korea	Therapy of Pleural Effusions-Breast Cancer	Bonn, Germany
Bone Marrow Ablation	Ulm, Germany Detroit, USA	Liver Cancer	Freemantle, Australia IAEA Sites

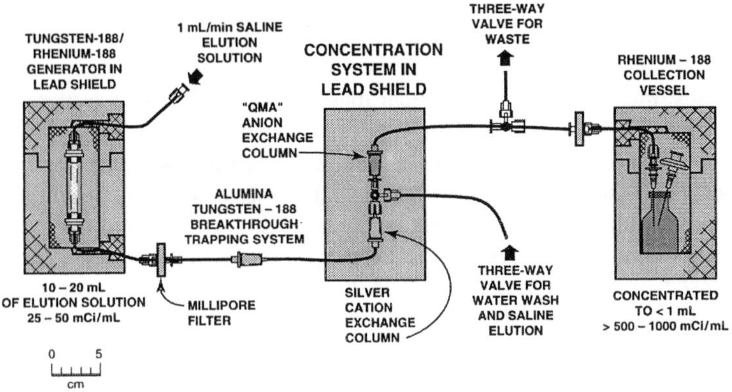

Figure 1. Schematic for set-up of tungsten-188/rhenium-188 generator system.

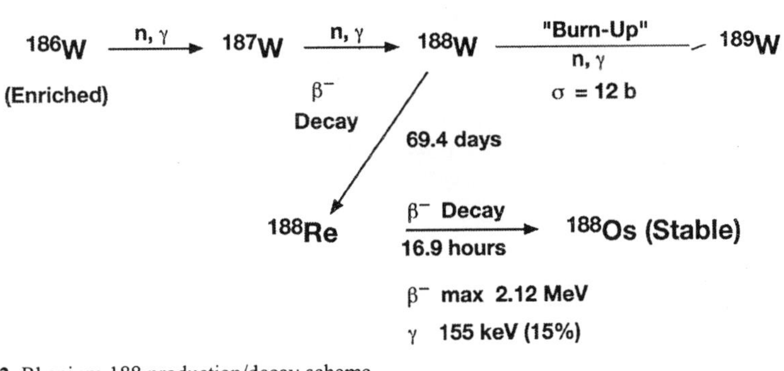

Figure 2. Rhenium-188 production/decay scheme.

Table 2. Examples of key research reactors for production of therapeutic radioisotopes.

Published Approximate Maximum Thermal Flux Values (x 10^{14} neutrons/cm^2/s)	Name of Reactor	Location	Country
30	HFIR	ORNL	U.S.A.
~30	SM3	Dimitrovgrad	Russia
12	BR2	Mol	Belgium
10	ATR	INEEL	U.S.A.
8	NRU	Nordion, Chalk River	Canada
8	HFBR	BNL	U.S.A.
5	MURR	University of Missouri	U.S.A.

irradiations at close to maximum flux conditions. The location of both the HT and PTP are shown in Figure 3.

Using the traditional enriched granular W-186 oxide targets, the maximum capacity of each rabbit is only about 700–800 mg. To increase W-188 production, the use of high-density pressed metallic W-186 targets has been evaluated, since the target density can be significantly increased. By pressing the tungsten powder under high pressure into a metallic cylindrical mass, the target density is increased by a least a factor of 3–4, which significantly increases the production yield per target from 4–5 Ci/target/cycle to as high as 12–15 Ci/target/cycle, resulting in a significant decrease in production costs. Initial tests with monolithic tungsten targets have highlighted the difficulties in achieving complete dissolution of the target, necessitating innovative approaches. The use of a pressed and sintered target, with a subsequent high-temperature oxidation step in the presence of atmospheric oxygen, has eliminated the dissolution problem.

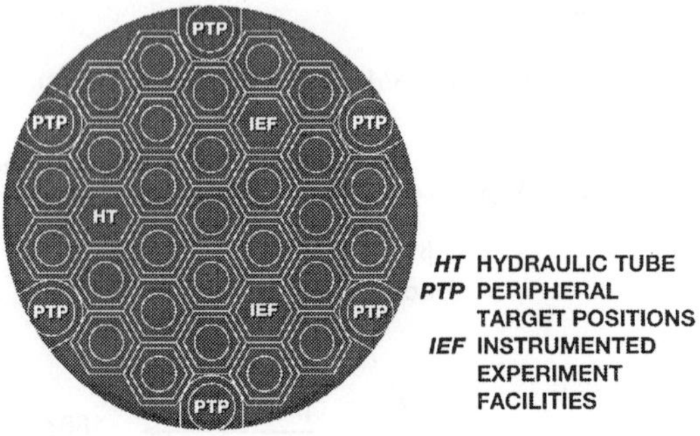

Figure 3. HFIR target unit assembly.

Pressed and sintered metallic targets have been prepared, irradiated for a full cycle in the HFIR, and successfully transformed to the tungsten oxide product, which is then dissolved in NaOH for analysis. Production yields of as high as 3.2mCi/mg of W-186 at the highest HFIR flux of 2.5 x 10^{15} neutrons/cm^2/s have been achieved. To date pressed metallic targets of up to 2.5 g have been successfully irradiated and processed. A larger target (8 g), consisting of a series of discs, has been prepared and provided to the BR2 Reactor in Mol, Belgium, for irradiation in one of their high-flux (1 x 10^{15} neutrons/cm^2/s) positions. The irradiated target was shipped to ORNL for hot-cell processing and analysis. Yields of 0.75 mCi/mg were achieved. The W-188 was then provided to MAP, Inc., in Finland, for preparation of W-188/Re-188 generators that were distributed to research sites funded through the International Atomic Energy Agency (IAEA) approved by the Department of Energy.

Based on the above results, the use of the pressed and sintered metallic targets is now becoming the "normal" mode for production of W-188. When higher-specific-activity material is required, the use of enriched W-186 oxide granular targets irradiated for two HFIR cycles will give specific activities of 7-9 mCi/mg of W-186. However, the use of metallic targets will permit large-scale production of W-188. This is expected to be an important requirement for future broader use of Re-188 and for bridging any downtimes in the HFIR operation schedule.

ABOUT AN OPPORTUNITY OF APPLICATION OF AN ISOTOPE LABEL ON THE BASIS OF STABLE ISOTOPES TO MARK VARIOUS PRODUCTS

ALEXANDRE POKIDYCHEV, MARINA POKIDYCHEVA
STABIS Co. Ltd., 10 Gamalei Street, Suite 54, 123098 Moscow, Russia
E-mail: ampokid@4unet.ru

The now existing relatively cheap technologies of stable isotopes production and quite simple techniques of isotope analysis allow to speak about long-range application of isotope labels to determine an origin of various products for a number of special (for example military) or routine customs, boundary, exchequer and other purposes. It is amazing, why isotope labels have not yet found wide application for reliable identification which would completely eliminate the possibility of copying, imitation or faking various products. Any stable isotopes rare in nature can be used as the isotope labels. Approximately 50 stable isotopes (from C-13, N-15, O-17 and O-18 up to rare isotopes of lead) can be used for these purposes. Owing to large number of chemical elements there is an opportunity to introduce any label into any substance or object (including liquids – lubricant materials, fuel etc.). These substances can be put as thin layers, are entered into painting substances, or are added directly in constructional materials. It is practically impossible to reproduce a label (even knowing its composition and having the necessary components) with the precisely same structure. More ample opportunities are given by using simultaneously 3-4 isotope-modified elements with a discrete variation of isotope concentration of each of isotopes of these elements. Theoretically there is an opportunity of assignment of an individual identification mark to each product issued in the world. Depending on a kind of a product, purpose of identification and used engineering, cost of a label and its drawing will make from several cents up to 100-1000 US dollars. It is possible to distinguish two essentially various basic fields of label's application:

a) For identification of mass parties of cheap products, valuable papers, banknotes, financial instruments, documents etc.;

b) Application of isotope labels for small-scale and special products, including military engineering.

1 Introduction

The now existing relatively cheap production technologies of stable isotopes of practically all polyisotope elements and procedure of isotopic analysis allow to speak about long-range opportunity of application of isotope labels for definition of derivation of different articles and goods. These labels can be used for the solution of many special problems (including military) or routine customs, boundary, exchequer procedures and other purposes.

It is amazing that isotope labels have not yet found application for reliable identification while working out the systems and/or protection marks which would completely eliminate a possibility of copying, imitation or fabrication of this or those articles. From the technical point of view such solution is apparent at present. There is all necessary development, procedures, technical and process equipment for its embodying.

Substances and materials, which include chemical elements with different content of stable isotopes, do not differ by chemical and macrophysical properties. The discrepancy is displayed on a microlevel only because of a different structure of nuclei of different isotopes of the same chemical element, for example, in electromagnetic fields of a particular configuration or at an irradiation by different kinds of ionizing radiation.

In accordance with this to detect, whether there are any material isotopic-modified substances of this or that kind in a composition, becomes possible only by using relevant techniques: NMR-spectrometry, mass-spectrometry, neutron-activation analysis etc. The range of serially produced instrument technique of such kind is rather wide now: from stationary precision devices up to field instruments for holding rapid analyses. Any ecological consequences connected with application of isotopic-modified substances miss. The stable isotopes, in contrast to radioisotopes, have no self radiation. They are natural substances and are a part of all existing materials.

2 The main principles of application of a label

From 1 up to 5 milligrams of isotopic-modified substance is applied on the article (frame, most important parts) or on its marking documents (label, excise marks, etc.). These substances can be sprayed in thin layers, be added in dye-stuffs, or be supplemented directly in materials (metals, plastics and so on), which the articles or their parts are produced from.

Any rare in nature stable isotopes, for example, the content of which in natural materials is less than 1 % at., can be used as an isotope label. About 50 isotopes from approximately 250 known and produced at present satisfy the requirements. Their range includes all diapason of chemical elements - from deuterium, carbon - 13, nitrogen - 15, oxygen - 17 and oxygen - 18 up to less-common isotopes of lead. Thus, there is an opportunity to add any label depending on a raised problem into any substance or object, including fluid - lubricant materials, fuel etc.

The existing analytical procedures and devices allow to define the isotope composition to within 3 significant figures at holding a rapid analysis in field requirements and up to 4-5 significant figures at holding the precision analyses in laboratory requirements.

At availability in a checked object of a label with exactly known isotope composition, it is possible to determine its authenticity or falsification precisely enough by results of the analysis. It is practically impossible to reproduce a label with absolutely the same composition even knowing its composition and having the necessary components.

The use of the labels containing simultaneously three - four polyisotope components with discontinuous variation of isotope concentrations of each component, gives considerably more ample opportunities.

So, using three isotopic-modified elements (for instance, carbon, nitrogen and oxygen while manufacturing banknotes or credit cards) and variation of concentration of each isotope in the range of 1-99,9 % at. at 100 steps of gradation (i.e. 1.0 %, 2.0 %, 3.0 %..., 99.9 %) the number of probable combinations exceeds 10^{10}.

It means, that the application of polycomponent isotope labels theoretically enables to assign individual identification code to each article released in the world, as the number of possible combinations exceeds by many times the total number of all articles of the world industry (from military engineering to food stuffs).

It is impossible to detect an isotope label even knowing about its presence without relevant instrument technique and knowledge about the place of its drawing. The plausible falsification of individual labels appropriated to each article of mass serial production is technically extremely complicated, expensive and practically unrealizable problem even if there is the information about the composition and places of drawing of isotope labels.

The cost of the label and its drawings amounts from several cents up to 100-1000 USD depending on the kind of an article, problem put by and applied technique. According to it the problem of identification can be parted into two basic directions.

3 Identification of bulk sets of cheap articles, securities, banknotes, credit cards, excise marks, documents, etc.

- The main purpose is the protection against falsification, determination of an origin source.
- The labels should be easily applied and be cheap.
- The analytical procedures should ensure a possibility of execution of the prompt continuous analysis of bulk sets by low-skilled staff or in a self-acting mode (for example - at the rummage or checkout of banknotes).
- Any cheap isotopes can be label materials: isotopes of hydrogen, boron, nitrogen, carbon, oxygen, and, probably, silicon, sulfur, iron.
- The labels are applied by addition of isotopic-modified substances in paints, paper, plastics.
- The quantity of isotope material in the label that is necessary for reliable rapid analysis amounts from several milligrams up to several tens milligrams.
- The cost of a label should not exceed 1-10 cents, that is quite reasonable at existing price level for isotopes of light elements.

4 Application of isotope labels for small-serial and special products

The basic requirement is the complexity of an unauthorized detection and elimination or replacement of a label.

The most suitable materials for labels are high-enriched compounds of rare isotopes ytterbium-168, osmium-187, palladium-102, tungsten-180 and some others.

The label is superimposed by spraying of a thin metal film by width 0,5-1 microns on the area of several square centimeters in any place of an article, for example, on the surface of the most important parts or a frame. It is desirable to superimpose the label under a lay of paint coat for protection from possible casual removal of a stratum of the label and difficulty of detection of its location.

A concrete place of location of a label can be individual for each article and correspond to its serial number.

If necessary identification, the thin scrape of skin with mass of 2-3 milligrams may be done from the known location this label. An external coat and its composition doesn't matter much and doesn't have any significant influence on precision of the analysis. The existing and widespread serial mass-spectrometers allow to carry out an isotopic analysis of a label within several hours.

For the most important details it is possible to add an isotope impurity directly in constructional materials during their manufacture. Suitable materials for this purpose are high-enriched compounds of iron-57, -58, tungsten-180, zinc-67, cadmium-106, -108 and some others rare isotopes. The application of isotopes of germanium, silicon, tin, silver is possible for units of electronic equipment.

The possibility of unauthorized detection of the label is the lowest and will in most cases be accompanied by the necessity of nonreversible disassembly of an article or complete destruction of its parts.

The value of the label will amount from 50 USD up to 1000 USD depending on both applied technology and required degree of reliability and protection.

Session M

Closing Plenary Session: "Isotope Production and Applications in the 21st Century"

"EXCESS" NUCLEAR MATERIALS HOLD KEYS TO NUCLEAR MEDICINE, RESEARCH, AND SPACE POWER

K. K. S. PILLAY

Nuclear Materials Technology Division, Los Alamos National Laboratory, Los Alamos, NM 87545 USA, USAE-Mail: s_pillay@lanl.gov

The U.S. weapons complex at one time had the only facilities and resources in the free world to produce some of the uncommon and uniquely valuable nuclear materials. Those facilities are being eliminated altogether in the name of "mortgage reduction." Since there is no chance that such facilities will be rebuilt in the foreseeable future, it is appropriate to consider saving some of the unique resources they produced as "national treasures" for the future. The scientific community has an opportunity to participate in a thoughtful process of planning to preserve some of the valuable materials within the DOE complex for the future. The "excess materials" now within the DOE complex are invaluable resources that should be preserved for the future of medicine, research, and space exploration. The investments required to stabilize and store these valuable resources are minuscule compared to investments already made to create them.

1 Introduction

Management and disposition of nuclear materials accumulated in various physical and chemical conditions are major issues facing the US Department of Energy (DOE). These residual materials are derived not only from the DOE's Nuclear Weapons Program but also from all the nuclear technology programs developed under the auspices of all US governmental agencies and some of the materials produced by the civilian sector..

During the past six decades, radioactive isotopes have been used in almost all areas of basic and applied sciences. The US Nuclear Regulatory Commission has issued about 47,000 general licenses to various nongovernmental users. Of these, 22,000 licenses are held by medical, academic, and industrial users of nuclear medicines.

The use of radioactive isotopes in nuclear medicine, fundamental and applied research, as well as in a variety of power sources used in space explorations and terrestrial applications, will be seriously impacted by the current downsizing and closing of nuclear facilities in the US. These facilities also provided some of the rare and unusual radionuclides to the world community for research and developmental projects in numerous interdisciplinary areas.

This paper identifies some of the issues relevant to the excess nuclear materials and argues for a thoughtful process of planning to preserve some of the valuable materials within the US DOE complex for the future.

2 A Unique Source of Radioactive Isotopes

The U.S. weapons complex at one time had all the facilities to produce most of the radioactive isotopes known. In addition, it had the only facilities and resources in the free world to produce some of the uncommon and uniquely valuable nuclear materials. Also, the DOE complex was the sole source of transuranic elements used by universities and research institutions to study the beneficial applications of such elements. Since almost all the facilities for producing such valuable isotopes have been shut down and are being decommissioned, it is prudent to consider saving some of the unique resources they produced as "national assets" for the future. Several categories of radio isotopes may be considered for preservation as national assets. They include:

- Large beta-gamma radiation sources
- Actinides in RTGs and pacemakers;
- Other Radioisotopes

2.1 Large beta-gamma radiation sources

Prominent among large beta-gamma sources that are in industrial uses are radioactive sources of ^{137}Cs, ^{90}Sr, and ^{60}Co designed for use as intense beta-gamma sources for industrial and scientific applications. These sources exist in a variety of physical and chemical forms, mostly as fabricated sources. Cesium-137 and ^{90}Sr are fission products whereas ^{60}Co is a product of neutron capture in a stable nuclide, ^{59}Co. The distinct nuclear properties of these isotopes make them useful for a variety of industrial applications.

At Hanford, ^{137}Cs and ^{90}Sr were originally separated from wastes from chemical processes used to produce weapons-grade plutonium. Because the heat generated by these two fission products was a serious problem for storing liquid wastes in large tanks they were separated and were solidified as cesium chloride and strontium fluoride. The solids were encapsulated in stainless steel or Hastalloy containers. The sealed sources were then stored in large pools of water to remove heat generated by these sources. Of the nearly 1,600 cesium capsules and 640 strontium capsules produced at Hanford approximately 1,330 cesium capsules and 600 strontium capsules are still at Hanford. The curie content of cesium capsules ranges from 63,000 - 84,000 Ci/capsule, and the strontium capsules average about 74,000 Ci/capsule.[1] The high curie content of these sources makes them extremely hazardous and demands special handling, storage, and maintenance.

The US government encouraged both industry and government agencies to use these materials free of charge for a variety of beneficial uses, such as sewage treatment and sterilization. However, the sources were thermally cycled, causing stress beyond their design basis, and one of the capsules developed a leak. The consequences were extensive cleanup efforts by DOE and an outcry resulting from radioactive contamination incidents. Although sound scientific fixes are available to stabilize these sources, none was attempted or explored. As a result, almost all the ^{137}Cs, and ^{90}Sr sources were returned to Hanford for safe storage and disposition.

2.2 *Actinides in RTGs and Pacemakers*

RTGs are generally made of radioactive sources of high specific activity, such as ^{238}Pu or ^{90}Sr, which convert thermal energy from alpha or beta decay into electrical energy. The Department of Defense, the National Aeronautics and Space Administration, and other users are returning depleted and unused RTGs to DOE for disposition. A specialized group of RTGs used to power artificial hearts in animals is known as pacemakers. These are small thermoelectric sources fabricated from actinides such as ^{238}Pu used in pacemakers and other medical devices and deep space heat/power sources for extraterrestrial payloads and living quarters for humans and other animals.

Another practical example is the need for ^{238}Pu as a reliable source of thermoelectric power for extended space missions and a variety of terrestrial applications. Although the supply of ^{238}Pu in the US is extremely limited, the need for ^{238}Pu for future space mission can be partially solved by creatively using two other isotopes available in plenty within the DOE complex—^{237}Np and ^{241}Am—that are considered excess to national security. Both these isotopes can be readily converted into ^{238}Pu via neutron irradiation.

2.2.1 Other Radioisotopes

In addition to the two groups of materials identified above, smaller sources (lower curie-content) of all types are valuable in all areas of fundamental and applied research as well as in a wide spectrum of tracer studies.

At one time, the DOE complex was the sole source of transuranic elements used by universities and research institutions to study the beneficial applications of such elements. An example is that of the isotope ^{252}Cf, a spontaneous fission source and efficient source of neutrons. Since this is an isotope with a relatively short half-life, it is important to preserve the parent nuclides, such as isotopes of Am, Cm, and Bk, to produce this material in the future. These and other rare isotopes are also necessary in our search for the island of stability in the periodic table beyond presently known elements

The continuing growth in radioimmunotherapy and the rapid developments in the new field of brachytherapy for localized cancer treatment are areas that require special attention because most of the active ingredients used here are short-lived alpha-emitters. These, in turn, are derived from a variety of long-lived heavy isotopes, not generally available outside the DOE complex. Typical source materials required for the generation of these short-lived alpha-emitters include ^{233}U, ^{235}U, ^{232}Th, and ^{226}Ra.

During the past decade, shortages of radioisotopes for medical research, diagnosis, treatment, and industrial applications have been well documented in numerous reports and hearings before the US Congress. In the US alone, over 30,000 diagnostic procedures using ^{99m}Tc are performed each day. Maintaining a reliable supply of these isotopes became an issue in early 1990 as a result of the closing down of DOE facilities supported by other missions. Efforts by the scientific community and the US Congress eventually resulted in a decision to have the DOE develop a reliable domestic source of this medical isotope. [2]

DOE's decision to maintain a reliable source of a few radioisotopes is a good beginning to address this need. This effort by the DOE should become a model for maintaining reliable supplies of a number of other valuable nuclear materials identified during the NMI exercise that have beneficial uses to society.

3 Orphan Nuclear Materials - Neutron Sources

A unique example of orphan nuclear materials is the man-made neutron sources that are used extensively in defense projects, industries, universities and research organizations. Examples of spontaneous fission sources include ^{252}Cf, ^{244}Cm, and several other transplutonium isotopes. However, most of the neutron sources routinely used are fabricated from a homogeneous mixture of naturally occurring and manmade alpha-emitting radionuclides, such as radium, polonium, plutonium, and americium, with elements of low atomic number, such as boron, beryllium, and fluorine. According to current laws, the DOE is responsible for disposition of almost all the neutron sources in both defense and the civilian sectors in the US.[3]

Approximately 30,000 neutron sources that currently exist in the US are likely to come back to DOE for eventual disposal. Of these, nearly two-thirds are in the civilian sector. The vast majority (>80%) of sealed actinide neutron sources use ^{241}Am as the alpha emitter; the remainder use either ^{238}Pu or ^{239}Pu as the alpha source. A few neutron sources are made from other alpha emitters, such as polonium or radium. Approximately 3,000 neutron sources per year are still being made and distributed in the US.

So far, DOE has accepted only neutron sources containing ^{239}Pu for disposition, except on an emergency basis. These sources are processed to recover plutonium, and the residues are discarded as TRU waste. Since 1979, about 1,100 ^{239}Pu-bearing neutron sources have been accepted and processed at the Los Alamos National Laboratory, including a few ^{241}Am-bearing neutron sources. Unfortunately, the implementation of this elegant technology to address large-scale disposition of man-made neutron sources is in serious doubt.

4 Conclusions

Although the DOE is in the process of winding down operations of its nuclear facilities, most of the rest of the world is moving forward with the development and applications of nuclear technologies for beneficial uses. The isotopes that are in beneficial uses are the ones that are growing in popularity. Therefore, there will continue to be a demand for the use of these isotopes both in the U.S. and worldwide. The private sector of the US economy will probably fill the supply needs from foreign sources as appropriate. In addition to the current 22,000 licensed users of radioisotopes in the US, in recent years the NRC has been issuing approximately 300 new licenses for isotopes users each year and that trend is expected to continue. A preliminary but incomplete estimate made last year identified over 34,000 records of 72 million curies of miscellaneous nonactinide radioactive materials. The actual numbers may be many times those identified last year.[4]

A significant portion of the so-called "excess materials" now within the DOE complex are invaluable resources that should be preserved for the future of medicine, research, and space exploration. The investments required to stabilize and store these valuable resources are minuscule compared to investments already made by U.S. taxpayers to create them. The total expenditures incurred by the taxpayers for the weapons complex is estimated by the Brookings Institution at $5.5 trillion.[5] The current reality is that many facilities within the DOE complex will be closed in an expedited manner to reduce future mortgage costs. During the closing process, disposition of excess materials has been largely overlooked. A unique example is the need for safe disposition of spent neutron sources. A prudent management strategy must recognize that proper preservation and management of these resources is absolutely essential to the future of science and technology and to the future of the nation.

References

1. "Project Management Plan for WESF ^{137}Cs & ^{90}Sr Capsule Storage," HNF-SD-WM-PMP-025, B&W Hanford Company, Hanford, WA, 1997.

2. "Medical Isotope Production Project: ^{99}Mo and Related Isotopes—Record of Decision," (6450-01-P), US Department of Energy, Washington, DC, September 1996.
3. Public law (PL 99-240) " Low-Level Radioactive Waste Amendment Act of 1985."
4. "National Inventory of Sources and Standards: Development of Nonactinide Isotopes and Sealed Sources Database," A draft Predecisional Document, WSRC-RP-99-00320, Westinghouse Savannah River Company, Aiken, S.C (April 1999).
5. Schwartz, Stephen I., "Atomic Audit," Brookings Institution Press, Washington, DC, 1998.

THE BULK RADIOISOTOPE PORTFOLIO OF THE ATOMIC ENERGY CORPORATION OF SOUTH AFRICA LIMITED - TECHNICAL AND MARKETING DEVELOPMENTS

D G ROBERTSON, P A LOUW, P L VAN DER WALT

Atomic Energy Corporation of South Africa Ltd., PO Box 582, Building 3100, PRETORIA, 0001, SOUTH AFRICA, Tel.: +27 12 305 5027, Fax: +27 12 305 5149, E-mail.: dgr@aec.co.za

The Atomic Energy Corporation of SA Ltd was founded some 50 years ago and for a long period had as its primary concern and as a strategic imperative the successful, independent development of most elements of the nuclear fuel cycle and other nuclear technological capabilities. Since the late Eighties a vigorous programme of commercialisation of AEC technology to meet market needs has been embarked upon and many strategic programmes have been terminated or downscaled. The bulk isotope manufacturing capability of the AEC initially focused on the production of ^{99}Mo and the successful completion of this project provided a reliable weekly production capability of 2500 Ci, a significant proportion of which is shipped to users throughout the world. Deliveries of the product are made to an ever-increasing client base whose specifications are consistently met.

The challenge was then taken up to expand the bulk isotope product portfolio and projects were initiated to produce ^{131}I, ^{32}P and ^{35}S. Development of an ^{131}I production capability received highest priority as it complemented a program to benificiate the waste stream resulting from the ^{99}Mo process. This project has recently been completed and significant quantities of ^{131}I are being supplied to various customers. The ^{32}P and ^{35}S production programmes are at an advanced stage of development and these products will be available to the market before the year 2000.

Certain key technical and marketing issues surrounding bulk isotope supplies are discussed in this paper against the background of progress made in this field at AEC.

1 INTRODUCTION

A review of the available global isotope production capacity could lead to the conclusion that there is excess capacity and hence no room for new entrants to the market. In the case of ^{99}Mo, current capacity probably exceeds market demand by a factor of two. The critical issue is not the total capacity however, but rather the distribution of capacity among the various producers. Radioisotopes are strategic raw materials for the radiopharmaceutical and life science industries and a disruption in the supply of ^{99}Mo, for example, would have disastrous consequences for the nuclear medicine industry. It is therefore crucially important that sufficient production capacity should be distributed among a number of manufacturers to enable an emergency backup arrangement to be activated should one or more of these producers suffer any form of production failure. Ideally, users of isotopes should be in a position to adopt a multi-sourcing strategy to minimise the risk of supply disruption. Isotope manufacturers, on the other hand, should develop strategies to allow for backup supply arrangements in the event of emergency situations resulting from breakdown in supply.

The AEC believes that there will be a growing trend towards strategic alliances between the major producers of isotopes with a view to meeting the needs of the market for surety of supply. At the vanguard of this tendency, AEC has formed a strategic alliance with the Institut National Des Radioelements (IRE) of Belgium whereby customers are offered an assured supply of ^{99}Mo from two geographically and organisationally distinct production facilities with guaranteed access to the products from a number of reactors.

2 AEC BACKGROUND

The corporate headquarters, manufacturing plants and associated facilities of the AEC are located at the large Pelindaba complex, approximately 25 km west of Pretoria, South Africa.

The AEC was founded for the purpose of carrying out nuclear research and strategic programs, including development of the nuclear fuel cycle. The advent of political changes in the early 1990s and consequent lifting of economic sanctions rendered many of these programs redundant and they were terminated at short notice. In the late 1980s the AEC embarked upon a commercialisation program aimed at re-deploying existing facilities and resources (including technological and scientific skills and knowledge) into profitable commercial entities in keeping with market needs. Good progress has to date been made with this commercialisation program and the AEC now is organisationally divided into two groupings, namely Pelindaba Technology (the commercial arm of the AEC) and the Pelindaba Nuclear Institute (nuclear research and development plus facility management). The isotope program of the AEC is a major business undertaking resident within the Pelindaba Technology Division.

3 ISOTOPE PRODUCTION PROGRAM

Isotope production activities at the AEC date back to the early Seventies when ^{131}I was produced by the irradiation of Tellurium oxide. This product has been used in the manufacture of a range of therapeutic and diagnostic capsules supplied to the South African nuclear medical market for the past 20 years. In parallel with ^{131}I developments, the AEC embarked upon a program to develop a ^{99m}Tc generator which was introduced in 1975 and which provided the backbone for development of the local nuclear medicine industry. With the growth of nuclear medicine and the resultant increase in the demand for ^{99m}Tc generators, a program was initiated to manufacture fission ^{99}Mo to meet local requirements. Building of a ^{99}Mo production facility was commenced in 1988 with a planned capacity of 80 Ci per week.

Good progress with this program, together with increased emphasis being placed on the commercialisation drive and the availability of an excellent hot cell facility led to a decision to embark upon large-scale production of radioisotopes, predominantly for the international market.

In April 1993 the first domestically produced ^{99}Mo was loaded into a PELTEK-F generator. Thereafter local requirements have been met without disruption and in 1994 the first export contract was negotiated and supplies of South African produced ^{99}Mo to international markets have since then shown consistent growth.

4 RADIATION FACILITIES AND INFRASTRUCTURE

The SAFARI reactor forms the foundation of the isotope program at the AEC. In addition to the reactor, AEC owns and operates various other irradiation and support facilities which include a world class hot cell complex, fuel fabrication and waste disposal facilities. These facilities, together with high level expertise in fields such as reactor physics, radio-chemistry and hot cell design and operation placed the AEC in a position to successfully embark upon a significant isotope production program.

The AEC is unique amongst ^{99}Mo manufacturers in that it has under its control the entire production cycle, from manufacture of the target plates through to ultimate processing and disposal of waste materials.

5 THE ^{99}Mo PROGRAM

The ^{99}Mo production plant consists of two back-to-back lines of ten lead shielded hot cells. Production capacity currently stands at 2500 Ci per week (6-day pre-calibration) and revisions to Safari irradiation capacity will boost weekly capacity to in excess of 3000 Ci.

Since the first successful production run in 1993, efforts have not only been focused on increasing plant capacity but also refining and improving efficiency of the production process. The success of this program is illustrated by the fact that the yield per target plate has been increased by a factor of 2,5 since 1993.

6 THE ^{131}I PROGRAM

^{131}I is a fission product which is available in irradiated ^{99}Mo target plates. In the plate dissolution process, nuclides of a few elements are dissolved together with

Molybdenum and Iodine. Following dissolution, the product solution passes through purification steps which are designed specifically to ensure proper decontamination of Molybdenum and Iodine in separate production streams. Purification of Iodine is carried out by means of an anion exchange resin and a distillation step.

The Iodine production facility consists of 5 hot cells which are served by a dedicated ventilation system keeping the cells under a slight negative pressure relative to the surrounding areas.

7 ^{32}P AND ^{35}S PRODUCTION

The inherent bulk processing capability of the AEC's Hot Cell Facility lends itself well to the production of large quantities of ^{32}P and ^{35}S and two production lines have been introduced to manufacture these isotopes.

Three fully shielded hot cells and one glove box are used for processing irradiated sulphur and potassium chloride targets. Hot commissioning of these production lines has commenced and excellent results are being achieved. AEC's production capacity is equal to current world requirements for these isotopes and supplies from AEC will commence towards the end of 1999.

8 CONCLUSION

The AEC has successfully made the transition from institutional organization to a fully-fledged production and distribution center for bulk isotopes from its base in South Africa and is well positioned, technologically and commercially, to play an ever increasing role in the global radioisotope market.

As we enter the 21st Century it is foreseen that there will be a significant change in the dynamics of global isotope supply. There will be a move away from the vulnerability to single sources supply and the trend will be towards the formation by manufacturers of strategic alliances so as to provide customers with a comprehensive, assured supply of isotopes. To this end AEC has established alliances with international organizations and is constantly reviewing other alliance opportunities.

List of Participants

3ICI PARTICIPANTS LIST

Dr. Jorma Aaltonen
University of Helsinki,
Dept. of Chemistry, Lab. of Radiochemistry,
P.O. Box 55
FIN-00014 Helsinki, FINLAND
jorma.o.aaltonen@helsinki.fi

Dr. John Aberhart
Cambridge Isotope Laboratories Inc
50 Frontage Road
Andover MA
USA 01810-5413
cilmkt@isotope.com

Dr. Brian Abeysekera
MDS NORDION
4004 Wesbrook Mall
Vancouver BC
Canada V6T 2A3
babeyseker@mds.nordion.com

Dr. Mike Adam
TRIUMF
4004 Wesbrook Mall
Vancouver BC
CANADA V6T 2A3

Mr. Vladimir Afanassiev
Siberian Group of Chemical Enterprises
1 Kurchatov Street
Seversk Tomsk Region
RUSSIA 636070

Dr. Peter Airey
ANSTO
Private Mail Bag 1
MENAI NSW
Australia 2234
pla@ansto.gov.au

Prof. Jigang An
INET Tsinghua University
Haidian District
Beijing
CHINA 100084
liye@inet.tsinghua.edu.cn

Dr. Gunnar Antoni
Uppsala University PET Centre
UAS,
Uppsala, SWEDEN S-75185
gunnar.antoni@pet.uu.se

Mr. Vladimir Antsoupov
Siberian Group of Chemical Enterprises
1 Kurchatov Street
Seversk Tomsk Region
RUSSIA 636070

Dr. Tatsuo Aono
National Institute of Radiological Sciences
3609 Isozaki
Hitachinka Ibaraki
JAPAN 311-1202
t_aono@nirs.go.jp

Prof. Ramon Aravena
University of Waterloo
200 University Avenue West
Waterloo Ontario
Canada N2L 3G1
roaraven@sciborg.uwaterloo.ca

Mr. Gordon Ashfield
Nordon International
447 March Road
Kanata Ontario , Canada K1Y 1Y8
gashfield@mds.nordion.com

Dr. Gerd Asperger
Chemotrade GmbH
Chemiehandelsgesellschaft Brakestreet 27
Leipzig
GERMANY D-04347
asp.chemotrade@t-online.de

Ms. Clara Bajgar
Implant Sciences Corp.
107 Audubon Road, #5
Wakefield MA
USA 01880

Mr. Dirk Becker
German Cancer Research Center
Radiology IM Neuenheimer Feld 280
Heidelberg
GERMANY 69120
D.Becker@dkfz-heidelberg.de

Dr. Carlos Bekerman
Michael Reese Hospital & Medical Ctr
5471 Hyde Park Blvd, apt 2B
Chicago Ill
USA 60616

Mr. Neil Bennett
REVISS Services (UK) Ltd
6 Chiltern Court Asheridge Road
Chesham Buckinghamshire
UK HPS 2PX
neil_bennett@reviss.co.uk

Mr. Owen Blacklaws
MDS Nordion
4004 Wesbrook Mall
Vancouver BC
CANADA V6T 2A3

Mr. Rick Bloemhard
TRIUMF - ATG
4004 Wesbrook Mall
Vancouver BC
Canada V6T 2A3
brick@triumf.ca

Prof. Ove Boman
University of Uppsala
Inst. Of Chemistry Box 531
Uppsala SWEDEN S-75121
ove.bohman@kemi.uu.se

Mr. Fred Boyd
Canadian Nuclear Society
9 Sandwell Crescent
Kanata Ontario
Canada K2K-1V2
fboyd@sympatico.ca

Mr. Dev Britto
UBC Department of Botany
6270 University Blvd
Vancouver BC
Canada V6T 1Z4
dtbritto@interchange.ubc.ca

Dr. Darren Brown
Trace Sciences International
15 Wertheim Court, Suite 404
Richmond Hill Ontario
CANADA L4B 3H7
darren@isotopetrace.com

Mr. Kenneth Buckley
UBC/TRIUMF PET Program
4004 Wesbrook Mall
Vancouver BC, Canada V6T 2A3
buckley@triumf.ca

Mr. Leon Butler
Studsvik Nuclear AB
SE-611 82 Nykoping
SWEDEN

Dr. Victor Bykov
V.G. Khlopin Radioum Institute
28, 2nd Murinsky Av
St. Petersburg
RUSSIA 194021
bykov@atom.nw.ru

Mr. John Carty
US Dept of Energy, Isotope Program
19901 Germantown Road
Germantown, MD
USA 20878
John.Carty@HQ.doe.gov

Mr. Chris Chater
Urenco Nederland B.V.
P.O. Box 158
Almelo
The Netherlands 7600 AD
isotopes@urenco.nl

Prof. Amares Chatt
Dalhousie University
Slowpoke 2 Facility Dept of Chemistry
Halifax NS
B3H 4J3
a.chatt@dal.ca

Mr. Oberon Cheng
Trace Sciences International
15 Wertheim Court Suite 404
Richmond Hill Ontario
Canada L4B-3H7

Mr. Alexander Chikshov
Trace Sciences International
15 Wertheim Court Suite 404
Richmond Hill Ontario
Canada L4B-3H7

Dr. Grazyna Ciszewska
Novartis Pharmaceuticals Corp.
59 Route 10, Bldg 405/141
East Hanover, NJ
USA 07936
grazyna.ciszewska@pharma.novartis.com

Dr. David Clark
Ion Beam Applications, S.A.
Chemin du Cyclotron 3
B-1348 Louvain la Neuve
BELGIUM
ternier@IBA.BE

Mr. Glenn Coarson
Isotec Inc
3858 Benner Road
Miamisburg OH
USA 45342
Gcoarson@Mathesongas.com

Mr. Milo Conard
PAC sprl
av. des Moissonneurs, 43
Dion-Valmont
Belgium 1325
milo@skynet.be

Mr. Mario Cornacchia
Ontario Power Generation
700 University Avenue H4-D14
Toronto ON
CANADA M5G 1Z5
mario.cornacchia@ontariopowergeneration.com

Prof. Raymond Counsell
Univ. of Michigan, Medical School
Dept. of Pharmacology
1301 Medical Research Bldg III
Ann Arbor, MI USA 48109-0632
counsell@umich.edu

Dr. Jerry Cuttler
AECL & CNS
2251 Speakman Drive
Mississauga On
CANADA L5K 1B2
cuttlerj@aecl.ca

Prof. Song Daohai
Nuclear Power institute of China (NPIC)
P.O. Box 291-107 Chengdu
Sichuan 610005
CHINA
qbs@npic.ac.cn

Mr. Keith Darabos
Spectra Gases
3434 Rt. 22 West
Branchburg NJ
USA 08876
gdietzel@raytest.de

Mr. Bernard David
Inst. National des Radioelenents I.R.E.
Zoning Industriel
6220 Fleurus
BELGIUM
generalmail@ire.be

Mr. Robert David
Isotope Products Labs.
1800 North Keystone Street
Burbank California
USA 91601
rdavid@isotopeproduct.com

Dr. Elisabete De Nadai Fernandes
CENA
Avenida Centenário 303 Bairro São Dimas
Piracicaba São Paulo
Brazil 13416-000
lis@cena.usp.br

Mr. Antoine Debauche
I.R.E.
Zoning Industriel
6220 Fleurus
BELGIUM
generalmail@ire.be

Dr. Cathie Deeley
Reviss Services (UK) Ltd
6 Chiltern Court Asheridge Road
Chesham Buckinghamshire
ENGLAND HP5 2PX
cathie_deeley@reviss.co.uk

Mr. Jean Deflers
CHEMGAS
31 Bis Ave R. Schuman
Boulogne
FRANCE 92100
chemgas@chemgas.com

Dr. Morgan Dehnel
Dehnel Consulting
PO Box 201
Nelson BC
CANADA V1L 5P9
mdehnel@wkpowerlink.com

Dr. Guenter Dietzel
Raytest GmbH
Benzstr.4
Straubenhardt
GERMANY

Mr. Peter Dodwell
Cambridge Isotope Laboratories, Inc.
50 Frontage Road
Andover MA
USA 01810-5413

Mr. Michael Druce
Radiopharmaceuticals Div. ANSTO
PMBI
Menai NSW
AUSTRALIA 2234
mjd@ansto.gov.au

Mr. Stephen Dueker
UC Davis
Meyer Hall One Shields Avenue
Davis California
USA 95616
srdueker@ucdavis.edu

Ms. Maureen Duffy
Cambridge Isotope Laboratories, Inc.
50 Frontage Road
Andover MA
USA 01810-5413
maureen@isotope.com

Dr. John Duke
Slow Poke Reactor Facility
University of Alberta 3118 Dent/Pharm Bldg
Edmonton AB
CANADA T6G 2E3
John.Duke@Ualberta.ca

Mr. Johnny Dulku
SFU, 8888 University Drive
Burnaby BC
Canada V5A 1S6
jdulku@sfu.ca

Dr. Gerardo Dutto
TRIUMF
4004 Wesbrook Mall
Vancouver BC
V6K 1C4
dutto@triumf.ca

Mr. John Easter
Pharmacia & Upjohn
7265-300-367 301 Henrietta
Kalamazoo MI
USA 49007
John.A.Easter@am.pnu.com

Dr. David Ehst
Argonne National Laboratory
9700 S. Cass Ave., Bldg. 315
Argonne IL
USA 60439
ehst@anl.gov

Prof. Mohamed El - Absy
Atomic Energy Authority
Cairo
Egypt 13759

Dr. George Ellames
Sanofi Research
Alnwick Research Centre Willowburn Avenue
Alnwick Northumberland
UK NE 66 2JN
george.ellames.sanofi.com

Dr. Fontana Erminia
Pharmacia & Upjohn
vle Pasteur 10
Nerviano, Milan
ITALY 20014
erminia.fontana@eu.pnv.com

Ms. Elisabetta Fabbro
624 Cleveland Street
Davis California
USA 95616
eefabbro@ucdavis.edu

Mr. Ping Fang
National Laboratory of Nuclear Medicine
No. 20 Qianrong Road
Wuxi Jiangsu
P.R.China 214063
jsinm@public1.wx.js.cn

Mr. Jack Faught
Spectra Gases
3434 Rt. 22 West
Branchburg NJ
USA 08876

Dr. Vladimir Feodorov
V.G. Khlopin Radium Institute
28, 2nd Murinsky Av.
St. Petersburg
RUSSIA 194021
fedvv@atom.nw.ru

Dr. Gunter Firnau
Hamilton Health Sciences Corp.
1200 Main Street West
Hamilton Ontario
Canada L8N 3Z5
gfirnau@yahoo.com

Dr. Christian Foged
Novo Nordisk A/S
Novo Nordisk Park
DK-2760 Malov
DENMARK
cf@novo.dk

Ms. Jennifer Follet-Dueker
UC Davis
36999 Russell Blvd
Davis California
USA 95616

Dr. Klaus Freyer
Centre of Environmental Research
Leipzig-Halle, Dept of Analytical Chemistry
Permoserstrasse 15, Leipzig
GERMANY D-04318
freyer@ana.ufz.de

Dr. Ryoko Fujiyoshi
Hokkaido University
Kita13, Nishi8, Kita-ku
Sapporo, Hokkaido
Japan 060-8628
ryoko@ccms1.hucc.hokudai.ac.jp

Prof. Michiko Fukushima
Ishinomaki Senshu University
Minamisakai
Ishinomaki Miyagi
Japan 986-8580
fukusima@isenshu-u.ac.jp

Ms. Shiori Furihata
TRIUMF
4004 Wesbrook Mall
Vancouver BC
Canada V6T 2A3
furihata@triumf.ca

Mr. Keith Garnes
SmithKline Beecham Pharmaceuticals
Mailcode UW 2830 709 Swedeland Road
King of Prussia PA
USA 19406
Keith_T_Garnes@sbphrd.com

Prof. Helen Garnett
Australian Nuclear Science Technology
Private Mailbag 1 Mena 1
Sydney NSW
AUSTRALIA 2234
helen.garnett@anstro.gov.au

Dr. Zev Gelbart
TRIUMF
4004 Wesbrook Mall
Vancouver BC
CANADA V6T 2A3
gelbart@triumf.ca

Dr. Pierre Girouard
AECL
2251 Speakman Drive
Mississauga Ontario
CANADA L5K 1B2
giriuard@aecl.ca

Dr. Oleg N. Godissov
Sci & Technol. Center, Centrotech-EChZ
Stachek Ave 47
Saint-Petersburg
RUSSIA 198096
godisov@neva.spb.ru

Dr. Om Goel
Parke-Davis Co
2800 Plymouth Road
Ann Arbor Mi
USA 48105
goelo@aa.wi.com

Mr. Thomas Golembeski
NEN Life Science Products
331 Treble Cove Rd
N Billerica MA
USA 01862
golembeski@nenlifesci.com

Mr. Clive Goodall
Nycomed Amersham
White Lion Road
Amersham Labs
Amersham, Buckinghamshire UK HP23 4DX
clive.goodall@uk.nycomed-amersham.com

Dr. Galen Greer
Information, Science & Technology
3rd Floor - 563 Superior Street
Victoria BC
CANADA V8W 9V1
galen-greer@gems8.gov.bc.ca

Prof. Henry Griffin
Univ. of Michigan
Chemistry 1055 Univ. of Michigan
Ann Arbor MI
USA 48109-1055
hcg@umich.edu

Dr. Gennadi Grigoriev
IMP RRC "Kurchatov Institute"
123182 Kurchatov sq.1
Moscow
Russia
ggrig@imp.kiae.ru

Dr. Terry Grimm
U of Michigan
3558 Benner Rd
Miamisburg Ohio
USA 45342
dhemmig@mathesongas.com

Dr. Surendra Gupta
American Radiolabeled Chemicals Inc
11624 Bowling Green Dr.
St. Louis MO
USA 63146
arcinc@arc-inc.com

Dr. Kazuharu Hamada
C/DIN Isotopes
88 Leacock St
Pointe-Claire Quebec
CANADA H9R 1H1
rad.chemie.tu-muenchen.de info@cdniso.com

Dr. Donald Hamlin
University of Washington
2121 N 35th Street
Seattle WA
USA 98103-9103
dkhamlin@u.washington.edu

Dr. Robert Hamm
AccSys Technology Inc.
1177 A Quarry Lane
Pleasanton CA
USA 94566
rhamm@linacs.com

Ms. Mary Hansen
Amersham Pharmacia Biotech
800 Centennial Avenue P.O. Box 1327
Piscataway NJ
USA 08855
Mary.Hansen@amapbiotech.com

Dr. Clarence Hardy
Australian Nuclear Association
P.O. Box 85
Peakhurst NSW
AUSTRALIA 2210
cjhardy@ozemail.com.au

Mr. Herbert Hegener
Chemotrade GMBH
Harless Str. 1A
D-40239 Dusseldorf
GERMANY
chemotrade@t-online.de

Dr. David Hemmig
Isotec Inc
3558 Benner Rd.,
Miamisburg Ohio
USA 45342
dhemmig@mathesongas.com

Mr. Leonard Hendrickson
Isotope Products
1800 North Keystone
Burbank California
USA 91504
lhendrickson@isotopeproducts.com

Dr. Richard A. Henkelmann
Tech. Universitat Muenchen, Radiochemie
Walter-Meissner Strasse 3
D-85740 Garching, Bavaria
GERMANY

Prof. Yoshio Homma
Kyoritsu College of Pharmacy
1-5-30, Shibakoen, Minato-ku
Tokyo
JAPAN 105

Mr. Nobumi Horikawa
Technol Suport System Corp.
5F, Kashiwaya-building, 3-40-11 Hongo,
Bunkyo-ku, Tokyo
JAPAN 113-0033
m6876nf277h@asahi-net.or.jp

Mr. Toshikazu Hosoda
Chiyoda Technol Corp.
1-7-12, Yushima,
Bunkyo-ku, Tokyo
JAPAN 113-8681
hosoda-t@c-technol.co.jp

Dr. JI Hu
Chinese Inst. Of Atomic Energy
Beijing
P.R. CHINA

Dr. Yaer Hu
Shanghai Second Medical University
280 South Chongquing Road
Shanghai
CHINA 200025
hyx@shsmu.edu.cn

Dr. Che Huang
Parke-Davis Pharmaceutical Research
2800 Plymouth Road
Ann Arbor MI
USA 48105
che.huang@wl.com

Mr. Rod Huggins
MDS Nordion
447 March Road
Kanata Ontario
CANADA K2K 1X8
rhuggins@mds.nordion.com

Dr. Don Hunter
TRIUMF
4004 Wesbrook Mall
Vancouver BC
CANADA V6T 2A3
hunter@triumf.ca

Dr. Duncan Hunter
University of Western Ontario
Chemistry Department
London Ontario
Canada N6A 5B7
dhunter@julian.uwo.ca

Ms. Tamara Hurtado
TRIUMF/PET Programme
4004 Wesbrook Mall
Vancouver BC
Canada V6T 2A3
magic@triumf.ca

Dr. Ahmed Hussein
UNBC
3333 University Way
Prince George BC
Canada V2N 4Z9
hussein@unbc.ca

Dr. Arslan Iliassov
RITVERC GmbH
28, 2nd Murinsky Ave
St. Petersburg
RUSSIA 194021
ritverc@atom.nw.ru

Dr. Teruo Inoue
Daiichi Radioisotope Lab. Ltd.
17-10, Kyobashi 1-Chome Chuo-Ku
Chuo-Ku Tokyo
JAPAN 104-0031
iteruo@dri.co.jp

Ms. Kaeko Iwai
Chiyoda Technol Corp.
1-7-12, Yushima,
Bunkyo-ku, Tokyo
JAPAN 113-8681
iwai-k@c-technol.co.jp

Mr. David Jamriska
Los Alamos National Laboratory
MS J514, Los Alamos NM
USA 87545
djamriska@lanl.gov
pbesp@sar.nnov.ru

Dr. Zhigang Jian
Monsanto
645 Hyacinth Place
Highland Park Illinois
USA 60035
zhigang.jian@monsanto.com

Dr. Jian Jin
National Laboratory of Nuclear Medicine
No.20 Qianrong Avenue
Wuxi Jiangsu
P.R.China 214063
jsinm@public1.wx.js.cn

Ms. Salma Jivan
TRIUMF/PET Program
4004 Wesbrook Mall
Vancouver B.C.
Canada V6T 2A3
salma@triumf.ca

Prof. John Jones
University of Surrey
Chemistry Department
Guildford Surrey
UK GU2 5XH
J.R.Jones@surrey.ac.uk

Prof. Alexei K. Kaliteevskii
Centrotech-EChZ
Stachek Ave 47
Saint-Petersburg
RUSSIA 198096

Dr. Bo-Chan Kao
Rhone-Poulenc Rorer
500 Arcola Road
Collegeville PA
USA 19426
bochan.kao@rp-rorer

Mr. Rudolph Karin
DuPont Pharmaceuticals Co.
331 Treble Cove Road
N. Billerica MA
USA 01862
rudolph.w.karin@dupontpharma.com

Mr. Peter Kavitov
RFIAC-VNIIEF
Bessarabenkc Street 9,
ap 69 Schelernodoroschnaja street 22
Sarov Nizhny Novgorod Region
RUSSIA 607180

Dr. Jerry Klein
Oak Ridge National Laboratory
P.O. Box 2008
Oak Ridge Tennessee
USA 37831-6182
kleinja@ornl.gov

Mr. George Klody
Nat'l Electrostatics Corp.
P.O. Box 620310
Middleton WI
USA 53562-0310
nec@pelletron.com

Dr. George Klonda
Nat. Inst. of Standards & Tech.
Route 270 & Quince Orchard Road /
Building 222, Room B150
Gaithersburg Maryland USA 20899-8372
george.klonda@nist.gov

Dr. Russ Knapp
Oak Ridge National Laboratory
Post Office Box 2008
Oak Ridge, TN
USA 37831-6229
jkp@ornl.gov

Dr. Kathryn Kolsky
Brookhaven Nat'l Laboratory
Medical Department
Upton NY
USA 11973
kolsky@bnl.gov

Dr. Jozsef Kornyei
Institute of Isotopes Co.
29-33 Konkoly Thege U
Budapest
HUNGARY H-1121
kornyei@izotop.kfkipark.hu

Ms. Ekaterina Kouzova
UNICON
26, Staromonetniy per
Moscow
RUSSIA 109180

Dr. Krish Krishman
Canadian Nuclear Society
2251 Speakman Drive
Mississauga Ontario
Canada L5K-1B2
krishan@aecl.ca

Mr. Boris Kudelin
V.G. Khlopin Radium Institute
28, 2nd Murinsky Ave
St. Petersburg
RUSSIA 194021
ritverc@atom.nw.ru

Mr. Yuri Kulinech
Electrochemical Plant
Energetikov 2-189
Zelenogozsk, Zelenogozsk Region
RUSSIA 663690
taifun@ecp.kts.ru

Mr. Alexander Kupfer
Freie Universitat Berlin
Forbeckstr 34136
Berlin GERMANY 14135

Dr. Rostislav Kuznetsov
Research Institute of Atomic Reactors
Dimitrivgrad 10
RUSSIA

kar@orip.niiaz.simbizsk.su

Dr. Steve Laflin
Int'l Isotopes Idaho Inc.
2325 W. Broadway, Suite D
Idaho Falls, Id
USA 83402
slaflin@srv.net

Dr. Jacques Laizier
CEA-DAMRI
BP No 52 91193
Gif sur yvette Cedex
France
Jacques.laizier@cea.fr

Prof. Sheldon Landsberger
University of Texas
Pickle Research Campus R-9000 Building #159
Austin Texas
USA 78712
s.landsbergeremail.utexas.edu

Dr. Sheldon Landsberger
University of texas
Prickle Research Campus R 9000, Bldg 159
Austin Texas
USA 78712

Mr. David Lapointe
Université de SHerbrooke
3001 12 Avenue Nord
Sherbrooke Quebec
Canada J1H 5N4
dlapoint@tep.crc.usherb.ca

Mr. Phil Larabie
MDS Nordion Inc
4004 Wesbrook Mall
Vancouver BC
Canada V6T 2A3
plarabie@mds.nordion.com

Dr. Doug Latornell
MDS Nordion
4004 Wesbrook Mall
Vancouver BC
CANADA V6T 2A3
dlatornell@mds.nordion.com

Dr. Juergen Leonhardt
IUT GmbH
Rudower Chaussee 5
12489 Berlin
GERMANY 433510

Dr. Iddo Levin
NRCN
PO Box 9001
Beer Sheva
ISRAEL 84149
iddole@internet-zahav.net

Mr. Kris Lewett
CNL International Inc.
433 Town Center, #488
Corte Madera California
USA 94925
kristjan@csi.com

Prof. Dewi Lewis
Nycomed - Amersham
White Lion Road
Amersham Bucks
United Kingdom MP7 9LL
dewi.lewis@uk.nycomed-amersham.uk

Mr. Robert Lewis
DuPont Medical Imaging
331 Treble Cove Road
N. Billerica MA
USA 01862

Dr. Wuu-Jyh Lin
Institute of Nuclear Energy Research
Post Office Box 3-27
Lung Tan Tao-Yen
Taiwan 325
bblin@usa.net

Dr. Xilei Lin
Institut fuer Radiochemie
Walther-Meissner-Str. 3
Garching bei Munich Bavaria
Germany 85748
lin@rad.chemie.tu-muenchen.de

Dr. Yumei Lin
UC Davis
One Shields Avenue
Davis California
USA 95616
ymlin@ucdavis.edu

Dr. Jeanne Link
Univ. of Washington,
UW Medical Center, Rm. NW053
1959 NE Pacific St. Box
Seattle WA 356004 , USA 98195-6004
jeanne@u.washington.edu

Mr. Hans-Joachim Linke
Welischmiller Dresden
Rossendorfer Ring 42
Schoenfeld-Weissig
GERMANY
waelisch-miller_dresden@t-online.de

Dr. Sueo Machi
Int'l Atomic Energy Agency
Vienna International Center, PO Box 100,
Wagramerstr. 5 Schoenfeld-Weissig
Vienna, AUSTRIA A-1400
S.Machi@iaea.org

Dr. Andrej Matthies
Waelischmiller Dresden
Rossendorferring 42
GERMANY
WAELISCHMILLER_DRESDEN@T-ONLINE.DE

Mr. William Litschgi
Monsanto Co.
700 Chesterfield Pkwy Mail Zone: BB46
St. Louis Missouri
USA 63198
wjlits@ccmail.monsanto.com

Ms. Janet MacIntosh
MDS Nordion Inc
4004 Wesbrook Mall
Vancouver BC
CANADA V6T 2A3
jmackintosh@mds.nordion.com

Dr. Leonard Mausner
Brookhaven National Laboratory
Medical Dept. Bldg 801 PO Box 5000
Upton NY
USA 11973-5000
mausner@bnl.gov

Dr. Yu-Ying Liu
Hoffmann LaRoche Inc.
340 Kingsland St.
Nutley NJ
USA 07110
yu-ying.liu@roche.com

Mr. Martin Magerl
Nycomed Amersham
3350 N. Ridge
Arlington Heights IL
USA 60004
marty.magerl@us.nycomed-amersham.com

Mr. Everaldo Maximo
CENA/USP
Avenida Centenário, 303 Cx.Postal 96
Piracicaba São Paulo
Brazil 13400-970
evmaximo@cena.usp.br

Mr. Peter Louw
A.E.C. of S.A. Ltd.
P.O. Box 582
Pretoria
Rep. of South Africa 0001
palouw@aec.co.za

Dr. Yasunori Mahara
CRIEPI
1646 Abiko
Abiko Chiba
JAPAN 270-1194
mahara@criepi.denken.or.jp

Dr. Brad Maxwell
Monsanto Company
Mail Zone BB5G
700 Chesterfield Parkway North
St. Louis Missouri, USA 63366
brad.d.maxwell@monsanto.com

m
JML / TRIUMF
4004 Wesbrook Mall
Vancouver BC
Canada V6T 2A3
lulu@triumf.ca

Dr. Verinder Mahindroo
G.D. Searle (Monsato)
4901 Searle Parkway
Skokie Illinois
USA 60077
verinder.k.mahindroo@monsato.com

Mr. Chris McCooeye
C/D/N Istotopes Inc
88 Leacock Street
Pointe Claire Quebec
CANADA H9R 1H1

Dr. Stephane Lucas
IBA
Chemin du Cyclotron 6
Louvain-la-Neuve
BELGIUM 1348
lucas@iba.be

Mr. Sviatoslav Makarovski
AO "Techsnabexport"
26 Staromonetniy per
Moscow
RUSSIA 109180
info@cdniso.com

Dr. Alexander McEwan
Cross Cancer institute
11560 University Ave
Edmonton Ab
CANADA T6G 1Z2
sandymce2cancerboard.ab.ca

Ms. Elena Lukashina
Moscow State University
Radiochemistry Div., Chemistry Dept.,
Vorob'evy Gory, Moscow
Russia 119899
lena@inorg.chem.msu.ru

Mr. Serguei Maltsev
PA MAYAK, Radioisotope Plant
31 Lenin Street
Ozyorsk, Chelyabinsk Region
RUSSIA 456780
riplant@mayak1.chel-65.chel.su

Mr. Richard McKannay Jr
Isoflex USA
PO Box 29475
San Francisco CA
USA 94129
isoflex.usa@att.net

Mr. Rolf Lundstrom
Studsvik Nuclear AB
SE-611 82 Nykoping
SWEDEN
rolf.lundstrom@studsvik.se

Prof. Gunter Marx
Freie University of Berlin
Buggestr 5
Berlin
GERMANY 12163

Dr. Paul McNamara
Schering-Plough Research Institute
2015 Galloping Hill Rd
Kenilworth New Jersey
USA 07033
pail.mcnamara@spcorp.com

Dr. David Melillo
Merck Research Laboratories
PO Box 2000
Rahway,NJ
USA 07065
dave_melillo@merck.com

Mr. Vladimir Meshkov
AO "Techsnabexport"
26 Staromonetniy per
Moscow
RUSSIA 109180

Dr. Viatcheslav Michine
RSC Applied Chemistry
Leningradsky Region

RUSSIA 198663

Dr. Nikolaj Mikhajlov
Inst of Geological Sciences
National Academy of Sciences of Belarus Kuprevich str.7, Minsk
BELARUS 220141
mihailov@ns.igs.ac.by

Mr. Mark Moore
ANSTO
Mail Box M34 Camperdown
Sydney NSW
AUSTRALIA 2050
mark@ansto.gov.au

Dr. Lutz Moritz
TRIUMF
4004 Wesbrook Mall
Vancouver BC
CANADA V6T 2A3
moritz@triumf.ca

Prof. Serag Morsy
Atomic Energy Authority, Cairo
Abu Zaabel

Cairo Egypt

Dr. Bhaskar Mukherjee
ANSTO, Safety division
PMB-1
Menai, NSW
AUSTRALIA 2234
muherjee@ieee.org

Prof. Boris Myasoedov
V.I. Vernadsky Inst of Geochemistry
19 Kosygin Street
Moscow
RUSSIA 117975

Dr. Iouri Mykhov
Russian Scientific Center, Applied Chemistry
14, Dobrolubova Pr. St. Petersburg
Leningradsky Region
RUSSIA 197198

Prof. Takashi Nakamura
Tohoko University
Cyclotron and Radioisotope Center Aoba, Aramaki, Aoba-ku
JAPAN 980-8578
nakamura@cyric.tohoku.ac.jp

Mr. Dimitry
OAO GAZ
Kosmicheskaja Street 56, Apt. 96
Sarov Nizhny Novgorod
RUSSIA 603 018

Ms. Kathy Nielsen
Slowpoke-2 at RMC/CMR Chem & Chem Eng.
P.O. Box 17000 Stn Forces
Kingston Ontario
CANADA K7K 7B4
nielsen-k@rmc.ca

Radioissotope Centre POLATOM
Otwock-Swierk
Otwock
POLAND 05-400
banko@cxl.cyf.gov.pl

Dr. Kennedy O'Brien
Nen Life Science Products
10 Britton Place

Wilmington DE
USA 19805
obrienkj@nenlifesci.com

Dr. Raymond Oekonomopoulos
Hoechst Marion Roussel Deutschland GmbH
Chemical Development
Radiosynthesis Griesheim
D-65926 Frankfurt/Main GERMANY
raymond.oekonomopulos@hmrag.com

Ms. Inna Okhotina
Ministry of Atomic Energy MINATOM
ul.b.Oedynka, 24/26

Moscow

Mr. Jozef Orzechowski
TRIUMF
4004 Wesbrook Mall
Vancouver B.C , Canada V6T 2A3
joski@triumf.ca

Dr. Jim Ounsworth
C/D/N/ Isotopes Inc
88 Leacock Street
Pointe Claire Quebec
CANADA H9R 1N1
jim@cdniso.com

Dr. Allan Pashkluski
ISOFLEX Russia

Mr. Dariusz Pawlak
Radioistope Centre POLATOM
Otwock-Swierk
Otwock
POLAND 05-400
pawlak@cxl.cyf.gov.pl

Ms. Ann Pedder
Nycomed Amersham Imaging
Amersham Laboratories White Lion Road
Amersham Bucks
United Kingdom HP7 9LL
ann.pedder@uk.nycomed-amersham.com

Dr. Yuan Ji Pei
Nat'l Synchrotron Radiation Lab, USTC
96 Jinghai Road, Hufei Anlmi 42 South Road, Hezuo Hua Ave
Hefei Anhui Province
P.R.CHINA 230029
yjpei@ustc.edu.ln

Dr. Jeffrey Perine
Norvatis Crop Protection Inc
PO Box 18300
Greensboro NC
USA 27419-8300
jeff.perine@cp.nova

Dr. Luiz Carlos Ruiz Pessenda
Centre for Nuclear Energy in Agriculture
Piracicaba São Paulo
Brazil 13416-000
pessenda@cena.usp.br

Dr. Gene Petersen
Los Alamos National Laboratory
Post Office Box 1663 MS J514
Los Alamos New Mexico
USA 87544
ejp@lanl.gov

Dr. Dennis Phillips
Los Alamos National Laboratory
MS J514
Los Alamos NM
USA 87544
drp@lanl.gov

Dr. K.K.S. Pillay
Los Alamos Nat'l Lab
NMT/DO Mail Stop: E.500
Los Alamos NM
USA 87544
S-Pillay@lanl.gov

Dr. Ulrich Pleiss
Bayer Ag
Aprather Weg 18a
Wuppertal
GERMANY 42096
Ulrich.Pleiss.UP@bayer-ag.de

Mr. Andrey Pogorelov
REViss Services (UK) Ltd
6 Chiltern Court, Asheridge Road
Chesham Buckinghamshire
UK HP5 2PX
andrey-pogorelov@reviss.co.uk

Dr. Marina Pokidycheva
STABOS Co. Ltd.
10 Gamalei Street, Suite 54
Moscow 123098
RUSSIA
ampokid@4unet.ru

Mr. Leonid Polyakov
FKP
Lsenoy
Russia
CANADA V6R 2H5

Mr. Bernard Ponsard
SCK.CEN/BR2 Reactor
Boeretang 200
MOL
BELGIUM B-2400
bponsard@sckcen.be

Mr. Ivan Pulnikov
Electrochemical Plant/TENEX
centralnaya 5
Zelenogozsk, Krasnoyazsk Region
RUSSIA 663690
taifun@ecp.kts.ru

Dr. Shirish Raja
Abbott Laboratories
Dept 46V, Bldg Ap9 100 Abbott Park Road
Abbott Park IL
USA 60064-6114
shirishn.raja@abbott.com

Drs. Hubert Rakhorst
Urenco Nederland BV
PO Box 158
7600 AD Almelo
THE NETHERLANDS
mail@urenco.nl

Dr. Srikantam Rao
Bhaba Atomic Research Centre
Trombay
Mumbai Maharashtra
INDIA 400 085
smrao@magnum.barc.ernet.in

Dr. J. Kent Rinehart
Bristol-Myers Squibb
PO Box 4000
Princeton New Jersey
USA 08543
rinehart@bms.com

Dr. Peter Roberts
Inst. Of Geological & Nuclear Sciences
P.O. Box 31-312
Lower Hutt
New Zealand
p.roberts@gns.cri.nz

Mr. John H. Rolston
RCHEM Consultants
4263 West 10th Avenue
Vancouver, BC

Mr. Gary Rotert
Abbott Laboratories
D46V, AP9 100 Abbott Park Rd
Abbott Park IL
USA 60064
gary.a.rotert@abbott.com

Dr. Lenny Roytvarf
Spectra Gases
80 Industrial Drive
Alpha New Jersey
USA 08865
lennyr@sgi-east.spectra-gases.com

Dr. Tom J. Ruth
TRIUMF, PET Group
4004 Wesbrook Mall
Vancouver, BC
CANADA V6T 2A3
truth@triumf.ca

Dr. Tim Saarinen
Isotec Inc
3858 Benner Rd
Miamisburg Ohio
USA 45342
tsaarinen@mathesongas.com

Dr. Roger Saintonge
Isotec Inc
3859 Benner Rd
Miamisburg Ohio
USA 45342
Rsaintonge@mathesongas.com

Prof. Ken-ichi Sasaki
Rikkyo University
3 Nishi-Ikebukuro, Toshima-ku
Tokyo
JAPAN 171-8501
Ksasaki@mail.com

Dr. M.A. Sattar
Bangladesh Inst of Nuclear Agriculture
PO Box 4
Mymensingh 2200 Bangladesh
Bina@bdmail.net

Mr. John Scheigetz
Merck Frosst Canada Co
16711 Trans Canada Hwy
Kirkland Quebec, Canada, H9H 3L1
John_scheigetz@merck.com

Dr. Robert Schenter
Pacific Northwest National Laboratory
MS-P7-52 Post Office Box 999
Richland Washington
USA 99352
re_schenter@pnl.gov

Dr. Frederic Schmitz
IOM Beam Applications (IBA)
3 Chemin du Cyclotron
Louvain-la- Neuve
BELGIUM 1348
schmitz@iba.be

Dr. Peter Schreck
UFZ-Centre for Environmental Research
Dept. of Hydrogeology Th. Lieser Str 4
Halle/Saale
GERMANY D-06120
schreck@hdg.ufz.de

Dr. Ronald Schubank
TRIUMF
4004 Wesbrook Mall
Vancouver, BC
CANADA V6T 2A3
schub@triumf.ca

Mr. Yann Seimbille
Université de Sherbrooke
3001, 12e avenue Nord
Sherbrooke Québec
Canada J1H 5N4
Seimbill@tep.crc.usherb.ca

Dr. Svetlana Selivanova
Universite de Sherbrooke
Dept. of Nucl. Med. 3001, 12 Ave. Nord
Sherbrooke Quebec
CANADA J1H 5N4
svetlana@hermes.usherb.ca

Dr. Yair Shamai
Soreq Nuclear Research Centre
Yavne, ISRAEL 81800
shamai@ndc.soreq.gov.il
npdept@hcm.vnn.vn

Mr. David Shepherd
Information Science & Technology Agency
560 Superior Street, 3rd Floor
Victoria BC, Canada
dshepherd@genis5.gov.bc.ca

Mr. Shell Shetti
Bioscan, Inc.
4590 MacArthur Blvd., NW
Washington DC
USA 22193
Shell@Bioscan.com

Mr. Ineiy Shipilov
OAO GAZ
Djakonora Street 20, ap 38 Prospect Lenina,
Sarov Nizhny Novgorod
RUSSIA 603 065
Jean.St.Pierre@meca.polymtl.ca

Dr. Walter Shmayda
Ontario Hydro
800 Kipling Avenue
Toronto Ontario
Canada M8Z-1C4
shmaydaw@oht.hydro.on.ca

Dr. Marco Skrinjar
Astra Draco
Medicinal Chemistry
PO Box 34
S-221 00 Lund SWEDEN
Marco.Skrinjar@draco.se.astra.com

Mr. Thomas Sloan
Chemical Abstracts Service Div
2540 Olentangy River Rd PO Box 3012
Columbus Ohio
USA 43210
tsloan@cas.org

Dr. Leonid Solin
V.G. Khlopin Radium Institute
2nd Murinsky Av 28
St. Petersburg
RUSSIA 194021
solin@Nuk.convey.ru

Mr. Phan Son Hai
Nuclear Research Institute
01 Nguyen Tu Luc Str. Dalat, Vietnam
Dalat Lan-dong
VIETNAM

Dr. Akinari Sonoda
SNIRI, 2217-14 Hayashi
Takamatsu KAGAWA
JAPAN 761-0395
sonoda@sniri.go.jp

Dr. Suresh Srivastava
Brookhaven National Laboratory
Medical Department, Building 801 PO Box 5000
Upton NY
USA 11973-5000
suresh@bnl.gov

Mr. Jean St.Pierre
Ecole Polytechnique
PO Box 6079
Montreal Quebec
CANADA H3C 3A7

Dr. Colyn Steeves
MDS Nordion
447 March Road
Kanata Ontario
CANADA K2K 1X8

Dr. Nigel Stevenson
Theragenics Corporation
5203 Bristol Industrial Way
Buford, GA 30518

stevensn@theragenics.com

Mr. Wayne Stolle
Pharmacia & Upjohn
301 Henrietta St Unit 7265-300-369
Kalamazoo MI.
USA 49001
wayne.t.stolle@am.pnu.com

Mr. Peter Strom
Astra Draco
Medicinal Chemistry
PO Box 34
S-221 00 Lund SWEDEN
peter.strom@draco.se.astra.com

Dr. Bruce Surber
Abbott Laboratories
D46V, AP9 100 Abbott Park Road
Abbott Park IL
USA 60064-6114
Bruce.Surber@Abbott.com

Dr. Alexander Susan
International Isotope Clearing House Inc
9110 Lee Boulevard
Leawood, KS
USA 66206
iichkansas@worldnet.att.net

Dr. Sandor Takacs
Institute of Nuclear Research
Bem ter 18/c
Debrecen
Hungary H-4026
takacs-s@atomki.hu

Dr. Hidelei Takano
JAERI
Shirakata,Tokai-mura-Naka-gum Ibaraki-ken,
Tokai Ibaraki ,
JAPAN 319-1195
takano@cens.tokai.jaeri.go.jp

Dr. Chou Tok Tan
Isotec Inc
3858 Benner Rd
Miamisburg Ohio
USA 45342
ctan@mathesongas.com

Dr. Frank Tang
Merck
126 East Lincoln Ave
Rahway NJ
USA 07065
Frank_tang@merck.com

Mr. Sergey Tarasov
Deputy Director of RFIAC-VNIIEF
RFIAC-VNIIEF Murrykov Street 18, ap 125
Sarov Nizhny Novgortod Region
RUSSIA 607190
otesp@sar.nnov.ru

Ms. Tatjana Tarasova
RFJAC/VNIIF
Muzrukov Street 18, ap 125 Prospect Mira, 37
Sarov Niznny Novgorod Region
RUSSIA 607 190
pcotrive@cena.usp.br

Dr. Alexy Tatarinov
EKP
Lesnoy
Russia
JAPAN 113-8941

Mr. Menashe Tidhar
Rotem Ind. Ltd.
Post Office Box 9046
Beer-Sheva
Israel 84190

Dr. Yuri Toporov
Research Institute of Atomic Reactors
Dimitrivgrad 10
RUSSIA 433510
kar@orip.niiar.simbirsk.su

Mr. Jose Torres
MDS Nordion
447 March Road
Kanata Ontario
CANADA K2K 1X8
jtorres@mds.nordion.com

Dr. Patrick Towers
Amersham Pharmacia Biotech
Forest Farm Whitchucrh
Cardiff
UK CF4 7YT
pat.towers@nycomed-amersham.com

Dr. Guennauiv Trembovetski
Isoflex Russia
RUSSIA

Dr. Hans-Christian Treutler
Centre for Environmental Research
Leipzig Halle Dept. of Analytical Chemistry
Permoserstrasse 15
Leipzig GERMANY D-04318
treutler@ana.ufz.de

Dr. Paulo Trivelin
CENA/USP
Av. Centenario 303 CX. P. 96
Piracicaba Sao Paulo
BRAZIL 13400-970

Dr. Hirokazu Umezawa
Japan Radioisotope Association
28-45 Honkomagome 2-Chome Bunkyo-ku
Tokyo
umezawa@jrias.or.jp

Prof. Turan Unak
Ege University
Faculty of Science,
Department of Chemistry, Division of
Nuclear Chemistry, Bornova
Izmir TURKEY 35100
unak@egenet.com.tr

Dr. Jacob Stenmann Valsborg
Novo Nordisk A/S
Novo Nordisk Park
Malov
DENMARK DK-2760
jstv@novo.dk

Mr. Rene van den Elzen
TRIUMF
4004 Wesbrook Mall
Vancouver BC
Canada V6T 2A3
elzen@triumf.ca

Dr. Alain Vanhove
Euriso-Top
Parc des Algorithmes - Batiment Homere
Route de L'Orme
Saint-Aubin, FRANCE 91194
vanhove@eurisotop.fr

Dr. Bertrand Vieillard-Baron
Framatome
Tour Framatome
Paris La Defense-Cedex
FRANCE 92084
bvieillardbaron@framatome.fr

Dr. Rolf Voges
Novartis Pharma AG
Lichtstr. Building WSJ-507.9.51
Basel
SWITZERLAND CH-4002
Rolf.Voges@pharma.novartis.com

Mrs. Maria Volskaia
AO "Techsnabexport"
26 Staromonetniy per
Moscow
RUSSIA 109180

Dr. Roland von Gogh
AEC of S.A. Ltd
PO Box 582
Pretoria
SOUTH AFRICA 0001

Prof. Manfred Wahren
University of Leipzig
Institute of Organic Chenistry
Permoserstr. 15
Leipzig
GERMANY D-04303

Dr. Shyh-Jen Wang
Veterans General Hospital
160, Sec 3, Taichung Harbor Road
Taichung Taiwan
TAIWAN 40705
sjwang@vghtc.vghtc.gov.tw

Mr. Bangun Wasito
Nat'l Nuclear Research Centre,Yogyakarta
Jl. Babarsari PO Box 1008
Yogyakarta
INDONESIA 55010
P3TM@indo.net.id

Dr. Leonard Wassenaar
Environment Canada
11 Innovation Blvd
Saskatoon SK
Canada S7N3H5
len.wassenaar@ec.gc.ca

Mr. Brian Webster
MDS Nordion Inc
4004 Wesbrook Mall

Vancouver BC
CANADA V6T 2A3
ytakao@drl.co.jp

Dr. Ken Westaway
Laurentian University
Dept of Chemistry
Sudbury Ontario
CANADA P3E 2C6
kwestawa@nickel.laurentian.ca

Prof. Leonard Wiebe
Noujaim Institute, University of Alberta
3117 DP University of Alberta
Edmonton Alberta
Canada T6G-2N8
leonard.wiebe@ualberta.ca

Dr. D. Scott Wilbur
University of Washington
2121 N 35th Street
Seattle WA
USA 98101-9103
dswilbur@u.washington.edu

Mr. Alfred Wong
TRIUMF
4004 Wesbrook Mall
Vancouver B.C. Canada V6T 2A3
awong@triumf.ca

Mr. Ray Wong
ATG,TRIUMF
4004 Wesbrook Mall
Vancouver BC
Canada V6T 2A3
rwong@triumf.ca

Ms. Zhifang Wu
INET, Tsinghua University
Haidian District
Beijing
P.R.China 100084
liye@inet.tsinghua.edu.cn

Dr. Masatoshi Yamada
National Insitute of Radiological Sci.
Research Center for Radioecology, Isozaki 3609
Hitachinaka Ibaraki
JAPAN 311-1202
m_yamada@nirs.go.jp

Mr. Takao Yamamoto
Daiichi Radioisotope Lab. Ltd
453-1, Shimo-okura,

Matsuo-machi, Sanbu-gun
Chiba , JAPAN 289-1592

Dr. Albert Yang
U.S. Geological Survey
Box 25046, Federal Centre MS-421
Lakewood Co
USA 80225
ayang@usgs.gov

Dr. Taufik Yassine
Atomic Energy Commission
P.O. Box 6091
Damascus
SYRIA
aecs@syriatel.net

Dr. Vivek Yelgaonkar
Bhabha Atomic Research Centre
Isotope Applications Division Hirup, BARC,
Mumbai, Maharashtra
INDIA 400085
yelgaon@magnum.barc.ernet.in

Mr. Huang Yong
Nuclear Power Institute of China
PO Box 291-107 Chengdu
Sichuan 610005, CHINA
qbs@npic.ac.cn

Dr. Sydney Yu
Singapore General Hospital
Dept. of Nuclear Medicine Outram Road
Singapore
SINGAPORE 169608
gnmywk@sgb.gov.sg

Dr. Kathleen Zandi
Pfizer
Eastern Point Road
Groton Connecticut
USA 06340
kathleen_s_zandi@groton.pfizer.com

Dr. Jan Rijn Zeevaart
AEC of SA
PO Box 582
Pretoria
South Africa 0001
radchem@aec.co.za

Dr. Stefan Zeisler
German Cancer Research Center DKFZ
Radiochemistry E0304 IM neuenheimer Feld
280
Heidelberg
GERMANY D-69120`
zeisler@dkfz-Heidelberg.de

Dr. Hongyan Zhang
Chinese Inst. Of Atomic Energy
P.O. Box 275 (12)
Beijing 102413
P.R. CHINA

Dr. Zhang Xuelian
Institute of Archaeology
Chinese Academy of Social Sciences
27 Wangfujing Street, Beijing, 100710
P. R. China
liu.jg@history.cass.net.cn

Mr. Juergen K. Ziegler
BEBIG GmbH
Robert Rossle Str. 10
Berlin-Buch
GERMANY D-13125

Conference Photos

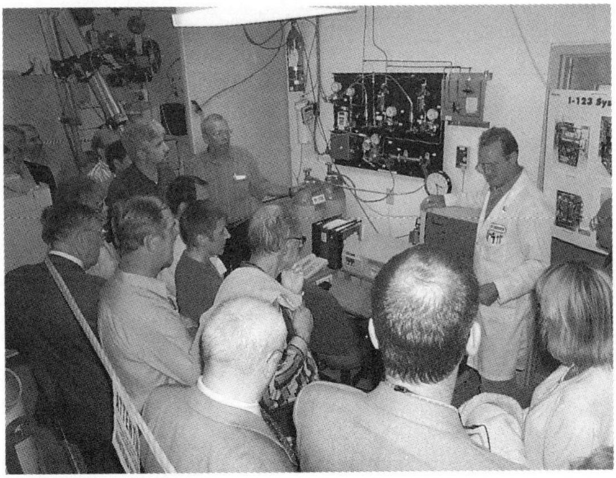
I-123 Sys

EXIT

Cambridge Isotop